# Polymers of Biological and Biomedical Significance

A C S  S Y M P O S I U M  S E R I E S **540**

# Polymers of Biological and Biomedical Significance

**Shalaby W. Shalaby,** EDITOR
*Clemson University*

**Yoshito Ikada,** EDITOR
*Kyoto University*

**Robert Langer,** EDITOR
*Massachusetts Institute of Technology*

**Joel Williams,** EDITOR
*Becton Dickinson Research Center*

Developed from a symposium sponsored
by the Division of Polymer Chemistry, Inc.,
at the 204th National Meeting
of the American Chemical Society,
Washington, DC,
August 24–28, 1992

American Chemical Society, Washington, DC 1994

**Library of Congress Cataloging-in-Publication Data**

American Chemical Society. Meeting (204th: 1992: Washington, D.C.)
   Polymers of biological and biomedical significance / Shalaby W.
Shalaby . . . [et al.], editor.

      p.    cm.—(ACS symposium series, ISSN 0097–6156; 540)

   "Developed from a symposium sponsored by the Division of
Polymer Chemistry, Inc., at the 204th National Meeting of the American
Chemical Society, Washington, DC, August 24–28, 1992."

   Includes bibliographical references and index.

   ISBN 0–8412–2732–2

   1. Polymers in medicine—Congresses.

   I. Shalaby, Shalaby W. II. American Chemical Society. Division of
Polymer Chemistry. III. Title. IV. Series.

R857.P6A427   1992
610'.28—dc20                                          93–34697
                                                              CIP

The paper used in this publication meets the minimum requirements of American National Standard for Information Sciences—Permanence of Paper for Printed Library Materials, ANSI Z39.48–1984. ∞

Copyright © 1994

American Chemical Society

PRINTED IN THE UNITED STATES OF AMERICA

# Foreword

THE ACS SYMPOSIUM SERIES was first published in 1974 to provide a mechanism for publishing symposia quickly in book form. The purpose of this series is to publish comprehensive books developed from symposia, which are usually "snapshots in time" of the current research being done on a topic, plus some review material on the topic. For this reason, it is necessary that the papers be published as quickly as possible.

Before a symposium-based book is put under contract, the proposed table of contents is reviewed for appropriateness to the topic and for comprehensiveness of the collection. Some papers are excluded at this point, and others are added to round out the scope of the volume. In addition, a draft of each paper is peer-reviewed prior to final acceptance or rejection. This anonymous review process is supervised by the organizer(s) of the symposium, who become the editor(s) of the book. The authors then revise their papers according to the recommendations of both the reviewers and the editors, prepare camera-ready copy, and submit the final papers to the editors, who check that all necessary revisions have been made.

As a rule, only original research papers and original review papers are included in the volumes. Verbatim reproductions of previously published papers are not accepted.

*M. Joan Comstock*
Series Editor

# Contents

Preface .......................................................................................... xi

TOPICAL REVIEWS

1. **Polymers in Pharmaceutical Products** ........................................ 2
   H. Park and Kinam Park

2. **Tissue Engineering Using Synthetic Biodegradable Polymers** ..... 16
   C. A. Vacanti, J. P. Vacanti, and R. Langer

3. **Interfacial Biocompatibility** ................................................... 35
   Yoshito Ikada

4. **Polymeric Devices for Transcutaneous and Percutaneous
   Transport** ......................................................................... 49
   Joel L. Williams

5. **Microcellular Foams** ............................................................ 58
   S. L. Roweton and Shalaby W. Shalaby

SYNTHESIS, SURFACE ACTIVATION, AND CHARACTERIZATION
OF BIOMATERIALS

6. **Surface Biolization by Grafting Polymerizable Bioactive
   Chemicals** ......................................................................... 66
   Y. Ito, K. Suzuki, and Y. Imanishi

7. **Hydrophilic, Lipid-Resistant Fluorosiloxanes** ............................ 76
   G. Friends, J. Künzler, R. Ozark, and M. Trokanski

8. **Synthesis of Poly(ether urethane amide) Segmented
   Elastomers** ........................................................................ 87
   A. Penhasi, M. Aronhime, and D. Cohn

9. **Interaction of Water with Polyurethanes Containing
   Hydrophilic Block Copolymer Soft Segments** ............................. 103
   N. S. Schneider, J. L. Illinger, and F. E. Karasz

10. Formation and Reactivity of Surface Phosphonylated
    Thermoplastic Polymers .................................................................. 116
    K. R. Rogers and Shalaby W. Shalaby

11. Surface Modification of Polymeric Biomaterials
    with Poly(ethylene oxide): A Steric Repulsion Approach ............. 135
    Mansoor Amiji and Kinam Park

12. Ascorbic Acid as an Etchant–Conditioner for Resin
    Bonding to Dentin........................................................................... 147
    James E. Code, Gary E. Schumacher, and
    Joseph M. Antonucci

13. Salt Partitioning in Polyelectrolyte Gel–Solution Systems .......... 157
    Yu-Ling Yin and Robert K. Prud'homme

14. Ring-Opening Polymerization of a 2-Methylene Spiro
    Orthocarbonate Bearing a Pendant Methacrylate Group ........... 171
    Jeffrey W. Stansbury

15. Ring-Opening Dental Resin Systems Based on Cyclic Acetals.... 184
    B. B. Reed, Jeffrey W. Stansbury, and Joseph M. Antonucci

16. Synthesis of Novel Hydrophilic and Hydrophobic
    Multifunctional Acrylic Monomers ............................................... 191
    Joseph M. Antonucci, Jeffrey W. Stansbury, and
    G. W. Cheng

17. Effect of Structure on Properties of Absorbable Oxalate
    Polymers ......................................................................................... 202
    Russell A. Johnson and Shalaby W. Shalaby

18. Fluorescent Cure Monitoring of Dental Resins ........................... 210
    Spurgeon M. Keeny III, Joseph M. Antonucci,
    Francis W. Wang, and John A. Tesk

BIOLOGICAL EFFECTS RELATED TO SPECIFIC
PHYSICOCHEMICAL FACTORS

19. Activation of Leukocytes by Arg–Gly–Asp–Ser-Carrying
    Microspheres ................................................................................. 220
    Keiji Fujimoto, Y. Kasuya, M. Miyamoto, and
    H. Kawaguchi

20. Poly(vinyl alcohol) Hydrogels Prepared under Different
    Annealing Conditions and Their Interactions with Blood
    Components ........................................................................ 228
    Keiji Fujimoto, Masao Minato, and Yoshito Ikada

21. Phase Transition's Control of Collagenous Tissue Growth
    and Resorption Including Bone Morphogenesis ........................... 243
    David Gilbert Kaplan

22. pH-Sensitive Hydrogels Based on Hydroxyethyl Methacrylate
    and Poly(vinyl alcohol)–Methacrylate ..................................... 251
    Y. J. Wang, F. J. Liou, S. W. Tsai, and G. G. C. Niu

23. Kinetic Model for Degradation of Starch–Plastic Blends
    with Controlled-Release Potential ........................................... 258
    Liu Zhang, John J. Harvey, and Michael A. Cole

24. Cross-Linking and Biodegradation of Native and Denatured
    Collagen ........................................................................... 275
    K. Tomihata, K. Burczak, K. Shiraki, and Yoshito Ikada

SYNTHETIC BIOACTIVE CHAIN MOLECULES
AND POLYMERS FOR CONTROLLED TRANSPORT
OF BIOACTIVE AGENTS

25. Poly(methacrylic acid) Hydrogels as Carriers of Bacterial
    Exotoxins in an Oral Vaccine for Cattle ................................... 288
    T. L. Bowersock, W. S. W. Shalaby, M. Levy,
    M. L. Samuels, R. Lallone, M. R. White, D. Ryker,
    and Kinam Park

26. Dye-Grafted, Poly(ethylene imine)-Coated, Formed-in-Place
    Class Affinity Membranes for Selective Separation
    of Proteins ....................................................................... 297
    Y. Li and H. G. Spencer

27. Preparation of Insulin-Releasing Chinese Hamster Ovary
    Cell by Transfection of Human Insulin Gene: Its
    Implantation into Diabetic Mice ............................................ 306
    H. Iwata, N. Ogawa, T. Takagi, and J. Mizoguchi

28. **pH and Ionic-Strength-Dependent Permeation through Poly(L-lysine-*alt*-terephthalic acid) Microcapsule Membranes** ................................................... 314
    Kimiko Makino, Ei-ichi Miyauchi, Yuko Togawa,
    Hiroyuki Ohshima, and Tamotsu Kondo

INDEXES

**Author Index** ........................................... 327

**Affiliation Index** ........................................ 327

**Subject Index** ........................................... 328

x

# Preface

IMPRESSIVE ADVANCES IN BIOTECHNOLOGY AND MATERIALS with unique properties have increased the interest in polymers and materials of significance in biological and biomedical research and development in the past decade. Biomedical polymers have received a lot of attention from authors with diverse research and clinical interests, which has led to a series of papers that are "snapshot" displays of topical discussions with varying degrees of technical depth. At the same time, the subject has received limited attention from authors and editors of the polymer community. This situation prompted the editors of this book to organize a broad-based symposium as a technical forum for integrated discussion of polymers that are significant to both the biological and biomedical communities.

More than two-thirds of the papers presented at the symposium were selected for incorporation as chapters in this volume. Experts other than the symposium participants were also invited to contribute to the book to provide comprehensive coverage of all key areas, allowing for the inclusion of the most recent research activities and topical reviews of fast-growing research areas. These reviews constitute the first five-chapter section of the book and present a bird's-eye view of key areas as well as concise introductory notes to several chapters of the book. The second section includes a broad range of topics with interrelated coverage of current activities on the synthesis of new materials, unique approaches to surface activation, and skillful implementation of known and emerging methods for the characterization of biomaterials. Six chapters on biological effects related to specific physicochemical factors are presented in the third section of the book and stress the importance of understanding the interaction of polymers with the biological environment. The last section focuses on the controlled transport of bioactive agents. Proteins as bioactive agents received a great deal of attention by the authors of most of the constituent chapters.

We hope that this book, devoted to biomaterials, biotechnology, and advanced materials for pharmaceutical and biomedical applications, will be valued by academic and industrial researchers, research managers, and industrial strategic planners, as well as graduate students. It will provide readers with access to state-of-the-art coverage of primary research and concise reviews of key areas, presented by authorities in diverse areas of biological and biomedical research.

We wish to express our appreciation to the American Chemical Society's Polymer Chemistry Division, Inc., for sponsoring the symposium. We also thank Anne Wilson of the American Chemical Society Books Department and Susan Roweton, a bioengineering graduate student at Clemson University, for their conscientious efforts to ensure a timely review and editing of the book.

SHALABY W. SHALABY
Clemson University
Clemson, SC 29634–0905

YOSHITO IKADA
Kyoto University
Kyoto, Japan

ROBERT LANGER
Massachusetts Institute of Technology
Cambridge, MA 02139

JOEL WILLIAMS
Becton Dickinson Research Center
Research Triangle Park, NC 27709

July 8, 1993

# TOPICAL  REVIEWS

# Chapter 1

# Polymers in Pharmaceutical Products

**H. Park and Kinam Park**

**Department of Industrial and Physical Pharmacy, Purdue University, West Lafayette, IN 47907**

Polymers have played indispensable roles in the preparation of pharmaceutical products. Their applications range widely from material packaging to fabrication of the most sophisticated drug delivery devices. This chapter reviews various polymers used in pharmaceutics based on their applications.

Of the many materials used in the pharmaceutical formulations, polymers play the most important roles. The use of polymers ranges from manufacturing of various drug packaging materials to the development of dosage forms. The important application of polymers undoubtedly resides in the development of sophisticated controlled release drug delivery systems. In conventional dosage forms, polymers are usually used as excipients, adjuvants, suspending agents, or emulsifying agents. In controlled release dosage forms, polymers are used mainly to control the release rate of drugs from the dosage forms. The presence of numerous polymers which are able to control the drug release profiles has been the basis for the explosive advances in the development of controlled release dosage forms during 80's. It would not be an overstatement that the future development of more sophisticated dosage forms entirely depends on the appropriate use of existing polymers and synthesis of new polymers.

One way of classifying polymers in pharmaceutical applications is to divide them into three general categories according to their common uses: (1) polymers in conventional dosage forms; (2) polymers in controlled release dosage forms; and (3) polymers for packaging.

## Polymers in Conventional Dosage Forms

Despite the well known advantages of controlled release dosage forms, conventional dosage forms are still most widely used probably because they cost less to manufacture. More than three quarters of all drug formulations are made for oral administration. Oral dosage forms such as tablets, capsules, and liquids are still most popular. Since tablet is one of the most widely used dosage forms and its preparation requires incorporation of polymers, we will focus on polymers used in tableting process.

0097–6156/94/0540–0002$06.00/0

**Tablet Excipients.** One of the important goals in tableting process is to produce tablets which are uniform in weight and strong enough to withstand the rigors of processing and packaging. It is also equally important that the tablets break down and dissolve upon administration for the release of the drug. Tablets with such desirable pharmaceutical properties can be prepared by using various types of polymeric excipients. It is the excipients that determine the compressibility, hardness, hygroscopy, friability, lubricity, stability, and dissolution rate of the prepared tablets.

The most commonly used polymeric excipients can be divided into four categories, such as binder, diluent, disintegrant, and lubricant. Table I lists examples of polymers used in each category. In many cases where drug powders alone have poor compressibility, binding agents need to be incorporated into the formulation to form suitable hard tablets. Starch is known as one of the best general binders (*1*). Starch is insoluble in cold water and in alcohol but gelatinizes in hot water to release insoluble amylopectin and soluble amylose. The amylopectin fraction is primarily responsible for the binding properties of starch (*1, 2*). Some powdered gums, such as agar, guar, karaya, and tragacanth, are used as binders due to their pronounced adhesiveness when wet. However, concentration of these substances in the tablets are restricted because they retard disintegration (*1*). Carbopol, high molecular weight poly(acrylic acid), is also frequently used as a binder in tablets. The adhesive polymers tend to resist breakup of the tablet and thus can be used in the formulation of prolonged action tablets. Binders are typically dispersed or dissolved in water or a water-alcohol mixture and sprayed, poured, or admixed into the powders to be agglomerated. The compaction properties of pharmaceutical powders can be quantitatively evaluated by measuring the bonding index, brittle fracture index, and strain index (*4*). When only a small amount of drug is required for pharmacological action, drug is diluted with inert materials to produce tablets of a reasonable size. Most of the diluents listed in Table I are those which are also used as binders.

Disintegrants are those which induce dissolution of tablets in the presence of water by overcoming the cohesive strength introduced into the mass by compression. Disintegrants break tablets through absorption of water by either capillary effect or by swelling (*1*). Microcrystalline cellulose, despite its water insolubility, induce water uptake into the tablet by the capillary effect (*5*). It is widely used in tablet manufacture as a binder and as well as a disintegrant in directly compressible formulations. Microcrystalline cellulose has been in more than 2000 pharmaceutical products (*5*). Starch is also known to induce water uptake by capillary effect, since the degree of hydration of starch at body temperature is negligible (*6*). On the other hand, alginic acid, cellulose derivatives, crosslinked Povidone, and gelatin disintegrate tablets by swelling action. To be effective as disintegrants polymers should not produce a sticky, gelatinous mass on swelling or increase the viscosity around the dispersing tablet mass.

When drug molecules are hydrophobic and consequently poorly wetted by aqueous media, a small amount of wetting agent such as poloxamer (PEO-PPO-PEO triblock copolymer), polysorbate, polyoxyl stearate, and polyoxyl oleyl ether (*3*), can be added to enhance disintegration and dispersion. Lubricants are used to prepare defect-free tablets during tableting process. As shown in Table I, some excipients have multiple uses. For example, microcrystalline cellulose as well as starch can be used as either a binder, a diluent, or a disintegant depending on the concentration used (*1*).

**Tablet Coating.** Many tablet dosage forms are coated to mask or minimize the unpleasant taste or odor of certain drugs, to protect the drug against decomposition, and to enhance the appearance. Traditionally, tablets were coated with sucrose by repeated application of a syrup followed by forced drying in hot air. The sugar

Table I. Some Polymers Used as Tablet Excipients

| | |
|---|---|
| A. Binder | Agar |
| | Alginic acid |
| | Poly(acrylic acid) (Carbopol) |
| | Carboxymethylcellulose, sodium |
| | Cellulose, microcrystalline (Avicel) |
| | Dextrin |
| | Ethylcellulose |
| | Gelatin |
| | Guar gum |
| | Hydroxypropyl methylcellulose |
| | Karaya gum |
| | Methylcellulose |
| | Polypvinylpyrrolidone (Povidone) |
| | Starch, pregelatinized |
| | Tragacanth gum |
| B. Diluent | Cellulose, microcrystalline (Avicel) |
| | Cellulose, powdered |
| | Dextrin |
| | Starch |
| | Starch, pregelatinized |
| C. Disintegrant | Alginic acid |
| | Cellulose, microcrystalline (Avicel) |
| | Gelatin |
| | Povidone, crosslinked |
| | Sodium starch glycolate |
| | Starch |
| | Starch, pregelatinized |
| D. Lubricant | Poly(ethylene glycol) |

SOURCE: Adapted from ref. 3.

coating may take several days to obtain the required thickness and appearance. On the other hand, polymer coating is fast, simpler, and results in a relatively thin layer, approximately 20 to 100 μm, on the solid dosage forms (7). A polymer coating also gives better coated layer by sealing the pores and smoothing the rough texture of the core surface (8). The stability of drugs can be greatly improved if the dosage units are coated with polymers which are hydrophobic enough to block the penetration of moisture.

The coating polymers can be divided into the two major groups: enteric and non-enteric. Enteric coating polymers are polyacids which are water-insoluble in low pH environments, such as in the stomach, and dissolve in alkaline conditions such as those found in the intestine through ionization of the acid groups. They are used to protect acid-labile drugs which may be destroyed by the acidic gastric juice and to improve tolerability of drugs which irritate the stomach. Since the enteric coated tablets release drugs only in the intestine, the enteric coating polymers allow delayed release as well as targeted release in the intestine. The most widely used polymers for enteric coating are listed in Table II along with their pH values above which the polymer coating dissolves (dissolution pH). The dissolution pH depends on the relative hydrophobicity of the polymer and the number of acid groups. For exmaple, the dissolution pH of poly(methacrylic acid-co-methyl methacrylate) ranges from 5.5 to 7 depending on the relative ratio of hydrophilic methacrylic acid and hydrophobic methyl methacrylate. Thus, mixing different copolymers makes it possible to achieve fine differentiation in the dissolution pH. The same goal can be achieved by mixing two different types of enteric coating polymers, such as cellulose acetate trimellitate and cellulose acetate phthalate or Shellac. It is preferred that enteric coating polymers have high resistance to moisture permeability and high stability to hydrolysis. Cellulose acetate phthalate possesses low resistance to moisture permeability and are easily hydrolyzed by the moisture in the air during storage (10). Although the majority of enteric coated films have been applied as solutions in organic solvents, the use of aqueous film coating is in a growing trend due to the environmental and health-related concerns. It is noted that disintegration of the enteric coating polymers depends not only on pH but also on other factors such as coating method, coating thickness, the plasticizer used, and the test method (11).

Non-enteric polymers are those which are water-soluble regardless of pH of the environment. Of the synthetic polymers, polyvuylpyrrolidone (PVP or Povidone), polyethylene glycol (PEG, Carbowax), and poly(ethylene oxide)-poly(propylene oxide) block copolymers (Pluronics and Poloxamers) are widely used as coating materials. One of the factors which determine the property of the coating is the molecular weight of the polymer. For example, high molecular weight PEG's give tough coating whereas low molecular weight PEG's are mainly used as plasticizers which prevent the film coat from becoming too brittle (12). Plasticizers cause a decrease in both the modulus of elasticity and the glass transition temperature of the polymer (13). Plasticizers could be released from the film and contaminate the drug. This can be prevented by the method of internal plasticization, in which the glass transition temperature of the polymer is lowered through copolymerization (7).

## Polymers Used in Controlled Release Dosage Forms

One of the most important applications of polymers in modern pharmaceutics is the development of new, advanced drug delivery systems, commonly known as controlled release drug delivery systems. Controlled release formulations attempt to alter drug absorption and subsequent drug concentration in blood by modifying the drug release rate from the device. This leads to reduced fluctuations in the plasma drug concentration, sustained drug effects with less side effects, and increased patient compliance. Controlled release products consist of the active agent and the

Table II.  Some Commonly Used Polymeric-Film Coating Materials

| A.  Enteric Coating Materials | Dissolution pH |
|---|---|
| Cellulose acetate phthalate (CAP) | 6.0 |
| Cellulose acetate trimellitate (CAT) | 5.2 |
| Hydroxypropyl methylcellulose phthalate (HPMCP) | 4.5 - 5.5 |
| Poly(methacrylic acid-co-methyl methacrylate) | 5.5 - 7.0 |
| Poly(vinyl acetate phthalate) (PVAP) | 5.0 |
| Shellac (esters of aleurtic acid) | 7.0 |

B.  Non-Enteric Water-Soluble Coating Materials

Alginate, sodium
Gelatin
Pluronic[a]
Poloxamer[a]
Poly(ethylene glycol) (PEG)[a]
Starch derivatives
Water-soluble cellulose derivatives
    Carboxymethylcellulose, sodium salt
    Hydroxyethylcellulose
    Hydroxypropylcellulose
    Hydroxypropylmethylcellulose
    Methylcellulose

SOURCE:  Adapted from ref. 9 and 14.
[a]Used as a co-component in film coating.

polymer matrix or membranes that regulate its release. Advances in controlled release technology in recent years have been possible as a result of advances in polymer science which allow fabrication of polymers with tailor-made specifications, such as molecular size, charge density, hydrophobicity, specific functional groups, biocompatibility, and degradability. Controlled release dosage forms revitalize old drugs by reducing pharmaceutical shortcomings and improving biopharmaceutical properties of the drugs. This is an alternative to the development of new drugs which is extremely costly. The controlled release dosage forms are also important in the delivery of newly developed protein drugs. Currently, most protein drugs are administered by injection. Although protein drugs have exquisite bioactivity, its success in treating chronic illness largely depends on the development of new delivery systems for the routine administration other than injection.

The mechanisms of controlled drug delivery can be classified into the following five mechanisms: (1) diffusion; (2) dissolution; (3) osmosis; (4) ion-exchange; and (5) polymeric prodrug. The schematic descriptions of the release mechanisms are shown in Figure 1. Although any therapeutic system may employ more than one mechanism, there is a predominant mechanism for each delivery system. In all cases, polymers function as a principal component which controls the transport of drug molecules and the way this process is utilized in the device determines the primary mechanism for each drug delivery system.

**Diffusion-Controlled Systems.** In general, two types of diffusion-controlled systems have been used: reservoir and monolithic systems. In reservoir systems, the drug is encapsulated by a polymeric membrane through which the drug is released by diffusion. Polymer films used in the diffusion-controlled reservoir system are commonly known as solution-diffusion membranes (Table III). The polymer membrane can be either nonporous or microporous. Drug release through nonporous membranes is governed mainly by the diffusion through the polymer. Thus, the drug release rate through the solution-diffusion membrane can be easily controlled by selecting a polymer showing desirable drug solubility and diffusivity in the polymer matrix. In case of microporous membranes, which have pores ranging in size from 10 Å to several hundred mm, the pores are filled with a drug-permeable liquid or gel medium. Thus, the diffusion of the drug through the medium in the pores will dominate the drug release process. Microporous membranes are useful in the delivery of high molecular weight drugs such as peptide and protein drugs. One of the recent, commercially available diffusion-controlled reservoir devices is the Norplant system, which is a subdermal contraceptive implant delivering levonorgestrel for more than 5 years. Levonorgestrel is filled inside of the nonporous silicone medical grade tubing with both ends sealed.

In monolithic devices, the drug is dissolved or dispersed homogeneously throughout the water-insoluble polymer matrix. The polymer matrix can also be non-porous or microporous. Unlike the reservoir devices, the drug release from the monolithic devices rarely provide zero-order release. The zero-order release can be achieved by adjusting the physical shape of the device such as a half sphere (*15*).

**Dissolution-Controlled Systems.** Polymers used in the design of dissolution-controlled dosage forms are usually water-soluble, but water-insoluble polymers can be used as long as they absorb significant amount of water and disintegrate the dosage. The dissolution-controlled systems also have both reservoir and matrix dissolution systems. In reservoir systems, the drug core particles are coated with water-soluble polymeric membranes. The solubility of the polymeric membrane, and thus the drug release, depends on the thickness of the membrane and the type of the polymer used. Drug release can be achieved in more controlled fashion by preparing a system with alternating layers of drug and polymeric coats or by

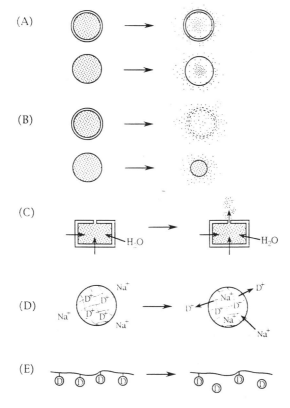

Figure 1. Mechanisms of controlled release drug delivery. Diffusion-controlled (A), dissolution-controlled (B), osmotically controlled (C), ion-exchange controlled (D), and degradation-controlled (polymeric prodrug, E) systems.

preparing a mixture of particles which have different coating characteristics. Contac capsules contain numerous small beads which have different coating thicknesses. Matrix dissolution devices are generally prepared by compressing powder mix of drug and a water-soluble or water-swellable polymer. They can also be made by casting and drying of a polymer solution containing a suitable amount of dissolved or dispersed drug. A variety of other excipients may optionally be included to aid formulation properties.

In an early stage of the dissolution process, the polymer starts swelling as a result of water penetration. During swelling, outer portion of polymer matrix forms a mucilaginous barrier which retards further ingress of water and acts as a rate-controlling layer to drug release. Thus, diffusion of the drug through this barrier also contribute to the drug release rate in addition to the dissolution kinetics of the polymer matrix. Various polymers used in the dissolution-controlled drug release systems are listed in Table IV. The most widely used polymers are cellulose ethers, especially, hydroxypropylmethylcellulose (HPMC) (16). Devices with coarser particles of HPMC result in fast drug release, since they minimize the gelling effect by swelling in water without losing the integrity of the cellulose fibers. On the other hand, finer particles of HPMC hydrate and form gel on the surface fast and this inhibits rapid penetration of water into the dosage form. Thus, the drug release is close to that of the diffusion-controlled devices (17, 18).

The term "dissolution" generally describes the physical disentanglement of polymer chains in the presence of excess water without involving any chemical changes. The cleavage of polymer chains into smaller fragments by either chemical or biological process and subsequent release into the medium is known as (bio)degradation. Biodegradable delivery systems can be considered as a special case of the dissolution-controlled system since they show the same physical phenomenon. Commonly used biodegradable polymers are poly(lactic acid) (19), poly(glycolic acid) (19), polyacrylstarch (20), dextran derivatives (21), polyanhydrides (22), polyamino acids and their copolymers (23), polyorthoesters (24) and polyalkylcyanoacrylate (25, 26). Biodegradable polymers are useful in the development of targetable drug delivery systems to specific cells or organs in the body where the removal of the polymeric systems becomes almost impossible.

**Osmotic Delivery Systems.** The osmotic device comprises a core reservoir of drugs, with or without osmotically active salt, coated with a semi-permeable polymer membrane. The presence of salt or drug molecules creates an osmotic pressure gradient across the membrane and the diffusion of water into the device gradually forces the drug molecules out through an orifice made in the device. The mechanical strength of the semi-permeable membrane should be strong enough to resist the stress building inside the device during the operational life time of the device. The drug release rate from the osmotic devices, which is directly related to the rate of external water diffusion, can be controlled by the type, thickness, and area of the semi-permeable polymer membrane. Table V lists some of the polymers which are used to form semi-permeable membranes. Table V also lists the rate of water vapor transmission (WVTR) through the membranes which can be used as a estimate of water flux rate. Substituted cellulosic polymers, such as cellulose acetate derivatives, have been most commonly used as a semipermeable membrane. Osmotic devices, such as elementary osmotic pump system (OROS system) for oral administration and Alzet osmotic pump for implant, are developed by Alza.

**Ion-Exchange Systems.** The ion-exchange systems are useful in the controlled release of ionic (or ionizable) drugs. Polyelectrolytes are crosslinked to form water-insoluble ion-exchange resins. The drug is bound to the ionic groups by salt

Table III. Some Polymers Used in Diffusion-Controlled Drug Release Formulations

Cellulose
Chitin
Collagen
Nylon
Poly(alkylcyanoacrylate)
Polyethylene
Poly(ethylene-co-vinyl acetate)
Poly(hydroxyethyl methacrylate)
Poly(hydroxypropylethyl methacrylate)
Poly(methyl methacrylate)
Poly(vinyl alcohol-co-methacrylate)
Poly(vinyl chloride)
Polyisobutene
Polyurethane
Silicone rubber

Table IV. Some Polymers Used in Dissolution-Controlled Formulations

Albumin
Carboxymethylcellulose, sodium salt, crosslinked
Cellulose, microcrystalline
Chitosan
Gelatin
Hydroxypropylmethylcellulose
Poly(ethylene glycol)
Poly(glycolic acid)
Poly(lactic acid)
Poly(vinyl alcohol)
Polyvinylpyrrolidone, crosslinked
Starch, thermally modified
Xanthan gum

Table V. Some Polymers Used in Making Semipermeable Membranes

| Film | WVTR[a] |
|------|---------|
| Poly(vinyl alcohol) | 100 |
| Polyurethane | 30-150 |
| Cellulose acetate | 40-75 |
| Ethylcellulose | 75 |
| Cellulose acetate butyrate | 50 |
| Poly(vinyl chloride), cast | 10-20 |
| Poly(vinyl chloride), extruded | 6-15 |
| Polycarbonate | 8 |
| Poly(vinyl fluoride) | 3 |
| Ethylene vinyl acetate | 1-3 |
| Polyesters | 2 |
| Cellophane, polyethylene coated | >1.2 |
| Poly(vinylidene fluoride) | 1.0 |
| Polyethylene | 0.5-1.2 |
| Ethylene propylene copolymer | 0.8 |
| Polypropylene | 0.7 |
| Poly(vinyl chloride), rigid | 0.7 |

SOURCE: Adapted from ref. 10.
[a]Water vapor transmission value expressed in g/100 $in^2$/24 hr/1 mil thick film.

formation during absorption and released after being replaced by appropriately charged ions in the surrounding media.  Cationic drugs form complexes with anionic charges in ion-exchange resin such as sulfonic and carboxylic groups of poly(styrene sulfonic acid) and poly(acrylic acid), respectively.  Hydrogen ions and/or other cations such as sodium or potassium ions activate the release of cationic drugs by replacing them from the drug-resin complex.  For the delivery of anionic drugs, one can utilize cationic ion-exchange resins which contain basic groups such as amino or quaternary ammonium groups of poly(dimethylaminoethyl methacrylate) (27). Sometimes the ion-exchange resins are additionally coated with a polymer film, such as acrylic acid and methacrylate copolymer or ethylcellulose to regulate the swelling of the resin and to further control the drug release (28).  The Pennkinetic system delivers dextromethorphan from the ethylcellulose-coated poly(styrene sulfonate) resins by this mechanism.

**Polymeric Prodrugs.**  Many water-soluble polymers possess functional groups to which drug molecules can be covalently attached.  Polymer backbone, which itself has no therapeutic effect, serves as a carrier for the drug.  The drug molecules are gradually released from the polymer by hydrolytic or enzymatic cleavage.  If the cleavage occurs by chemical hydrolysis, the drug release depends on the nature of the covalent bonds and pH of the environment.  In the body, this occurs very slowly. If the drug molecules are released by enzymatic hydrolysis, the release is mainly dependent on the concentration of enzymes.  Thus, the exact release profile depends on the in vivo condition not on the delivery system itself.  To be a useful drug carrier, a polymer needs to possess certain features (29).  The polymer should remain water-soluble even after drug loading.  The molecular weight of the polymer should be large enough to permit glomerular filtration but small enough to reach all cell types.  The drug-carrier linkages should be stable in body fluid and yet degradable once captured by the target cells.  This can be achieved by making the linkage degradable by lysosomal enzymes (29).  It is preferred that the polymer itself can be degradable by lysosomal enzymes to be eliminated from the body after releasing drugs.  The polymeric carrier, of course, has to be non-toxic, non-immunogenic, and biocompatible.  Poly(hydroxypropyl methacrylamide) (30), PVP, starch derivatives (31), dextran (32), and poly(amino acids) (33) have been used as polymeric drug carriers.

### Polymers for Drug Packaging

Many polymers are used as packaging materials for pharmaceutical products.  The properties of the plastic packaging materials, such as flexibility, transparency, and gas permeability, can be easily controlled.  Flexible packages are made from thin and flexible polymer films.  When they are wrapped around a product, they can easily adapt their shape to conform to the shape of the contents.   The thin, flexible films are usually produced from cellulose derivatives, polyethylene, polypropylene, PVC, polystyrene, polyamide (nylon), polyesters, polycarbonate, poly(vinylidene chloride), and polyurethanes.  These polymeric materials are generally heat sealable and are also capable of being laminated to other materials (34).  A tight package can be prepared by wrapping an article with these polymer films followed by a brief heat treatment.
     Rigid packages, such as bottles, boxes, trays, cups, vials, and various closures, are made from materials of sufficient strength and rigidity.  Widely used polymers are high-density polyethylene, polypropylene, polybutene, poly(vinyl chloride), acrylic copolymers, polycarbonate, nylon, and polyethylene terephthalate (PET). Biodegradable PET is preferred due to environmental concerns but it is expensive. The closure (or cap) of the container is typically made of polypropylene or

polyethylene. The cap is usually lined with moisture-resistant materials such as low-density polyethylene, poly(vinyl chloride), or poly(vinylidene dichloride) (*35*). Most of the containers for parenteral products have a closure made of rubber elastomers. Polyisoprene, ethylene propylene/dicylopentadiene copolymer, styrene/butadiene copolymer, polybutadiene, silicone elastomers, and natural rubber are used as polymeric ingredients of the rubber stopper (*36*). Although the primary polymer component of the end products are the same, their physical, chemical, and mechanical properties of the products can be varied greatly depending on the molecular weight, crystallinity, and alignment of the polymer chains of the polymer component.

One of the important requirements of any packaging material is that it should not release any component into the drug product. Preparation of containers free of any leachables such as monomeric component is especially important for the containers of ophthalmics, parenteral products, and any liquid products. It was shown that di(2-ethylhexyl) phthalate was released from the PVC bags and that caused haziness of the taxol solution (*37*). USP/NF offers the protocol of chemical, spectral, and water vapor permeation tests and tolerances for plastic containers (*38*). Among those, chemical test is designed to give a quantitative assessment of the extractable materials in both organic solvents and water.

When drugs are stored in the polymeric containers or in contact with polymer surfaces, the drug loss by adsorption to the polymer surface should be considered. It was reported that significant portions (ranging from 23% to 55%) of drugs, such as diazepam, isosorbide dinitrate, nitroglycerin, and warfarin sodium, were lost during the 24-hour storage in the PVC bags (*39*). Caution should be exercised in the use of polymeric containers and other devices for the delivery of protein drugs, since proteins readily adsorb and sometimes denature on the hydrophobic polymer surface. It was shown that more than 90% of tissue necrosis factor was lost when delivered using a buret mixing set simply due to the adsorption onto the surface of the delivery device (*40*). It has been proposed that insulin molecules aggregate in the bulk solution as a result of adsorption to the container surface and subsequent denaturation (*41*). Clearly, the loss of protein drugs by adsorption to the polymeric surface can be significant (*42, 43*).

## Conclusion

Polymeric substances are in contact with drugs not only as ingredients in final dosage forms but also as processing aids or packaging materials. In conventional dosage forms, the majority of polymers used as excipients are natural polymers and many are included in the GRAS list as a result of a long history of pharmaceutical marketing. In pharmaceutical packaging, polyethylene, polypropylene, and poly(vinyl chloride) have been used most widely. There is a trend, however, to replace them with more environment-friendly, biodegradable polymers (*44*).

Of the numerous roles played by polymers in the production of pharmaceutical products, emphasis often has been placed on the use of polymers in the fabrication of controlled release drug delivery systems. The progress in the area of controlled drug delivery has been possible only as a result of incorporation of polymer science into pharmaceutics. Development of sophisticated pharmaceutical products requires the multidisciplinary efforts. Polymers with special or multiple properties are need to be developed to achieve self-regulated drug delivery, long-term delivery of protein drugs, and drug targeting to specific organs in the body. These cannot be achieved by elaborate device design alone. These require development of smart polymeric systems which recognize and respond to physiological and pathological processes in the body. Polymers will remain as the indispensable component in the development of new pharmaceutical products.

## Acknowledgments

This study was supported in part by the Zeneca Pharmaceuticals Group.

## Literature Cited

1. Marshall, K.; Rudnic, E. M. In *Modern Pharmaceutics*; Banker, G. S.; Rhodes, C. T. Eds.; Drugs and the Pharmaceutical Scieces; Marcel Dekker Inc.: New York, NY, 1990, Vol. 40; pp 361-374.
2. Schwarz, J. B.; Zelinskie, J. A. *Drug Dev. Ind. Pharm.* **1978**, *4*, 463.
3. USP XXII NF XVII, Reference Tables / Pharmaceutical Ingredients, United States Pharmacopeial Convention, Inc., Rockville, MD, 1990, pp 1857-1859.
4. Hiestand, E. N.; Smith, D. P. *Powder Technol.* **1984**, *38*, 145-159.
5. Lowenthal, W.; Wood, J. A.; *J. Pharm. Sci.* **1973**, *62*, 287-292.
6. Smolinske, S. C. *Handbook of Food, Drug, and Cosmetic Excipients*; CRC Press: Boca Raton, FL, 1992.
7. Pillai, J. C.; Babar, A.; Plakogiannis, F. M. *Pharm. Acta Helv.* **1988**, *63*, 46-53.
8. Porter, S. C.; Bruno, C. H.; Jackson, G. J. In *Pharmaceutical dosage forms*; Lieberman, H. A.; Lachman, L., Eds; Marcel Dekker, Inc.: New York, NY, 1982, Vol. 3; pp 92-117.
9. Thoma, K.; Bechtold, K. Information Brochure, *CAPSUGEL*, 1-16.
10. *Sustained Release Medications*; Johnson, J. C. Ed.; Chemical Technology Review No. 177; Noyes Data Corporation: Park Ridge, NJ, 1980; pp 14-21.
11. Plazier-Vercammen, J.; Van Molle, M.; Steppé, K.; Cherretté, I. *Eur. J. Pharm. Biopharm.* **1992**, *38*(4), 145-149.
12. Shah, N. H.; Stiel, E.; Infeld, M. H.; Railkar, A. S.; Malik, A. W.; Patrawala, M. *Pharm. Tech.* **1992**, *16*, 126-132.
13. Rowe, R. C. *J. Pharm. Pharmacol.* **1981**, *33*, 423-426.
14. Marshall, K.; Rudnic, E. M. In *Modern Pharmaceutics*; Banker, G. S.; Rhodes, C. T. Eds.; Drugs and the Pharmaceutical Scieces; Marcel Dekker Inc.: New York, NY, 1990, Vol. 40; pp 388-397.
15. Chang, N. J.; Himmelstein, K. J. *J. Controlled Rel.* **1990**, *12*, 201-212.
16. Melia, C. D. *Crit. Rev. Ther. Drug Carrier Sys.* **1991**, *8*(4), 395-421.
17. Alderman, D. A. *Int. J. Pharm. Technol. Prod. Manuf.* **1984**, *5*, 1-9.
18. Nakagami, H.; Nada, M. *Drug Design and Delivery* **1991**, *7*, 321-332.
19. Jalil, R. U. *Drug Dev. Ind. Pharm.* **1990**, *16*, 2353-2367.
20. Stjärnkvist, P.; Laakso, T.; Sjöholm, I. *J. Pharm. Sci.* **1989**, *78*, 52-56.
21. Crepon, B.; Jozafonvicz, J.; Chytry, V.; Rihová, B.; Kopecek, J. *Biomaterials* **1991**, *12*, 550-554.
22. Langer, R. *J. Controlled Rel.* **1991**, *16*, 53-60.
23. Shen, W. C.; Du, X.; Feener, E. P.; Ryser, H. J. *J. Controlled Rel.* **1989**, *10*, 89-96.
24. Heller, J.; Maa, Y. F.; Wuthrich, P.; Ng, S. Y.; Duncan, R. *J. Controlled Rel.* **1991**, *16*, 3-13.
25. Couvreur, P.; Fattal, E.; Alphandary, H.; Puisieux, F.; Andremont, A. *J. Controlled Rel.* **1992**, *19*, 259-268.
26. Gaspar, R.; Preat, B.; Roland, M. *Int. J. Pharm.* **1991**, *68*, 111-119.
27. Moldenhauer, M. G.; Nairn, J. G. *J. Pharm. Sci.* **1990**, *79*, 659-666.
28. Chen, Y., *Novel Drug Delivery Systems*; Marcel Dekker, Inc.: New York, NY, 1992; pp 155-157.
29. Lloyd, J. B. In *Drug Delivery Systems, Fundamentals and Techniques*; Johnson, P.; Lloyd-Jones, J. G. Eds.; Ellis Horwood Ltd.: Chichester, UK, 1987; pp 95-105.

30. Seymour, L.; Ulbrich, K.; Strohalm, J.; Kopecek, J.; Duncan, R. *Biochem. Pharmacol.* **1990**, *39*, 1125-1131.
31. Laakso, T.; Stjärnkvist, P; Sjöholm, I. *J. Pharm. Sci.* **1987**, *76*, 134-140.
32. Larsen, C; Jensen, B. H.; Olesen, H. P. *Acta Pharm. Nord.* **1991**, *3*, 41-44.
33. Bennett, D. B.; Li, X.; Adams, N. W.; Kim. S. W.; Hoes, C. J. T.; Feijen, J. *J. Controlled Rel.* **1991**, *16*, 43-52.
34. Griffin, R. C. Jr.; Sacharow, S. *Drug and cosmetic packaging*, Noyes Data Corporation: Park Ridge, NJ, 1975; 136-148
35. Liebe, D. C. In In *Modern Pharmaceutics*; Banker, G. S.; Rhodes, C. T. Eds.; Drugs and the Pharmaceutical Scieces; Marcel Dekker Inc.: New York, NY, 1990, Vol. 40; pp 695-740.
36. *Extractables from elastomeric closures: Analytical procedures for functional group characterization/identification*; Technical Methods Bull. No. 1, Parenteral Drug Association, Inc.: Philadelphia, PA, 1980.
37. Waugh, W. N.; Trissel, L. A.; Stella, V. J. *Am. J. Hosp. Pharm.* **1991**, *48*, 1520-1524.
38. USP XXII NF XVII, General Tests and Assays, <661> Containers, United States Pharmacopeial Convention, Inc., Rockville, MD, 1990, pp 1570-1576.
39. Martens, H. J.; De Goede, P. N.; Van Loenen, A. C. *Am. J. Hosp. Pharm.* **1990**, *47*, 369-373.
40. Geigert, J. *J. Parent. Sci. Tech.* **1989**, 43, 220-224.
41. Arakawa, T.; Kita, Y.; Carpenter, J. F. *Pharm. Res.* **1991**, *8*, 285-292.
42. Tarr, B. D.; Campbell, R. K.; Workman, T. M. *Am. J. Hosp. Pharm.* **1991**, *48*, 2631-2634
43. D'Arcy, P. F. *Drug Intell. Clin. Pharm.* **1983**, *17*, 726-731.
44. Guise, B. *Manuf. Chem.* **1991**, June, 26-29.

RECEIVED June 2, 1993

Chapter 2

# Tissue Engineering Using Synthetic Biodegradable Polymers

C. A. Vacanti[1], J. P. Vacanti[2], and R. Langer[3]

[1]Department of Orthopaedic Surgery and Anesthesiology, Massachusetts General Hospital and Harvard Medical School, Boston, MA 02115
[2]Department of Surgery, Children's Hospital and Harvard Medical School, Boston, MA 02115
[3]Department of Chemical Engineering, Massachusetts Institute of Technology, Boston, MA 02139

A new approach to create new animal tissues utilizing biodegradable polymer is presented. The use of this approach to create cartilage, liver and other tissues are reviewed.

Murray and collaborators performed the first successful human organ transplant, that of a kidney from an identical twin into his brother(1). One of the most formidable obstacles to whole organ transplantation is the immunologic attack of the allograft. This biologic barrier was again breeched in humans at the Peter Bent Brighham Hospital under Murray's direction in the early 1960s(2). With the success of transplantation came explosive growth, and a new major problem has emerged, that of organ scarcity. Technical difficulty, complex labor-intensive care, and expense however, presented significant barriers to whole or partial organ transplantation. In 1985, Russell stated that if one could effectively replace only those important functional cellular elements of an organ, there would be many conceptual advantages over organ transplantation(3). In the ideal situation a cell sample could be obtained from the afflicted individual and the particular enzymatic defect might be corrected with genetic

0097–6156/94/0540–0016$06.00/0
© 1994 American Chemical Society

engineering. The disease free cells could then be grown in vitro and allowed to multiply. After a sufficient number of cells had been obtained, the cells could then be transplanted back into the original donor, and thus avoid problems associated with immunogenicity. This technology could ensure viability and function of the transplanted cells before the native organ was removed. In many instances, the native organ might be used to maintain other organ functions as long as a sufficient number of functioning cells survive to overcome the specific defect occurring in the individual.

**A. Cell Transplantation** There have been many attempts to replace organ function with cell transplantation. Islet cell and hepatocyte transplantation have been attempted for almost 20 years. Initial studies focused on the concept of injecting dissociated cells into other tissues, such as liver(4), spleen(5-7), or subcutaneous tissue(8-10). More recent studies have focused on using natural extracellular matrix (ECM) proteins such as collagen as structural supports for cell transplantation. Yannas and Burke designed an artificial dermis from collagen and glycosaminoglycans(11). Others created new blood vessels(12) and cartilage(13,14). Poor mechanical properties and variable physical properties with different sources of the protein matrices have hampered progress with these approachs. Concerns have recently arisen regarding immunogenic problems associated with the introduction of foreign collagen. Also, there are inherent biophysical constraints in collagen used as scaffolding. For example, cartilage grown from cells seeded on collagen must be confined in a rigid "well" such as underlying bone or cartilage, because of collagen's non-moldability. Demetriou et. al.(15,16) employed a hybrid matrix of inert plastic microcarrier beads coated with ECM proteins to transplant hepatocytes. Cells on beads were injected as a slurry into animals and were able to replace enzyme function in metabolically deficient rats.

This paper focuses on synthetic biodegradable polymers and their use in creating or engineering new tissue. As far back as the early 1970s, Greene predicted that with the advent of new synthetic biocompatible materials, transplantation of cells on such a material might result in the formation and successful engraftment of new functional tissue(17). Advances in material sciences have now enabled the production of biocompatible as well as biodegradable materials. We postulated that the synthetic polymer scaffolds might act as cell anchorage sites and give the transplanted complex intrinsic structure.

Employment of synthetic rather than naturally occurring polymers would allow exact engineering of matrix configuration so that the biophysical limitations of mass transfer would be satisfied. Synthetic matrices would also give one the flexibility to alter physical properties and potentially facilitates reproducibility and scale-up. The configuration of the synthetic matrix could also be manipulated to vary the surface area available for cell attachment as well as to optimize the exposure of the attached cells to nutrients. The chemical environment surrounding a synthetic polymer can be affected in a contolled fashion as the polymer is hydrolyzed. The potential exists to continuously deliver nutrients and hormones that can be incorporated into the polymer structure(18).

Much became known about the chemical environment surrounding these compounds as they degraded. The surface area to mass ratio can be altered or the porosity of differing configurations can be changed. The pore size of polymers can be altered to increase or decrease the intrinsic strength and elasticity of the polymer matrix, as well as compressability or creep recovery. We can also change the rate of degradation of the polymer matrices and the environment into which the cells are implanted by systematically altering the surface chemistry of the polymers, creating an acidic or basic environment as they degrade.

For instance, polyiminocarbonates cause a local basic environment, while polyglycolic acid and polyanhydrides cause a local acidic environment. Polyanhydrides and polyorthoesters show surface erosion; other polymers show bulk erosion. By manipulating design configurations one can increase and decrease the surface area of the polymers. Also, synthetic polymers offer the advantage of being able to be consistently reproduced, and thus varying quality is not a problem as it sometimes is with naturally occurring polymers. With synthetic polymers, one has the potential of adding side chains to the polymer structure. Thus one might potentially deliver nutrients and hormones to the cells as the polymer breaks down.

In our laboratory, we use synthetic biodegradable polymers as templates to which cells adhere and are transplanted. The polymers act as a scaffolding which can be engineered to allow for implantation of transplanted cells within only a few cell layers from the capillaries, and thus allow for nutrition and gas exchange by diffusion until successful engraftment is achieved. In this manner, we hoped to generate permanent functional new tissues composed of donor cells and recipient interstitium and blood vessels.

We have had encouraging results with several cell types: osteocytes, chondrocytes, hepatocytes, as well as intestinal mucosal and urothelial cells. Although many of the issues associated with cell transplantation are the same, unique problems associated with each cell type are very different. For example, while cartilage consists of only one cell type, the chondrocyte, the liver has several cell types organized in defined spatial relationships with each other. In any cell transplantation system, it is necessary to provide the right conditions for cell survival, differentiation, and growth.

Our approach was based on several observations:  1) Although isolated cells remodel to only a limited degree when placed as a suspension into the midst of mature tissue without intrinsic organization, they will however reform appropriate tissue structure if proper conditions are provided. For example, mammary epithelial cells will form acini which secrete milk(19), and capillary endothelial cells tend to form tubular structures when placed on the properly adhesive substratum in vitro(20). 2) Tissues undergo constant remodeling due to attrition and remodeling of constituent cells. 3) Transplanted cells will survive only if implanted within a few hundred microns from the nearest capillary(21).

We therefore undertook the development of a system whereby we could deliver a high cell density into a recipient, anchored to suitable template to provide intrinsic organization. This cell-template complex had to be transplanted into an animal as a unit with physical characteristics that placed each cell only a few cell layers from an intact capillary system to provide for adequate nutrition, and elimination by diffusion of waste until successful engraftment was accomplished. The template had to be easy to work with, have structural integrity and provide a very large surface area for cell attachment. There needed to be sufficient space to allow for cell proliferation. We felt that the optimal system would not only be biocompatible, but would also be biodegradable and thus leave only functional tissue as the synthetic polymer degraded and was replaced by natural matrix. The templates also had to have reproducible porosity, pore size distribution, and degradation rates.

**B. Cartilage and Bone.**  Cartilage was the first new functional tissue to be successfully engineered in our laboratory. Cartilage contains only one cell type, chondrocytes, which can be easily isolated in large numbers and are quite viable. Chondrocytes have a very low oxygen requirement, approximately 1/50 that of hepatocytes, and, *in vivo,* are nourished by diffusion, as there is no vascularization of cartilage. They also can be harvested from an animal for several days after sacrifice, and they readily multiply *in vitro*. We have demonstrated that chondrocytes are capable of multiplying and maintaining functional activity after being stored in a refrigerator at $4^0$ C in an appropriate culture media for up to 30 days. These properties suggested to us that chondrocytes would be an ideal candidate for transplantation and survival by diffusion until successful engraftment was achieved.

Cartilage is also a very important tissue to create since, like many other organ systems, it has a limited capacity to regenerate. Until recently, the primary approaches for repair of cartilage defects, were either synthetic material utilization (prosthetic devices), or cartilage transplants. A significant amount of adjacent normal cartilage and underlying bone must be excised to replace diseased cartilage with prosthetic material. The interface of the underlying bone and prosthetic material also presents a problem. Although various adhesives and techniques have been utilized, in time, all of the interfaces break down and the prosthesis loosens. Generally occurring at the same time, there is further resorption of underlying bone, and thus even more loss of unreplaceable tissue.

Another potential consequence of a synthetic prosthesis is the ever present possibility of infection. This generally results in loss of the prosthesis, and thus function of the joint. Although cartilage is considered to be an immuno-privileged tissue, in time, with the stress of weight bearing and mechanical insults, cells are eventually exposed to circulating antibodies like any other non-autologous tissue. This results in tissue rejection with cell death and loss of function.

In our laboratory we began transplanting chondrocytes seeded in vitro onto branching networks of unwoven synthetic biodegradable polymer fibers 14-15 microns of polylactic and polyglycolic acid. Bioabsorbable, biocompatible synthetic polymers of polyglactin 910 (Vicryl, Ethicon Inc.) and polyglycolic acid (Dexon, Davis & Geck) configured as fibers or as foams of varying porosity, have been seeded with chondrocytes and transplanted into animals.

We initially harvested cartilage from the articular surfaces of surfaces of newborn calf shoulders. Chondrocytes were isolated from the harvested cartilage under sterile conditions using a technique described by Klagsbrun(22). Next, braided threads of the biodegradable co-polymer polyglactin 910 (Vicryl, Ethicon Inc.) or polyester, a non-biodegradable polymer (Ethibond, Ethicon Inc.), were cut into pieces of approximately 17 mm in length. One end was knotted to prevent total disintegration of the fibers, while the other end was unbraided to expose multiple fibers, approximately 14 microns in diameter. Our preliminary work with chondrocytes and polymer fibers suggested that the density of cellular attachment was optimized when the distance between polymer fibers was on the order of 100-200 microns. Using the above configuration, we were able to create varying inter-fiber distances while keeping the unit intact. Additionally, an embossed non-woven mesh of polyglycolic acid with interfiber interspaces averaging 75 - 100 microns (Dexon, Davis & Geck) was divided into pieces of approximately 5mm X 15mm. One to two hundred microliters of cell suspensions containing up to $10^7$ chondrocytes have been placed on polymer complexes (experimental) in parallel with polymers incubated in a solution free of chondrocytes (controls). These constructs have been studied in vitro in 5% $CO_2$ at $37^0$ C for up to six weeks, and evaluated for the presence and the morphologic appearance of chondrocytes using phase contrast microscopy. They were then evaluated using hematoxylin and eosin stains and an aldehyde fuchsin-alcian blue stain.

Several hundred cell-polymer constructs have been implanted into animals. We have followed the fate of these implants for the formation of new cartilage. Initially, cell-polymer constructs were surgically implanted subcutaneously in nude mice for varying periods of time of up to one year. Specimens were excised and evaluated histologically using hematoxylin and eosin stains. Aldehyde fuchsin-alcian blue stains were also used for confirmation of strongly acidic sulfate mucopolysaccharides, i.e., chondroitin sulfates, of cartilage. In addition, several mice received subcutaneously injections in the same region with 200 ul suspensions, each containing 5 x $10^5$ chondrocytes, without attachment to polymers.

Using this system, we found several interesting results(23). We found that in culture the chondrocytes appeared to readily adhere to branching polymer fibers 14-15 microns in diameter in multiple layers, retain a rounded configuration and bridge small distances between fibers (**Fig. 1**). This is in contrast to standard culture systems of chondrocytes, where they grow in a monolayer, and become flattened, thereby lacking the ability to perform their differentiated function. Hematoxylin and eosin (H and E) staining of the experimental specimens *in vitro* for at least 18 days demonstrated a basophilic matrix. Aldehyde fuschin-alcian stains of the same specimens indicated the presence of chondroitin sulfate. In contrast, histologic evaluation using H and E stains showed no evidence of chondrocytes or chondroitin sulfate in the *in vitro* control fibers.

All polymer fibers *in vitro* (controls and experimentals) began to dissolve by 4 weeks, which is comparable to their anticipated time for dissolution. When evaluating implants from *in vivo* experiments, on gross examination, the experimental biodegradable polymer fibers were progressively replaced by cartilage, until only cartilage with very little evidence of polymer remained(**Fig. 2**). Histologic examination of these specimens using H and E stains revealed evidence of cartilage formation (**Fig. 3**) in greater than 90% of the experimental bioabsorbable implants with all specimens having been implanted for at least 4 weeks appearing very similar in appearance to normal human fetal cartilage. In contrast, less than two thirds of the polyester control implants resulted in cartilage formation(24). Aldehyde fuschin-alcian staining of the experimental specimens suggested the presence of chondroitin sulfate.

The size of the cartilage formed in the experimental bioabsorbable implants increased over a period of about 7 weeks after which time it remained stable. The growth appeared histologically to be at the expense of the fibrous tissue initially seen and associated at least temporally with a decrease in neovascularization and resolution of the mild inflammatory response originally noted. There was a decrease in inflammatory response, as evidenced by decreases in the number of polymorphonuclear leukocytes and giant cells, which correlated with the disappearance of the polymers. Very little evidence of either inflammatory response or polymer remnants were seen after 7 weeks when using bioabsorbable material, while moderate inflammation and minimal cartilage formation was noted in the nonabsorbable implants after this period of time.

Implantation of polymer fibers not seeded with chondrocytes or injection of chondrocytes alone did not result in cartilage formation grossly or histologically using the hematoxylin and eosin stain. These studies suggested that it was possible to grow chondrocytes *in vitro* on biocompatible polymer fibers in a cell density sufficient to allow implantation of the cell-polymer construct into animals with successful engraftment and formation of new cartilage. Nonabsorbable polyester fibers function less satisfactorily for this purpose. There was a decrease in neovascularization associated with the maturation of the cartilage formed which may reflect the production of an angiogenesis inhibitory factor by the newly formed cartilage. We then undertook studies to determine the usefulness and ultimate fate of cartilage engineeered in this fashion.

In New Zealand White Rabbits, we demonstrated this technology to be useful in resurfacing the joint surfaces surgically denuded of articular cartilage(25). Hyaline cartilage was harvested from the articular cartilage on the joint surface of the distal femur of New Zealand White Rabbits. Chondrocytes were isolated as described above and seeded onto synthetic sheets of polyglycolic acid. The distal femoral grooves of the contralateral knee were then surgically denuded

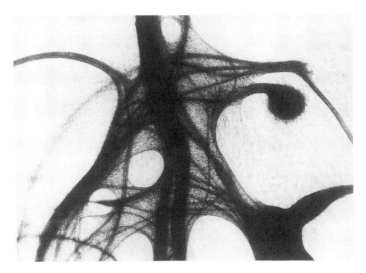

Figure 1. Photomicrograph (4X) of polymer construct seeded with chondrocytes. Note the rounded configuration of the cells and the cell attachment to the polymer in multiple layers.

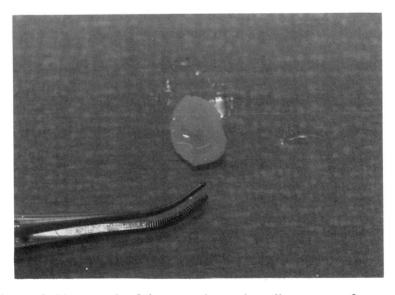

Figure 2. Photograph of tissue engineered cartilage grown from synthetic biodegradable polymer mesh seeded with chondrocytes and then implanted for 7 weeks in a nude mouse.

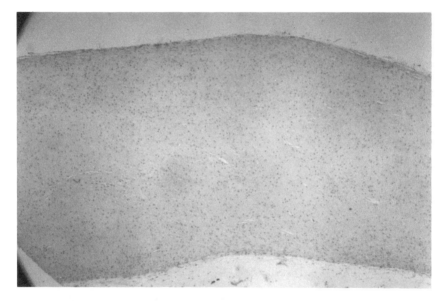

Figure 3. Photomicrograph demonstrating the histology using hematoxylin and eosin stains of the new cartilage.

of their articular cartilage and defects covered with cell-polymer constructs, blank polymers, or nothing at all.

Seven weeks later, all joint surfaces were examined grossly and histologically. Joint defects covered with cell-polymer constructs showed evidence of new cartilage formation, while there was very little evidence of repair in control defects. Prior to implantation, rabbit chondrocytes were labelled in vitro with a thymidine analogue, BrdU. That the cartilage present in the defects repaired with cell-polymer constructs was generated by the implanted cells as evidenced by the presence of BrdU labelled chondroctyes in the new cartilage. Experiments implanting chondrocytes isolated from different sources, i.e. hyaline cartilage, fibrocartilage in ribs, and fibroelastic cartilage of the ear, have demonstrated resultant growth of cartilage. We believe this is secondary to loss of cell types other than chondrocytes during our isolation techniques.

The generation of human cartilage has also been acheived in our laboratory. We have also shown that cartilage generated in this fashion is useful in the repair of large bone defects made in the craniums of rats. Of interest is the fact that cartilage created in a bone defect remains cartilage without undergoing morphogenesis to bone. Several of our studies have demonstrated that it is possible to grow tissue engineered cartilage in predetermined shapes, such as the shape of a human ear (26). We are also currently employing these techniques in the hopes of generating cartilagenous cylinders lined with tracheal epithelial cells for use in tracheal reconstruction.

**C. Bone**  Applying concepts derived from our experience with chondrocytes, we began to transplant similar polymer matrices seeded with osteocytes or osteoblasts isolated from different sources. Initial efforts at harvesting osteocytes from cortical bone, via a series of enzymatic digestions, were not fruitful. We then began to use cells shed from first specimens of cortical or membranous bone, and finally from the periosteum stripped from cortical and membranous bone.

Shed cells were nourished in Tissue Culture Media 199 (Gibco) with 10% fetal calf serum  and 5 ug/ml ascorbic acid with L-glutamine (292 ug/cc), penicillin (100 U/cc), streptomycin (100 ug/cc), ascorbic acid (5 ug/ml) and calciferal (40 ng/cc) was added to each well. The specimens were then allowed to incubate in vitro at $37^0$ C  in the presence of 5% $CO_2$ for approximately 2 weeks, until they multiplied to form a monolayer on the bottom of each well. At that time, a non woven mesh approximately 100 microns thick, composed of polymer fibers of polyglycolic acid 15 microns in diameter was cut into pieces  2 centimeters square. One polymer fiber was wiped on the bottom of each well, and the cells allowed to adhere to it. The cell polymer constructs were then kept in the incubator for an additional week to ten days, until most of the strands were coated with a multiple cell layer of osteocytes.

Supernatant from the culture media was analysed for the presence of osteocalcin, to confirm the presence of functioning osteoblasts, and cells from representative constructs were counted. We discovered that cells shed from periosteum multiplied well in vitro. Immunohistochemical techniques demonstrated osteocalcin in the supernatant of culture dishes containing perosteal cells, confirming the presence functioning osteoblasts in our cultures. Implantation of these cell-polymer constructs resulted in some significant findings. All implants containing periosteal cells appeared to result in the generation of hyaline cartilage in the first few weeks after implantation. This was determined by gross examination and histologically using hematoxylin and eosin stains. In time, all specimens matured to form new bone, containing the cellular elements of bone marrow (**Fig. 4, 5**) (27).

The rate of the morphogenesis from cartilage to bone was dependent on the site into which the cell-polymer construct had been implanted, and seemed to be proportional to the vascularity of the site. In contrast, polymers that had been seeded in vitro with chondrocytes remained cartilage, with no evidence of progression to bone, regardless of the site or duration of time in which they had been implanted.

**D. Hepatocyte transplantation**     Our initial polymer design was that of a small wafer of biodegradable polyanhydride. Hepatocytes were seeded in a monolayer onto the wafer in culture and then placed into the recipient animal while on the disc. We found that the cell number and cell density were inadequate for reliable successful engraftment when using this polymer design. Our preliminary studies indicated that for adequate hepatic function, we would need to implant at least 10% of the number of cells found in a normal liver(28). We estimate this number to be $10^9$ cells in a child.

To achieve delivery of this high cell number in close proximity to a capillary network, we needed to provide a scaffold with an immense surface area. In this context, we began to address the question of growth of multicellular organisms. The surface area of a mass of cells increases as the square of the radius, while the volume increases as the cube of the radius. In experiments involving several hundred animals, we employed branching networks of biocompatible, biodegradable polymers of several chemical compositions and physical characteristics as matrices onto which we seeded hepatocytes.

Figure 4. Photograph of tissue engineered bone after being implanted for 16 weeks in a nude mouse.

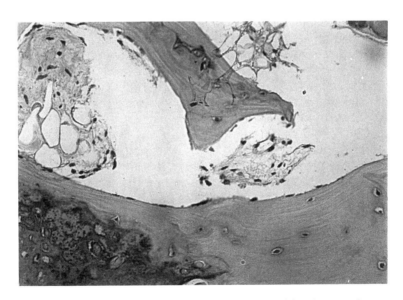

Figure 5. Photomicrograph demonstrating the histology using hematoxylin and eosin stains of the new bone. Note the trabeculated structure of the organized bone, and the presence of a hypocellular bone marrow.

Polyorthoesters, polyanhydrides, and combinations of polyglycolic and polylactic acid were all tested. Again, our most satisfactory results were obtained initially when the polymers consisted of fiber networks with a fiber diameter of 14-15 microns. We began to identify features of a substrate to which the cells were anchored which were important for maintaining cell function, and incorporate them into the polymer matrix used for transplantation.

It is known that when hepatocytes are freshly isolated and cultured under conventional conditions on plastic or collagen coated dishes, gene transcription is drastically depressed and liver specific mRNA declines while the mRNA of structure related genes increases many fold. By contrast, hepatocyte culture on extracellular matrix (ECM), rich in laminin, type IV collagen, and heparin sulfate proteoglycan exhibits increased longevity and maintenance of several liver-specific functions.

We have further characterized hepatocyte behavior on extracellular gel matrix and found that, unlike conventional cultures, the capability of gene transcription persists(29). Initial studies showed that hepatocyte shape and function could be controlled by varying the coating density of any of a variety of different naturally occurring extracellular matrix (ECM) substrates on bacteriologic plates(30).

Hepatocytes cultured on low density ECM exhibited a rounded morphology, while there was an increase in hepatocyte spreading and epithelial like morphology as ECM coating density increased. The ECM coating density also affected hepatocyte ability to enter the synthetic phase (S) and continue through the cell cycle, exhibiting the S phase in none of the low density ECM studies, while entering the S phase in 60% of the high density ECM studies(31).

The opposite was found when effects on differentiated function were studied. These results indicated that hepatocytes could be switched between programs of growth and differentiation simply by modulating the ECM coating density, thereby altering the substratum's ability to resist cell-generated mechanical load. These and other studies also suggest that cell shape may be important in determining cell function.

We felt that the physicochemical properties of the polymer might therefore be manipulated to alter the cell shape and thus the cell physiology. We next studied the effects of polymers of different physical and chemical configuration on cell attachment, viability and performance of differentiated function. The ability of the hepatocyte to maintain differentiated function was assessed by the rate of albumin secretion. We found that a suitable polymer was an uncoated 85:15 combination of polylactic/polyglycolic acid. This suggested to us that a simple substrate, unmodified by exogenous peptides or proteins, might fulfill the criteria for cell transplantation.

These findings have led us to conclude that the matrix to which the cells are attached in vitro would be one of the most important variables of our system as we continued to search for the ideal polymer matrix. The hepatocyte-polymer scaffolds, created as described above, were then implanted into the omentum, the interscapular fat pad, or the mesentery of the recipient animal for varying periods of time. We avoided surgical trauma as much as possible, because any fibrin clot or hematoma formation would decrease the ability to nourish the cells by diffusion. The rat was used as the animal model for hepatocyte transplantation, because there existed several models of metabolic deficiency states representative of human liver disease.

Routine histologic staining techniques (H&E) were used to evaluate the success of the transplants. Also Fisher rats were used for immunostaining for hepatospecific function, such as albumin secretion. Utilizing this system, we found that there was a large loss of cells in the six hours after implantation. However, clusters of hepatocytes remained adjacent to native tissue and blood vessel ingrowth, and retained differentiated function for up to six months.

We also performed studies where we transplanted hepatocytes from healthy Wistar rats into the Gunn rat model of UDP glucuronyl transferase deficiency, the enzyme needed for bilirubin conjugation. We have observed decreases in serum bilirubin levels of up to 50% and appearance of conjugated bilirubin products in the native bile up to nine weeks post transplantation, indicating metabolic activity in the transplanted cells. While promising, these results were difficult to consistently reproduce, and successful hepatocyte transplantation was limited by cell injury and hypoxia at the time of transplantation. Cell death prevented consistently adequate cell mass engraftment to replace function.

Since hepatocyte transplantation requires a highly vascularized environment for cell survival, we developed polymeric foams of polyvinyl alcohol and poly (lactic-co-glycolic acid) to induce vessel ingrowth into a noncompressable three dimensional matrix into which the cells are injected at a later time. The porosity and pore area per unit volume were tailored to allow tissue growth and maintain adequate space. Tissue ingrowth with a high capillary density occurred at a rate of 200um/day, and potential space for cell injection was maintained between the new tissue and polymer. We studied cell survival in foam discs with pore sizes of up to 200 microns, and porosity values up to 0.95. Porous structures with indwelling silastic cathethers were then placed into the mesentery of Fisher 344 rats. Hepatocytes for implantation were harvested from syngeneic donors. Between $5 \times 10^6$ and $1 \times 10^7$ cells were injected into the discs at $10^7$ cells/cc of Wm E medium after 5 days of prevascularization. Results have now been followed for 2 months.

Hepatocyte function has been quantitated by histology, morphometric analysis using an Image Technology 3000 Image analyzer and Northern slot blot analysis for albumin specific mRNA of implants. Sixty percent engraftment was seen by this technique compared to three percent engraftment seen before we used this approach(32).

It is clear from the transplant literature that graft survival requires portal blood flow and the hepatotrophic effect of hepatectomy is well established(33). To take full advantage of circulating hepatotropic factors, we created portocaval shunts (PCS) in the recipient rats to eliminate first pass clearance of these factors by the native liver. We therefore repeated the studies with prevascularized polymeric foams, but additionally created a PCS in one group of animals, a 70% hepatectomy in another group of animals, and both a PCS and hepatectomy in yet another group studied.

We found that the addition of either PCS or partial hepatectomy increased cell survival. The group of animals that received both a PCS and partial hepatectomy demonstrated a marked increase in overall cellularity with acinar and tubule formation evident. The increase in cellularity was evaluated using image analysis and morphometric quantification and demonstrated a 1-2 fold increase with hepatectomy alone, and a sixfold increase when hepatectomy was combined with PCS. We have made substantial progress in our efforts to restore liver function with our approach; however, to apply this technology as replacement therapy we may need to expand the cell population in vivo. We hope to stimulate transplanted cells with controlled release of hepatotrophic factors at the implantation site using previously described techniques (34).

**E. Endothelial cells: Urethelium, Intestinal Mucosa, and Tracheal Epithelium.** Urothelial cells obtained from New Zealand White Rabbits and Fisher syngeneic rats have been seeded onto polyglycolic acid in vitro and implanted after 1-4 days into the omentum or mesentery of host animals(35). These cell-polymer constructs were either implanted as flat sheets, or as tubes rolled around silicone cores.

After being implanted for up to 21 days, specimens were excised and evaluated. After 5 days, urothelial cells were seen randomly arranged in the polymer. After 10 days, single cell layers lined the polymers. After 15 days, urothelial cells lined the polymers in a 2 - 3 cell layer thickness. Cell continuity was seen at 20 days, and polymer degradation observed. This was true for both the tubular and flat polymers.

The epithelial origin of these cells was confirmed immunohistologically with AE1/AE3 anti-keratin antibodies. These findings suggest that it may be feasible to use reconfigured autologous urothelium in reconstructive procedures involving the ureter, bladder and urethra.

Similar studies using intestineal mucosal cells (36) and tracheal epithelial cells resulted in the formation of tubular conduits lined with these epithelial cells.

## F. Conclusions:

The potential exists to further manipulate the physical and chemical properties of synthetic polymers in order to optimize this system of cell delivery. With regard to polymer chemistry and design, there is already extensive data concerning the safety and biocompatability of many synthetic polymers. Our reseach is currently focused on the organ systems discussed above. It is our hope to develop a system to create functional tissue as an alternative to conventional organ transplantation and tissue reconstruction. Current advances in polymer design are allowing us to seed cells into polymers that can be molded into specific shapes such as a human ear, that have intrinsic flexibilty. With further study, this approach and related approaches may ultimately allow replacement therapy in many organ systems.

## References,

1. Murray J E, Merrill J P, Harrison J H. Renal homotransplantation in identical twins. Surg Forum **1955**; 6: 432-436.
2. Couch N P, Wilson R E, Hager E B, Murray J E. Transplantation of cadaver kidneys: experience with 21 cases. Surgery. **1966**; 59(2): 183-8.
3. Russell P S. Selective Transplantation. Ann Surg **1985**; 201:255-62.
4. Matas AJ, Sutherland DER, Steffes MW, Mauer SM, Lowe A, Simmons, RL, and Najarian JS. Hepatocellular transplantation for metabolic deficiencies: decrease of plasma bilirubin in Gunn rats. Science **1976**; 192: 892.
5. Vroemen JPAM, Buurman WA, Schutte B, Maessen JG, Van der Linden CJ, and Koostra G. The cytokinetic behavior of donor hepatocytes after syngeneic transplantation into the spleen. **1988**; Transplantation 45: 600-7.

6. Mito M, Ebata H, Kusano M, Onishi T, Saito T and Sakaamoto S. Morphology and function of isolated hepatocytes transplanted into rat spleen. **1979**; Transplantation 28: 499-505.
7. Maganto P, Traber PG, Rusnell C, Dobbins WO III, Keren D, Gumucio JJ. Long-term maintainance of the adult pattern of liver-specific expression for P-450b, P-450e, albumen, and alphafeto-protein genes in  intrasplenically transplanted hepocytes.**1990**; Hepatology 11: 585-93.
8. Gould MN, Biel WF, and Clifton KH. Morphological and quantitative studies of gland formation from innocula of mono-dispersed rat mammary cells. **1977**; Exp Cell Res. 107: 405-416.
9. Jirtle RL, Biles C, and Michalopoulos G. Morphologic and histochemical analysis of hepatocytes transplanted into syngeneic hosts. **1980**; Amer. J Path. 101: 115-26.
10. Strom SC, Jirtle RL, Jones RS, Novicki DL, Rosenberg MR, Novotny A, Irons G, McLain JR, and Michalopoulos G. Isolation, culture and transplantation of human hepatocytes. **1982**; J Nat Canc Inst 68: 771-78
11. Yannas IV. Regeneration of skin and nerve by use of collagen templates. **1988**; Collagen III, ME Nimni, ed., CRC Press, Boca Raton, FL.
12. Weinberg CB, and Bell E. A blood vessel model constructed from collagen and cultured vascular cells. **1986**; Science 231: 397-400.
13. Grande DA, Pitman ML, Peterson L, Menche D, Klein M. The repair of experimentally produced defects in rabbit articular cartilage by autologous  chondrocyte transplantation. **1989**; J Orthop Res. 7: 208 .
14. Wakitani S, Kimura T, Hirooka A, Ochi T, Yoneda M, Yasui N, Owaki H, Ono K. Repair of rabbit articular surfaces with allograft embedded in collagen gel. **1989**; J Bone Joint Surg. [Br] 63, 529.
15. Demetriou AA, Reisner A, Sanchez J, Levenson SM, Moscioni AD, and Roy-Chowdhury J. Transplantation of microcarrier-attached hepatocytes into 90 % partially hepatectomized rats. **1988**; Hepatology 8: 1006-1009.
16. Moscioni AD, Roy-Chowdhury J, Barbaour R, Brown LL, Rot-Chowdhury N, Compeetiello PL, and Demetriou AA. Human liver cell transplantation: prolonged function in athymic Gunn and athymic analbumenemic rats. **1989**; Gastroent. 96: 1546-1551.
17. Green Jr. W.T. Articular cartilage repair: Behavior of rabbit chondrocytes during tissue culture and subsequent allografting. **1977** Clinical Orthopaedics and Related Research  124, 237.

18. Langer, R New Methods of Drug Delivery. **1990**; Science 249: 1527-33.

19. Bissell MJ, and Barcellos-Hoff MH. The influence of extracellular matrix on gene expression: is structure the message? **1987**; J Cell. Sci., Suppl., 8: 327-43.

20. Folkman J, and Haudenschild C. Angiogenesis in vitro. **1980**; Nature 288: 551-6.

21. Folkman J, and Hochberg MM. Self-regulation of growth in three dimensions. **1973**; J Exper. Med. 138: 745-753.

22. Klagsbrun M. Large-scale preparation of chondrocytes. **1979**; Methods in Enzymology 58: 560-4.

23. Vacanti CA, Langer R, Schloo B, Vacanti JP. Synthetic Biodegradable Polymers Seeded with Chondrocytes Provide a Template for New Cartilage Formation In Vivo. Plast Reconstr Surg **1991**; 87: 753-9

24. Vacanti CA, Langer R, Schloo B, Vacanti JP. Engineered new cartilage formation from cell-polymer constructs: A comparison of biocompatible materials. Transactions of the 36th Annual Meeting, Orthopoedic Research Society. **1990**; 15: 25.

25. Vacanti CA, Schloo B, Vacanti JP. Joint Relinement with Cartilage Grown from Cell-Polymer Constructs. Annual Meeting of the American Orthopaedic Society for Sports Medicine, (Abstr.) **1992**.

26. Vacanti CA, Cima L, Ratkowski D, Upton J, Vacanti JP. Tissue Engineering of New Cartilage in the Shape of a Human Ear Employing Specially Configured Synthetic Polymers Seeded with Chondrocytes. Materials Research Society Symposium Proceedings. **1992**; 52: 367-73.

27. Vacanti CA, Kim WS, Upton J, Vacanti MP, Mooney D, Schloo B, Vacanti JP. Tissue Engineered Growth of Bone and Cartilage. Transplantation Proceedings, **1993**; (In Press).

28. Asonuma K, Gilbert JC, Stein JE, Takeda T, and Vacanti JP. Experimental quantitation of transplanted hepatic mass necessary to cure metabolic deficiency. J Ped. Surg.**1992**; 27(3): 298-301.

29. Bucher NLR, Aiken J, Robinson GS, Vacanti JP, Farmer SR. Transcription and translation of liver specific cytoskeletal genes in hepatocytes on collagen and EHS tumor matrices. **1989**; American Assoc. for the Study of Liver Disease

30. Mooney DJ, Langer R, Vacanti JP, Ingber DE. Control of hepatocyte function through variation of attachment site densities. **1990**; (Abstr.) American Institute of Chemical Engineers.

31. Mooney DJ, Hansen LH, Vacanti JP, Langer RS, Farmer SR, Ingber DE. Switching between growth and differentiation in

hepatocytes: control by extracellular matrix. 1990; J Cell Physiology. **1992**; 151: 497-505.

32. Stein JE, Gilbert JC, Hansen LK, Schloo B, Ingber D, Vacanti JP. Hepatocyte transplantation into prevascularized porous matrices. J Ped Surg **1991**; (Abstr.)

33. Niiya M, Hasumura S, Nagamori S. Consideration on the mechanism of liver regeneration on experimental results. Hum Cell. **1991**; 4 (3): 222-9.

34. Murray JB, Brown L, Langer R, and Klagsbrun M. A microsustained release system for epidermal growth factor. **1983**; In Vitro 19: 743-748.

35. Atala A, Vacanti J P, Peters C A, Mandell J, Retik A B, Freeman M R. Formation of urothelial structures in vivo from dissociated cells attached to biodegradable polymer scaffolds in vitro. J Urol. 1992; 148(2): 658-62.

36. Organ G M, Mooney D J, Hansen L K, Schloo B, Vacanti J P. Transplantation of enterocytes utilizing polymer-cell constructs to produce a neointestine. Transplant Proc. **1992**; 24 (6): 3009-11.

RECEIVED August 25, 1993

# Chapter 3

# Interfacial Biocompatibility

Yoshito Ikada

Research Center for Biomedical Engineering, Kyoto University,
53 Kawahara-cho, Shogo-in, Sakyo-ku, Kyoto 606, Japan

For a better understanding of the minimum requirements for
biomedical polymers it is important to distinguish biocompatibility
from non–toxicity and to divide biocompatibility into two subgroups;
mechanical and interfacial. This article presents an overview of
interfacial biocompatibility, a subject that can be itself further divided
into two groups non–stimulative and bioadhesive. The surface of all
currently available biomaterials stimulate, microscopically or
macroscopically, the host biological system when in direct contact. As
a result, complement activation, blood coagulation, thrombus
formation, encapsulation, and calcification take place. It has been
shown that surface modification by grafting techniques can markedly
reduce such stimulation, thus improving the interfacial biocompatibility
of currently used biomedical polymers. Further, it has been shown that
medical polymers can also attain  bioadhesive biocompatibility by
surface modifications.

Biomedical materials that are becoming increasingly important in medicine
and pharmaceutics include synthetic and natural polymers, metals, ceramics,
composites, and tissue–derived materials. These biomedical materials must meet
several requirements in addition to achieving their functionality. Table I summarizes
the minimum requirements necessary for biomedical materials. These materials can
be applied to the  permanent replacement of defective organs and tissues, temporary
support of defective or normal organs, storage and purification of blood, and control
of drug delivery.  An increasing number of high performance biomedical polymers
have been developed in recent years to more effectively fulfill their requirements.
In addition to their functionality, biomedical materials must be non–toxic and
capable of being sterilized as shown in Table I. These properties are essential for
biomaterials, although perhaps not required in industrial applications. This is
necessary as biomaterials are always in direct contact with living cells that are very
vulnerable to and readily killed by physical and chemical stimuli, for instance, by
toxic substance invasion. Toxic compounds associated with polymers are listed in
Table II.

0097–6156/94/0540–0035$06.00/0
© 1994 American Chemical Society

Table I. Minimum Requirements for Biomaterials

| Requirements | Examples |
|---|---|
| I. Non-toxicity | Non-pyrogenic, Non-hemolytic, Non-inflammatory (chronic), Non-tumorigenic, Non-allergenic |
| II. Functionality | Organ and tissue replacement, Tissue reconstruction, Internal organ support, Disposable medical devices, Drug delivery |
| III. Sterilizability | Radiation ( ray and electron beam), Ethylene oxide gas, Autoclave, Dry heating |
| IV. Biocompatibility | See Figures 1 and 4 |

Table II. Toxic Compounds Related to Polymers

| Compounds | Remarks |
|---|---|
| I. Monomers | Not polymerized, Depolymerized |
| II. Initiators | Intact and decomposed |
| III. Catalysts | For crosslinking, curing, and other reactions |
| IV. Additives | Antioxidants, Plasticizers, UV absorbents, Lubricants, Antistatic agents, Dyes, Pigments |
| V. Others | Byproducts, Degradation products, Stains |

## Biocompatibility

There is no authorized definition for the term biocompatibility, although it is generally agreed that biocompatibility is necessary for biomaterials. Often biocompatibility is used synonymously with the term non–toxicity. Indeed, it is difficult to rigorously define biocompatibility and differentiate non–toxicity from so–called "biocompatibility". In this article, however, biocompatibility will be considered as a concept separate from non–toxicity to avoid ambiguity. How is biocompatibility defined here? No attempt will be made to explain biocompatibility with a few sentences as this concept involves many aspects. Basically, biocompatibility can be divided into two groups, as represented in Figure 1. Mechanical compatibility is sometimes called bulk compatibility, whereas interfacial compatibility often as biological compatibility. Both compatibilities have little to do with the leachable or extractable substances which are the major cause of the toxicity of the biomaterials mentioned earlier. In the following it is assumed that all the biomaterials have neither leachables nor extractables, in other words, are not toxic in a narrow sense.

This chapter does not deal with mechanical biocompatibility though it is very important as a comprehensive discussion of this subject is presented in another chapter by Dr. S.W. Shalaby.

## Interfacial Biocompatibility

When a biomaterial comes in contact with a living body interactions occur between the foreign material and the living body, as shown in Figure 2. As a result the biomaterial may undergo detectable alteration in surface and bulk structure resulting in hydrolysis, deterioration, and fatigue, whereas the living body responds against the foreign material by invoking an immune reaction. The biological responses associated with biomaterials can be classified on the basis of toxic and foreign–body reactions(Figure 3). In some cases microscopic adhesion occurs between the surface of the biomaterial and the contacting tissue. The interfacially biocompatible materials, therefore, can be described as those in which these adverse biological responses occur at much reduced rates or in which the required tissue adhesion is promoted depending on the objective of the biomaterial, whether tissue adhesion is desirable or not.

The two large groups of interfacial biocompatibility, non–stimulative (the least foreign–body reaction) and bioadhesive, can be further divided into subgroups, as shown in Figure 4. It may be obvious from Figures 1 and 4 that the concept of biocompatibility cannot be summarized in a few sentences. Here, interfacial biocompatibility will be briefly explained according to the classification described in Figure 4 and results obtained in our laboratories will be presented as examples of interfacial biocompatibility.

**Non–stimulative Surfaces.** This biocompatible surface is often difficult to discriminate from non–toxic surface.

**Macroscopically Non–stimulative Surfaces.** When a biomaterial moves relative to the contacting tissue, the tissue surface is often damaged by mechanical abrasion resulting, for instance, in stenosis for the case of tubular tissues such as the urethra (1). Most of the inner and outer surfaces of soft tissues are lubricious because the surface is covered with a mucous layer though skin is one exception. Due to this lubricity serious damage to the tissue surfaces can be avoided even if they are brought in frictional contact with each other. In contrast to natural tissues, artificial materials generally do not possess such lubricious surfaces. A method to render the artificial surface lubricious is to coat it with a hydrophilic layer.

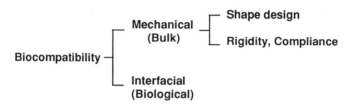

Figure 1.    Two major biocompatibilities.

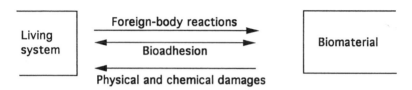

Figure 2.    Interactions between living system and biomaterial.

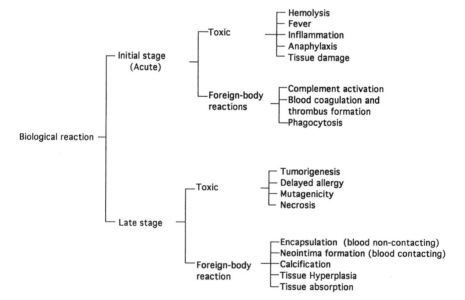

Figure 3.    Major biological reactions.

In fact, coating technology has been employed for lubrication of medical devices such as catheters, but it is not effective for long-term applications because of short life span of the coated layer. One of the better techniques to produce a lubricated surface is to covalently immobilize water-soluble polymers on the device surface. A model of a lubricious surface produced by grafting a thick layer of water-soluble polymer is illustrated in Figure 5a. Figure 6 presents the coefficients of friction observed when an ethylene-vinylacetate copolymer surface grafted with poly(N,N-dimethylacrylamide) was rubbed against a smooth glass plate in the presence of water (2). This surface exhibited permanent lubricity and induced no remarkable tissue damage on the rabbit urethra when repeatedly rubbed, in contrast with the ungrafted surface(3).

**Molecularly Non-stimulative Surfaces.**   This    type    of biocompatible surface has been most extensively studied in the field of biomedical polymers. Poor performance, low reliability, and the high failure rate of biomaterials both for short and long-term applications have been ascribed to molecular stimulation of the host defense system by the biomaterial surface. In other words, deleterious host reactions such as complement activation, blood coagulation, thrombus formation, encapsulation, and calcification have been thought to be due to a lack of interfacial biocompatibility. As a result, many studies have been devoted to the elucidation of such adverse biological reactions. These have found that the reactions are usually triggered by the material surface molecularly stimulating the body's living immune systems.

The mechanism of the biological reactions of the host to the biomaterial surface is not known in detail but it is very likely that the molecular stimulation of the biological systems is initiated on contact between the biomaterial surface and the molecular and cellular components of the living body. These components include plasma proteins, especially those of the complement and blood coagulation systems, and cells, especially those of the immune and clotting systems. At present, it is widely accepted that the initial process that takes place upon contact with the living structure is plasma protein adsorption onto the biomaterial surface. If it is adsorbed exclusively and permanently by the serum albumin, no adverse reaction to the biomaterial will be evoked by the host biological system, as the foreign surface will be covered with a layer of the abundant bioinert serum albumin. In general, the material surface is not entirely covered with serum albumin, but albumin desorption followed by adsorption of other bioactive proteins, such as fibrinogen and complements, will take place causing cell adhesion. These events are schematically depicted in Figure 7.

It is, therefore, reasonable to assume that surface modifications which prevent any protein adsorption and, hence, cell adhesion, will lead to the minimization of the adverse reactions, that is, insignificant stimulation of the living system being in direct contact with the biomaterial will occur. Indeed, in the past two decades a great number of studies have been performed to try and delineate the mechanism of protein adsorption on many material surfaces. One conclusion that can be drawn from these studies is that protein adsorption is minimized as the interfacial free energy between the material surface and the surrounding protein solution approaches zero. A surface with an electric charge similar to that of the protein will also lead to minimal protein adsorption because of the ionically repulsive interaction. However, there exist many kinds of proteins with positive and negative surface charges mixed in the living plasma. It follows that a surface with immobilized, non-ionic water-soluble chains may be very promising as it could thermodynamically reject adsorption of proteins, irrespective of the electric charge. A model of such a surface is schematically represented in Figure 5b.

A material with such a surface can be most easily produced by physical adsorption (4) or by physical entrapment of polymers into the base polymer surface

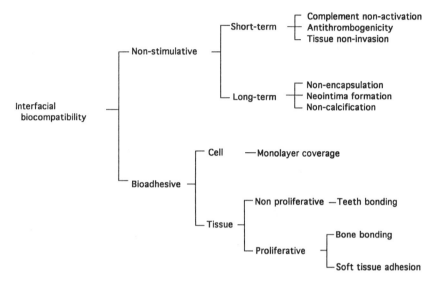

Figure 4.    Classification of interfacial biocompatibility.

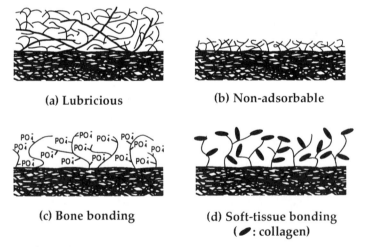

(a) Lubricious                    (b) Non-adsorbable

(c) Bone bonding               (d) Soft-tissue bonding
                                    (✐: collagen)

Figure 5.    Model of polymer surfaces with grafted chains.

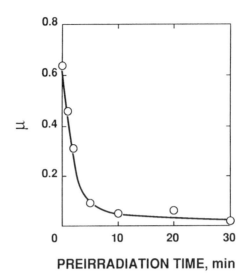

**PREIRRADIATION TIME, min**

Figure 6.    Coefficients of friction (μ) as a function of UV preirradiation time for ethylene–vinyl acetate copolymer films surface–grafted with poly(N,N–dimethylacrylamide).
Source: Reproduced with permission from reference 2.
Copyright 1991 Butterworth-Heinemann Ltd.

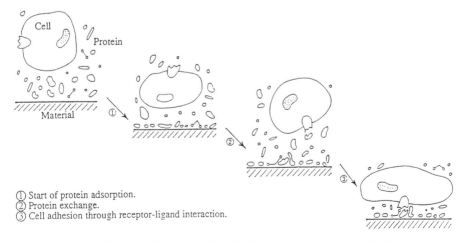

① Start of protein adsorption.
② Protein exchange.
③ Cell adhesion through receptor-ligand interaction.

Figure 7.    Schematic illustration of cell adhesion to protein–adsorbed surface.

(5). Promising polymers for this purpose are ethylene oxide homopolymer and block or graft copolymers consisting of poly(ethylene glycol) units and hydrophobic units. Presumably, the physically entrapped polymers will be desorbed under certain environmental conditions. A more reliable method to permanently fix a non–ionic water–soluble polymer on the biomaterial surface is through a chemical coupling reaction or graft polymerization. Both the methods are feasible. We have covalently immobilized poly(ethylene glycol) onto the surface of cellulose hollow fibers used for hemodialysis through esterification between the hydroxyl groups of cellulose and the terminal carboxyl groups of poly(ethylene glycol) (6). It was found that not only complement activation but also thrombus formation on the cellulose surface were markedly reduced by this surface modification (7).

However, the polymer coupling method is not always applicable to biomaterials, because their surface must possess functional groups for immobilization, otherwise, sophisticated reaction techniques are required (8,9). An alternative means for the polymer immobilization is graft polymerization of water–soluble monomers onto the material surface. It is generally for this reason that extensive studies have been conducted on radiation–induced graft polymerization onto polymer surfaces. However, it is difficult for the radiation technique to localize the graft polymerization on only the material surface without any change to the bulk properties resulting in dense grafted chains unless the combination of monomer and substrate polymer and the polymerization condition are correctly selected. To localize the formation of polymeric radicals and peroxides capable of initiating graft polymerization onto the surface region of polymer, we have been utilizing low temperature plasmas (10, 11), UV (12), and ozone (13) as the means of oxidation. Figure 8 shows peroxide production onto a polyethylene film when exposed to an argon plasma (10). With the use of these peroxides, graft polymerization of various water–soluble monomers can be performed and be restricted to the substrate polymer surfaces. Protein adsorption as well as platelet adhesion were found to be significantly reduced by the surface modification with graft polymerization. UV– and ozone–induced graft polymerization also gave similar results. For example, suppression of protein adsorption and platelet adhesion is shown in Figure 9 for a polyurethane film. The surface was modified by ozone–induced graft polymerization (13). A major problem yet to be solved in surface graft polymerization is a lack of analytical means available to characterize the grafted surface in detail. ATR–FTIR and XPS provide information on the chemical composition and functional groups in the dried surface region, but not on the surface density and molecular length of the polymer chains grafted on the material surface.

Encapsulation of biomaterials by collagenous connective tissue is the most common response of the living system to isolate implanted foreign materials from the surrounding living body for its protection. However, such encapsulation is, in many cases, not desirable for the biomaterial as its performance is often greatly affected by the thick fibrous capsule. A well–known example is the silicone mammary implant which can deform due to contraction of the newly–formed collagen fibers. It has been shown that collagen molecules are biosynthesized and excreted by fibroblasts which are anchor–dependent cells. Therefore, encapsulation must be largely dependent on the site where the biomaterial is implanted and on the material surface to which the fibroblasts are attached. If a biomaterial is implanted in a site that contains few fibroblasts and has a surface that undergoes little cell adhesion the thickness of the collagenous tissue encapsulating the implanted biomaterial will be very thin. On the other hand, a thick collagenous capsule will be formed if a biomaterial with a surface favorable for cell adhesion is implanted in a tissue where collagen is abundant and a remarkable encapsulation will be observed due to a high affinity of collagen to the fibroblast. Table III gives the capsule thickness formed on various polymer films subcutaneously implanted in rats for 16 weeks (14). This table shows that encapsulation occurred on more or less every surface. However,

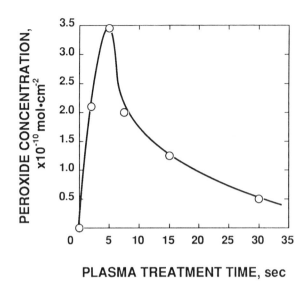

Figure 8.    Formation of peroxides on polyethylene surface exposed to argon plasma.

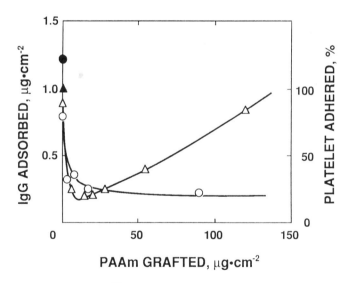

Figure 9.    Adsorption of [125]I–labeled IgG and adhesion of platelet to polyurethane surface grafted with polyacrylamide.

△ , IgG (virgin, ▲ );   ○ , platelet (virgin, ● )

Table III.  Thickness of the Fibrous Capsule Formed around Surface-modified Silicones Implanted in the Subcutaneous Tissue of Rabbits (n=5)

| Surface | Capsule thickness (um) | | | |
|---------|------|------|------|------|
|         | 2w   | 4w   | 8w   | 16w  |
| Virgin | 63 ± 1.3 | 76 ± 0.8 | 88 ± 2.5 | 90 ± 1.5 |
| Corona-treated(5min) | 53 ± 3.3 | 44 ± 1.9 | 77 ± 3.7 | 91 ± 2.6 |
| Corna-treated (10min) | 59 ± 1.1 | 68 ± 2.7 | 83 ± 2.5 | 91 ± 2.4 |
| Corona-treated (15min) | 58 ± 1.1 | 53 ± 2.8 | 64 ± 2.1 | 83 ± 2.3 |
| AAm-grafted | 50 ± 2.9 | 36 ± 1.3 | 60 ± 1.7 | 94 ± 3.1 |
| AA-grafted | 48 ± 2.5 | 50 ± 1.1 | 72 ± 2.1 | 50 ± 3.2 |
| DMAA-grafted | 29 ± 2.2 | 40 ± 0.9 | 79 ± 2.2 | 87 ± 2.4 |
| AMPS-grafted | 72 ± 2.5 | 69 ± 2.2 | 77 ± 2.6 | 92 ± 2.0 |
| Nass-grafted | 76 ± 1.1 | 74 ± 1.5 | 75 ± 3.5 | 80 ± 2.9 |
| DMAPAA-grafted | 80 ± 0.9 | 47 ± 2.8 | 55 ± 1.5 | 89 ± 2.6 |
| HA-fixed | 69 ± 2.6 | 46 ± 4.2 | 63 ± 2.7 | 88 ± 2.5 |
| Collagen-immobilized | 89 ± 1.1 | 42 ± 2.7 | 65 ± 1.8 | 63 ± 2.1 |
| Gelatin-immobilized | 85 ± 1.6 | 73 ± 2.3 | 81 ± 2.6 | 82 ± 0.9 |
| Fibronectin-immobilized | 76 ± 1.4 | 52 ± 2.3 | 45 ± 1.5 | 87 ± 2.7 |

AAm, acrylamide; AA, acrylic acid; DMAA, dimethylacrylamide; AMPS, 2-acrylamide-2-methylpropane sulfonic acid; Nass, styrene sulphonic acid sodium salt; DMAPAA, N,N-dimethylaminopropyl acrylamide; HA, hyaluronic acid

Source: Reproduced with permission from reference 14.

when polymer rods were intraperitoneally implanted for 3 months, little encapsulation occurred around polytetrafluoroethylene and poly(vinyl alcohol) hydrogel rods (15).

Infection has been reported to often occur when a biomaterial is implanted in a body or inserted in tubular organs as a catheter. Such infection seems to be a result of adhesion of infectious bacteria onto the biomaterial surface. If bacteria adhesion is similar to adhesion of other cells like fibroblasts, occurrence of infection through the material surface may be reduced by modifying the material surface so as to minimize cell adhesion. Such a surface can be obtained by surface grafting of water–soluble polymer chains. However, producing a hydrophilic biomaterial surface by plasma treatment only leads to enhanced cell adhesion as shown in Figure 10 (16).

Calcification is also an undesirable response of the host to an implanted material, because the biomaterial becomes hard and brittle as exemplified in biological heart valves and artificial hearts. The mechanism of such calcification is not clear, thus no effective method to avoid calcification has been proposed. However, if a hydrophobic biomaterial is free of adherent cells, calcification will be greatly reduced, as calcification is thought to be due, in part, to these dead cells. On the other hand, it seems also probable that calcium and phosphate ions sorbed within the material may provide the nucleus for calcification. If this is the case, care should be taken when hydrogel–type polymers are implanted in the body.

**Bioadhesive Surfaces.**   Bioadhesion is defined here as microscopic bonding between a biomaterial and a biological structure. Well–known examples include composite resins for dentistry and bioactive ceramics for bone replacement. The biological structures associated with such bioadhesion include living tissues and cells.

**Tissue–adhesive Surfaces.** Bioadhesion of biomaterials has been studied for three different tissues; teeth, bones, and soft connective tissues. Of these three tissues the bioadhesiveness of teeth has been most intensively studied. Restorative dentistry requires polymeric materials that adhere strongly to teeth tissue especially to dentin to seal and fill caries fissures and decay. Dentin is composed of an inorganic material, hydroxyapatite, and an organic material, collagen. In the mouth the adhesion process should be quickly completed. Moreover, the bond should have excellent durability considering its exposure to a harsh environment where water is present and temperature varies over a wide range. The most widely used monomers for dental composite resins are methacrylate derivatives having a large molecular size with two or more double bonds and hydrophilic and hydrophobic side chains in one molecule. The hydrophilic portion is necessary to enhance the adhesive strength, whereas the hydrophobic side chain contributes to high mechanical strength of the cured resins.

In contrast to teeth, bone and soft tissues can proliferate at a high rate when damaged. The bone–bonding materials currently being studied are not polymers but rather mostly ceramics such as hydroxyapatite, glass, and their composites. However, research on bone–bonding polymers has recently been published (17). If polymers that are adhesive to bones are successfully developed they will be applicable to the replacement of bone, tendon, and ligament. It appears probable that calcification mediates adhesion between the organic polymer and the contacting bone. If this is the case, a polymer surface grafted with polymer chains having phosphate groups may be promising as bone–bonding surface. The illustration is given in Figure 5c.

Polymers which can adhere to soft tissues are also necessary. They could be used as part of a percutaneous device to prevent tunnel infection through the dead space between the skin tissue and the material surface. We have demonstrated that

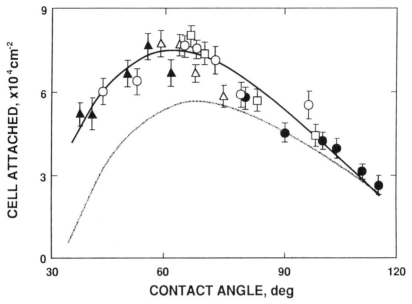

Figure 10.    Dependence of L cell adhesion on the water contact angle of
polymer surfaces with and without plasma treatment.
(———) with plasma treatment, (————) without plasma treatment
( O ), polyethylene; ( ● ), polytetrafluoroethylene;
( △ ), poly(ethylene terephthalate); ( ▲ ), polystyrene; ( □ ), polypropylene.
Source: Reproduced with permission from reference 16.
Copyright 1993 Butterworth-Heinemann Ltd.

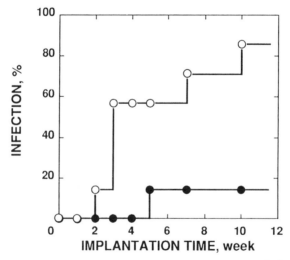

Figure 11.    Infection rate for polyethylene sponge percutaneously implanted in
rabbits. ( O ), without collagen immobilization; ( ● ), with collagen immobilization.
Source: Reproduced with permission from reference 18.
Copyright 1988 American Chemical Society.

**Table IV. Tumorigenesis in Rats by Implanted Porous Polyethylene with and without Immobilized Collagen**

| Polyethylene | Tumorigenesis rate | |
|---|---|---|
| | Rat based | Material based |
| Without collagen | 9/12 (75%) | 11/24 (46%) |
| With immobilized collagen | 1/12 (8%) | 1/24 (4%) |

Source: Reproduced with permission from reference 19.

Copyright 1993 Butterworth-Heinemann Ltd.

covalent immobilization of collagen onto polymer surfaces, as shown in Figure 5d, leads to microscopic adhesion of the surrounding connective tissue and disappearance of the dead space. Figure 11 shows how immobilized collagen reduces the infection rate for a percutaneous device (18). Collagen immobilization onto a silicone surface has been achieved using a water–soluble carbodiimide after surface graft polymerization of acrylic acid was performed to introduce carboxyl groups onto the material surface. It is interesting to note that surface immobilization of collagen greatly reduced the sarcoma formation when a porous polyethylene was subcutaneously implanted in rats (19). Table IV describes these results.

Bioadhesive polymers are also required for sustained drug delivery through mucous membranes of tissues such as the vagina and gastrointestine. It has been demonstrated that polymer surfaces with anionic groups or moieties capable of hydrogen bonding are promising for adhesion to the mucous layer of living tissues (20, 21).

**Cell–adhesive Surfaces** If a biomaterial is entirely covered with a monolayer of cells when in contact with blood or tissues and, in addition, if this layer is permanent the biomaterial could acquire excellent interfacial biocompatibility. However, a long–term cell covering is very difficult to achieve on man–made materials. For instance, interfacial biocompatibility of large–caliber vascular grafts is currently obtained by neointima formation but is inadequate partly due to the poor coverage of neointima, though still widely used in surgery. No endothelium cell is seen in the middle of the implanted vascular grafts made of poly(ethylene terephthalate) and polytetrafluoroethylene. It should be stressed that mechanical biocompatibility has a large effect on performance of vascular grafts.

If a technique is developed to readily and securely cover a polymer surface with a layer of adequate cells, hybrid–type artificial organs for use in vascular reconstruction and extracorporeal liver support, for example, may be possible. For cell–culture substrates several polymer surfaces are already available which distinctly favor cell covering. Cationic surfaces or immobilization of cell–adhesive proteins such as fibronectin and collagen have been utilized in these cases.

**Reference**
(1) Uyama, Y.; Ikada, Y. *Lubricating Polymer Surfaces*; Technomic Publishing: Lancaster, PA, 1993.

(2)    Uyama, Y.; Tadokoro, H.; Ikada, Y. *Biomaterials* 1991, *12*, 71.
(3)    Tomita, N.; Tamai, S.; Hirao, Y.; Okajima, E.; Ikeuchi, K.; Ikada, Y. submitted to *J. Appl. Biomater.*
(4)    Shen, M.S.; Hoffman, A.S.; Feijen, J. *J. Adhesion Sci. Technol.* 1992, *6*, 995.
(5)    Desai, N.P.; Hubbell, J.A. *Biomaterials*, 1991, *12*, 144.
(6)    Kishida, A.; Mishima, K.; Corretge, E.; Konishi, H.; Ikada, Y. *Biomaterials* 1992, *13*, 113.
(7)    Akizawa, T.; Kino, K.; Koshikawa, S.; Ikada, Y.; Kishida, A.; Yamashita, M.; Imamura, K. *Trans. Am. Soc. Artif. Organs* 1989, *35*, 333.
(8)    Park, K.D.; Piao, A.Z.; Jacobs, H.; Okano, T.; Kim, S.W. *J. Polym. Sci. Part A; Polym. Chem. Ed.* 1991, *29*, 1725.
(9)    Tseng, Y.–C.; Park, K. *J. Biomed. Mater. Res.* 1992, *26*, 373.
(10)   Suzuki, M.; Kishida, A.; Iwata, H.; Ikada, Y. *Macromolecules* 1986, *19*, 1804.
(11)   Iwata, H.; Kishida, A.; Suzuki, M.; Hata, Y.; Ikada, Y. *J. Polym. Sci., Part A: Polym. Chem. Ed.*, 1988, *26*, 3309.
(12)   Uyama, Y.; Ikada, Y. *J. Appl. Polym. Sci.* 1988, *36*, 1087.
(13)   Fujimoto, K.; Takebayashi, Y.; Inoue H.; Ikada, Y. *J. Polym. Sci., Part A: Polym. Chem. Ed.*, 1993, *31*, 1035.
(14)   Okada, T.; Ikada, Y. in press *J. Biomed. Mater. Res.*
(15)   Inoue, K.; Fujisato, T.; Gu, Y.G.; Burczak, K.; Sumi, S.; Kogire, M.; Tobe, T.; Uchida, K. Nakai, I.; Maetani, S.; Ikada, Y. *Pancreas*, 1992, *7*, 562.
(16)   Tamada, Y.; Ikada, Y. *Polymer* 1993,*34*,2208.
(17)   Blitterswijk, C.A. van; Grote, J.J.; Hesseling, S.C.; Bakker, D. *Interfaces in medicine and mechanics–2*; Williams, K.R.; Toni, A.; Middleton, J.; Pallotti, G., Eds. 1991, 1–10.
(18)   Okada, T.; Ikada, Y. *Polym. Mater. Sci. Eng. (ACS)*, 1988, *59*, 548.
(19)   Kinoshita, Y.; Kuzuhara, T.; Kirigakubo, M.; Kobayashi, M.; Shimura, K.; Ikada, Y. *Biomaterials* 1993, *14*, 546.
(20)   Park, H.; Robinson, J.R. *J. Control. Rel.*, 1985, *2*, 47.
(21)   Chickering, D.; Jacob, J.; Panol, G.; Mathiowitz, E. *Proceed. Intern. Symp. Control. Rel. Bioact. Mater.*, 1992, *19*, 88.

RECEIVED May 25, 1993

## Chapter 4

# Polymeric Devices for Transcutaneous and Percutaneous Transport

Joel L. Williams

Becton Dickinson Research Center, 21 Davis Drive,
Research Triangle Park, NC 27709

Development of control release polymeric drug devices has continued a strong growth since the 1960's. During the past three decades, extensive progress has been made in polymer synthesis and fabrication of various sustained drug delivery devices, and a few of the more specific developments are given in the appropriate section of this book. It is not the purpose of this topical overview to review the extensive patents, articles, and recent reviews on transdermal or subcutaneous polymeric devices but to provide the reader with an early historical background and references for further reading in these emerging areas of drug delivery (*1-10*).

The rationale behind controlled or sustained release has always been the need for a therapeutic drug level that is sufficient without the overdose level that is generally found with injection or other forms of bolus medicant delivery. Oftentimes, there are undesirable side effects associated with bolus drug delivery and, in the case of certain drug therapies, a second drug must be administered to prevent side effects such as nausea. In theory, a properly designed controlled delivery device should overcome or minimize problems associated with pulsed drug delivery.

### Encapsulated Drug Devices

In recent years, the transdermal patch device has been extremely popular for delivering drugs for motion sickness, nicotine, and nitroglycerine, to name a few. Some examples of commercial products are given in Table I. The drug release device, however, can take on many forms (see Table II) other than a patch, and the selection of a particular device is highly dependent on the molecular weight of the drug and the rate required for optimum therapeutic effectiveness.

0097-6156/94/0540-0049$06.00/0
© 1994 American Chemical Society

## Table I.  Transdermal Products

| Product Name | Active Agent | Polymer | Company |
|---|---|---|---|
| Transderm® Scop | Scopolamine | Ethylene vinyl acetate | Alza, Ciba |
| Transderm® Nitro | Nitroglycerin | Polypropylene | Alza, Ciba |
| Nitrodue® | Nitroglycerin | Polyvinyl alcohol Polyvinyl pyrrolidone | Key |
| Nitro Disk® | Nitroglycerin | Silicon | G. D. Searle |
| Catapres TTS (Therapeutic Transdermal System) | Clonidine | | Boehringer-Inglehein |
| Nitroderm™ | Nitroglycerin | Polyvinyl chloride copolymer | Hercon Div. Health-Chem. |

The rate controlling membranes in encapsulated drug devices are most commonly made of dense polymeric materials.  In these devices, a solution-diffusion mechanism controls release rates.  Fick's First Law governs the specific release rate (q):

$$q = -D \frac{dc}{dx}$$

where D is the diffusion constant, c is the concentration of drug in the membrane, and x is the membrane thickness.

Crank and Park discuss the numerous solutions to this equation that exist with boundary conditions such as tube, sphere, and slab (11).  Baker and Lonsdale (12) have also made an excellent compilation of mechanisms and release rates for drug delivery systems.  It is evident from Fick's First Law that the main parameters governing release rates are membrane thickness, area, and drug concentration.  If the release rate desirable for a given membrane cannot be achieved by manipulating these variables, a different material must be selected.  Unfortunately, drug permeability has been measured for only a few polymer-drug combinations, and the release rate for a particular drug must often be determined experimentally.  Duncan and Kalkwarf (13) describe an inexpensive and convenient method for determining release rates by using a simple flow system.  An alternative to a solution-diffusion system is a pore-diffusion system in which the drug is virtually immobilized in a microporous barrier and is diffused primarily through micropores.  The use of microporous membranes as a medium for controlled drug release has also received considerable attention.

**Table II. Polymeric Systems for Sustained Release**

Reservoir systems with rate-controlling membrane
    Microencapsulation
    Macroencapsulation
    Membrane systems
Reservoir systems without rate-controlling membrane
    Hollow fibers (microporous)
    Poroplastic® and Sustrelle® ultramicroporous cellulose triacetate
    Porous polymeric substrates and foams
Monolithic systems
    Physically dissolved in nonporous polymeric matrix
        Nonerodible
        Erodible
        Environmental agent ingression
        Degradable
    Physically dispersed in nonporous polymeric matrix
        Nonerodible
        Erodible
        Environmental agent ingression
        Degradable
Laminated structures
    Reservoir layer chemically similar to outer control layers
    Reservoir layer chemically dissimilar to outer control layers
Matrix devices
    Release by diffusion
    Release by ingression of environmental agent
Chemical erosion of polymer matrix
    Heterogeneous
    Homogeneous
Biological erosion of polymer matrix
    Heterogeneous
    Homogeneous

**Single Component System.** In a single component system, the drug is encapsulated in its pure form and release rates are essentially zero order (*12*). Polydimethylsiloxane and polyethylene are the materials most often chosen for encapsulation. Table III shows some typical release rates reported by Kincl, et al. (*14*), for various steroids through silicone rubber. Clearly, when a solution-diffusion mechanism controls the drug release rate, drug permeabilities can be expected to vary widely.

Kincl (*14*) has also reported release rates for progesterone when various polymer materials are used (Table IV). Except for nylon, most materials studied have release rates a thousand times lower than that for silicone rubber. Silicone rubber, widely employed in drug-releasing devices because of its high permeability, has been accepted by the medical industry mainly for its excellent biocompatibility. Unfortunately, silicone rubber does not have sufficient tensile strength and elastic modulus to meet all requirements for drug devices. Other polymers having a wider range of permeabilities than silicone are copolymers of polydimethylsiloxane and polycarbonate and polyurethane rubbers, which have the proper elastic moduli, strength, and biocompatibility. Rubbery polymers have potentially higher release rates than conventional "glassy" or semicrystalline polymers because of their high internal free volume (i.e., space between molecules).

The drug-release rate for a specific drug is generally increased if the "free volume" of the barrier is increased. The following equation, which describes diffusion (D) in a polymeric film, illustrates the concept of free volume:

$$D = D_o \, e^{k \, v_f / v_d}$$

where $D_o$ and k are phenomenological coefficients, $v_f$ is polymer free volume, and $v_d$ is the volume of the drug molecule.

This equation implies, for example, that polyvinylchloride that has been plasticized with a low molecular weight material such as phthalic ester will release a given drug much faster than an unmodified polyvinylchloride membrane. It follows that the drug-release rate can often be increased by plasticizing the membrane. Since most drugs have a fairly high molecular weight (>300), rubbery or plasticized materials will probably have to be employed to achieve the necessary release rates using homogeneous membranes.

**Multicomponent Reservoir.** One disadvantage of the single component reservoir device is that tissue fluids can diffuse across the membrane and dilute the encapsulated drug in subcutaneous applications. It follows that the release rate would be time dependent since the concentration gradient would change as a function of time. To offset this difficulty, Pharriss, et al. (15), used a silicone fluid saturated with the drug that released at a constant rate for periods in excess of one year. In their device, a constant release rate was maintained until the encapsulated solution became unsaturated. Although the main advantage of a multicomponent reservoir system is that it achieves zero-order release rates, the sustained release of drugs at various rates can be achieved by simply varying the amount of drug dissolved in the solvent. This technique could offer certain advantages when prolonged drug release is not required such as when a local antibiotic is administered.

### Matrix Devices

Matrix devices do not release drugs at a constant rate; instead, the initial release rate is rapidly followed by an exponential decay with time. The exact kinetics for numerous matrices have been treated by several authors (12, 16, 17, 18) and will not be repeated here. In most cases, however, release rates over a long period of time are a linear function with the square root of time.

**Nondegradable Devices.** The popularity of matrix systems can be attributed to the simplicity of either diffusing the drug into the substrate or mixing it in the polymer and/or solution before making the device.

Table III.  Average Diffusion Rates of Various Steroids Across
Polydimethylsiloxane Membrane*

| Steroid | Diffusion Rate (g)** |
|---------|----------------------|
| 19 Norprogesterone | 1353 |
| Progesterone | 469 |
| Testosterone | 317 |
| Megestrol acetate | 236 |
| Norethindrone | 73 |
| Estradiol | 61 |
| Mestranol | 43 |
| Corticosterone | 21 |
| Cortisol | 6 |

\*   SOURCE:  Adapted from ref. 14.
\*\*  Diffusion rate:  100 mm$^2$/.1mm/24 hours.

Table IV.  Diffusion of $^3$H Progesterone Through
Membranes of Various Polymers*

| Polymer** | Frequency of Sample Collection, Hrs. | No. of Observations | Progesterone Diffused, % |
|-----------|--------------------------------------|---------------------|--------------------------|
| Polydimethylsiloxane (Silastic) | 6,24 | 40 | 100 |
| Acetylcellulose (cellophane) | 6,24 | 8 | 0.1 |
| Fluoroethylene (Teflon) | 6,24 | 8 | 0.1 |
| Polyester (Mylar) | 6,24 | 8 | 0.1 |
| Polycarbonate (Lexan) | 6,24 | 8 | 0.1 |
| Polyethylene | 6,24 | 10 | 0.1 |
| Polyamid (Nylon) | 24 | 4 | 1.0 |
| Polystyrene copolymer (Cr39) | 24 | 4 | 0.1 |

\*   SOURCE:  Adapted from ref. 14.
\*\*  Thickness of membrane from 0.127 to 0.254 mm.

Again, because silicone rubber will hold up to 20 percent or more of solids, has a high capacity to imbibe drugs, and is compatible with tissue, it has been preferred for most matrix devices (*17, 19*). Kalkwarf, et al. (*20*), however, found that ample release rates (14 μg/day) could be achieved for progesterone by dry mixing this drug with low density polyethylene followed by melt extrusion. Initial release rates, however, were in excess of 200 μg/day and dropped to less than 14 μg/day in six days.

Unfortunately, some drugs may denature or degrade when subjected to the temperatures required to mold or extrude most polymeric materials. Such drugs can still be loaded in matrix devices up to their solubility limit by simply diffusing them into the substrate after the device is made. The specific amount of drug imbibed can be easily adjusted by determining the sorption isotherm for a given polymer-drug system. Once the equilibrium isotherm is known, a device can be preloaded to a given drug level by allowing it to equilibrate with a known drug concentration in solution.

### Dissolution and Biodegradable Devices

**Dissolution Devices.**  Zipper, et al. (*21*), have established that the slow release of metal ions, such as copper or zinc, can improve the contraceptive efficacy of intrauterine devices. Zipper and his colleagues coupled 200 mm$^2$ of copper surface area with a Tatum-T device and reduced the pregnancy rate from 18/100 to 0.05-1.5/100 women years.  Also, Zipper (*22*) found that the contraceptive efficacy was roughly proportional to the available copper surface area up to 200 mm$^2$. These self-contained dissolution devices can release metal ions for long periods of time.  Most drugs in their pure form do not have the mechanical integrity of metals or dissolution rates necessary for making an effective drug delivery device. In the case of progesterone, however, Rudel (*23*) has studied its slow release from pelletized mixtures.

**Biodegradable Devices.**  Considerable attention in recent years has been given to the use of biodegradable polylactic acid materials for dispensing drugs (*24, 25*). Nilsson, et al. (*26*), found the release rate of norgestrel mixed with a polylactic acid solution before film formation to be similar to the hydrolysis rate of the polymer; that is, drug diffusion contributed little or nothing to the release rate.  The authors cited lack of tissue inactivation as an advantage of the biodegradable polylactates.

In a separate study, Jackanicz, et al. (*27*), found that norgestrel released at a constant rate of 80 days. In their study, polylactic acid broke down less rapidly than the release rate of norgestrel thus indicating that diffusion contributed substantially to the overall drug release.

It is safe to assume that for a given dissolution device, the ultimate drug delivery rate will be a combination of hydrolysis (dissolution) and drug diffusion out of the polymer.  Of the systems studied to date, the release rates for biodegradable devices are essentially zero-order.

The major obstacle of subcutaneous devices has been the removal of the device in the event there are adverse reactions to the drug.

## Table V.  Polymers for Sustained Release Devices

### Natural Polymers

| | | |
|---|---|---|
| Starch | Gelatin | Carboxymethylcellulose |
| Ethylcellulose | Arabinogalactan | Cellulose acetate phthalate |
| Methylcellulose | Proteins | Propylhydroxycellulose |
| Nitrocellulose | Shellac | Gum arabic |
| Waxes-paraffin | Zein | Natural rubber |
| Sucinylated gelatin | | |

### Synthetic Polymers

| | | |
|---|---|---|
| Polybutadiene | Polyisoprene | Neoprene |
| Polysiloxane | Styrene-butadiene rubber | Ethylene-propylene-diene terpolymer |
| Hydrin rubber | Chloroprene | Silicone rubber |
| Nitrile | Acrylonitrile | Butyl rubber |
| Thermoplastic elastomers | | |
| Polyacrylic acid | Polyvinyl alcohol | Polyethylene |
| Polypropylene | Polystyrene | Acetal copolymer |
| Polyurethane | Polyvinylpyrrolidone | Poly(p-xylylene) |
| Polymethylmethacrylate | Polyvinyl acetate | Chlorinated polyethylene |
| Polylactones | Polylactides | Polyester |
| Polyvinyl chloride | Polyacrylate | Epoxy |
| Polyacrylamide | Polyether | Polyurea |
| Ethylene vinyl acetate copolymer | Hexamethylene oxalate copolymer | Polyhydroxyethyl methacrylate |

## Polymer Selection

The polymer selection is dependent mainly on the device type. However, there are a few considerations that are common to the selection process:
1. Nontoxic in a biological environment,
2. Solubility and diffusion characteristics in vivo or in vitro,
3. Compatible with drug,
4. Polymer should not contain additives that leach into the body,
5. Ease of fabrication and stable to sterilization,
6. Cost.

Both natural as well as synthetic polymers are used in drug devices and some of the most common materials are shown in Table V. The reader is referred to specific papers in this book for the preparation and properties of special polymers for use in sustained release devices. Details of materials employed in commercial products can sometimes be determined from the patent literature but are not widely publicized.

## Literature Cited

1.  *Controlled Drug Delivery*; Bruck, S. D., Ed.; Clinical Applications, Vol. II; CRC Press, Inc.: Boca Raton, FL, 1983.
2.  *Controlled Release Polymeric Formulations*; Paul, D. R.; Harris, F. W., Eds.; ACS Symposium Series 33; American Chemical Society: Washington, DC, 1976.
3.  *Percutaneous Absorption*; Bronaugh, R. L.; Maibach, Howard I., Eds.; Marcel Dekker, Inc.: New York, NY, 1989.
4.  *Prediction of Percutaneous Penetration, Methods, Measurements, Modelling*; Scott, R.C.; Guy, R. H.; Hadgraft, J., Eds.; IBN Technical Services Ltd.: London, 1990.
5.  Chien, Y. W. *Novel Drug Delivery Systems: Fundamentals, Developmental Concepts, Biomedical Assessments*; Drugs and the Parmaceutical Sciences, Vol. 14; Marcel Dekker, Inc.: New York, NY, 1982.
6.  *Controlled Release of Biologically Active Agents*; Tanquary, A.C.; Lacey, R.E., Eds.; Advances in Experimental Medicine and Biology, Vol. 47; Plenum Press: New York, NY, 1974.
7.  *Pulsed and Self-Regulated Drug Delivery*; Kost, J., Ed.; CRC Press: Boca Raton, FL, 1990.
8.  *Medical Polymers: Chemical Problems*; Sedlacek, B.; Overberger, C. G.; Mark, H. F., Eds.; Journal of Polymer Science: Polymer Symposium 66; John Wiley & Sons: New York, NY, 1979.
9.  *Biological Approaches to the Controlled Delivery of Drugs*; Juliano, R. L., Ed.; Annals of the new York Academy of Sciences, Vol. 507; The New York Academy of Sciences: New York, NY, 1987.
10. Kydonieus, Agis F.; *Transdermal Delivery of Drugs*; Berner, B., Ed.; CRC Press: Boca Raton, FL, 1987, Vol. I.
11. *Diffusion in Polymers*; Crank, J.; Park, G.S., Eds.; Academic Press: London, 1968.
12. Baker, R.W.; Lonsdale, H.K.; *Controlled Release of Biologically Active Agents*; Tanquary, A. C.; Lacey, R.E.; Plenum Press: New York, NY, 1974, pp 15-71.
13. Duncan, G. W.; Kalkwarf, D.R.; *Human Reproduction: Conception and Contraception*; Hafez, E.S.E.; Evans, T. N.; Harper and Row: Hagerstown, MD, 1973, pp 483-504.
14. Kincl, F. A.; Benagiano, G.; Angee, I.; *Steroids.* **1968**, *11*, pp 673-680.
15. Pharriss, B. B.; Erickson, R.; Bashaw, J.; et al.; *Fertil. Steril.* **1974**, *25*, pp 915-921.
16. Roseman, T. J.; *Adv. Exp. Med. Biol.* **1974**, *47*, pp 99-115.
17. Flynn, G.; *Adv. Exp. Med. Biol.* **1974**, *47*, pp 72-98.
18. Lacey, R. E.; Cowsar, D. R.; *Adv. Exp. Med. Biol.* **1974**, *47*, pp 117-114.

19. Fu, J. C.; Kale, A. K.; Moyer, D. L.; *J. Biomed. Mater. Res.* **1973**, *7*, pp 71-78.
20. Kalkwarf, D. R.; Sikov, M. R.; Smith, L., et al.; *Contraception*, **1972**, *6*, pp 423-431.
21. Zipper, J.; Medel, M.; Prager, R.; *Am. J. Obstet. Gynecol.* **1969**, *105*, pp 529-534.
22. Zipper, J.; Tatum, H.; Medel, M.; et al.; *Am. J. Obstet. Gynecol.* **1971**, *109*, pp 771-774.
23. Rudel, H.; U.S. Patent 3,656,483, January 15, 1970.
24. Kulkarni, R. K.; Moore, E. G.; Hegyell, A. F.; et al.; *J. Biomed. Mater. Res.* **1971**, *5*, pp 169-181.
25. Yolles, S.; U.S. Patent 3,887,699, December 29, 1970.
26. Nilsson, C. G.; Johansson, E.D.B.; Jackanicz, T. M.; et al.; *Am. J. Obstet. Gynecol.* **1975**, *122* pp 90-95.
27. Jackanicz, T. M.; Nash, H. A.; Wise, D. L.; et al.; *Contraception*, **1973**, *8*, pp 227-234.

RECEIVED May 28, 1993

# Chapter 5

# Microcellular Foams

## S. L. Roweton[1] and Shalaby W. Shalaby[1,2]

[1]Department of Bioengineering and [2]Materials Science and Engineering
Program, Clemson University, Clemson, SC 29634–0905

Microcellular materials exist in many forms. Methods for production of these materials are as varied as potential applications. This chapter reviews the technology of one class of microcellular materials, microcellular foams, which are sought for biomedical applications. Included is a description of several methods of foam production, foam morphologies, and present uses for microcellular foam materials. New methods of microcellular foam production and potential uses for the resultant foam materials are important to those interested in biomaterials and contemporary biomedical applications. It is for this reason that advances in microcellular foam formation are emphasized in the final section of this chapter. Increasingly, it is becoming evident that microcellular foams can be used effectively in many medical applications, particularly polymeric foam materials which are being investigated in this laboratory. For this reason, the focus of this chapter pertains to possible biomedical applications of polymeric microcellular foams.

## Types of Microcellular Foams

The subject of cellular materials has been explored extensively by Gibson and Ashby (1988) who defined foam materials as three-dimensional cellular solids. Given this definition, a wide variety of materials can be considered foams, including many foods. If the faces of cells composing the foam are solid, preventing communication between cells, the foam is considered to be closed-cell. If the cells of the foam have open faces, the foam is considered to be open-cell. The latter type of foam can exist with a variety of characteristics and contains cells of varying sizes and geometries. Most materials can be made into foams including metals, glasses, ceramics, and polymers (1). Composite foams can also be produced.

Microcellular foams are solids composed of relatively small cells. It is difficult to identify a universally accepted definition of the term "microcellular." Defining this term as "relatively small" is hardly quantitative. Aubert (1988) has defined microcellular foams as "foams whose cell or pore size is much smaller than that obtained in conventional foams." This is a subjective definition. The assertion is that traditional methods of foam formation can be used to produce foams with cells approximately 50 to 100 μm in diameter. This range in cell size serves as Aubert's basis for comparison. Aubert then defines microcellular foams as foams with cell diameters smaller than this, i.e., in the range of 10 μm or below. One author's attempt at defining "microcellular" illustrates this term's nebulous character. To provide a

0097–6156/94/0540–0058$06.00/0

simple, more general definition, microcellular foams are defined here as foams composed of cells with diameters under 400μm.

## Methods of Foam Formation

Methods for foam production are numerous, depending to a great extent on the material used and the desired morphology. Metals, glasses, and ceramics all can be converted to foams with controlled physical characteristics by utilizing one of several foam production methods. Polymeric microcellular foams are emphasized in this chapter and methods for their production are addressed.

One common method for the production of polymeric foams entails the introduction of gas to a liquid polymerizing monomer or molten polymer *(1)*. Solidification of the polymer traps gas bubbles, resulting in a foamed material. Gases can be introduced into a polymer melt in a variety of ways including intensive stirring, injection of inert gases, and incorporation of additives which decompose to gases upon heating. Depending on the processing parameters, both open- and closed-cell foams can be formed.

Sintering is another method of polymeric microcellular foam production. Polymer beads can be partially melted to join the spheres at their surfaces, leaving behind a porous material. Fibers can be joined in the same fashion to produce a cellular solid *(1)*. Emulsions can be used to produce foams by incorporating monomers in the oil phase of a water-oil emulsion with subsequent polymerization. A foam results upon water removal. Instead of a gas or a solvent, water serves as the pore-former around which polymeric cell walls develop *(3)*.

Thermally-induced phase separation (TIPS) is a technique, applicable to many polymers, which is widely used to produce open-cell microcellular foams. TIPS, a process which utilizes freeze-drying in its final stages, has been used to produce polystyrene microcellular foams as well as polyacrylonitrile, carbon, cellulose, and dextran foams *(4)*. The versatility of the TIPS process is illustrated in its applicability to both aqueous and non-aqueous systems. Polymer solubility is the main requirement for TIPS *(4)*. If this is satisfied, foams can be produced with controlled densities and morphologies. The TIPS process begins with a single-phase, polymer solution at a high temperature. Upon cooling, polymer crystallization or liquid-liquid phase separation takes place. Polymer concentration determines whether one or both of these types of phase separation will occur. A liquid-liquid phase separation produces a concentrated polymer solution in equilibrium with a solvent phase. The formation of a polymeric microcellular foam from the concentrated polymer phase requires additional steps. Thus, vitrification or crystallization are used to insure preservation of the characteristic microstructure. Then, freeze-drying is pursued to remove the solvent from the concentrated polymer solution. Sublimation preserves the morphology that develops as a result of the induced phase separation, a process that can remove the solvent while leaving behind an open-cell polymeric microcellular foam *(4)*.

## Microcellular Foam Morphologies

The variety of foam production methods and controllable processing parameters provides a virtually endless supply of unique polymeric microcellular morphologies. Commercially-produced foams and experimental foams can be distinguished morphologically by their differing cell sizes. Commercial foams typically can have larger cells, in the range from 100 to 1000 μm in diameter, while laboratory-scale foams may have cells in the 10 μm diameter range *(3)*. As mentioned previously, polymeric foams are often produced by introducing expanding gases to molten polymers. This method of commercial foam formation results in the production of foams with little variation in cell size and shape.

Because of their methods of production, experimental foams usually do not have cells of uniform size and shape but, instead, contain voids which are difficult to characterize concisely. For these foams, it is difficult to quantitate "cell size" (3). Polymeric microcellular foams formed using TIPS fall into this category. Increasingly, it is becoming evident that for TIPS, resultant foam morphology is dependent on a variety of parameters including polymer type and diluent crystallization kinetics, polymer concentration, the presence or absence of nucleating agents, and the cooling scheme of the system. All factors interact intimately to play an important role in determining foam morphology

Emulsions can be used to produce microcellular foams with relatively uniform morphologies. Round cells, with a range of diameters less than 100 μm, are typical. Foams with ultrafine microstructures can be formed utilizing gels produced by chemically reactive systems. Extraction reveals cells in the 0.1 to 0.3 μm diameter range (3).

Several types of microcellular foam morphologies have been described by LeMay et al. (1990). A froth is a relatively symmetric foam morphology which LeMay et al. describe as "curvilinear polyhedra sharing faces." Capillary forces are responsible for shaping the cells, as in the foam of soap. The froth morphology is characterized by closed cells and is likely to occur in blown foams. Syntactic, reticulate, inverse emulsion, and spinodal are terms used in describing other common foam morphologies. Syntactic foams are typically characterized by hollow, connected spheres and can contain both open and closed cells. Reticulate foams are open-cell foams characterized by straight, sturdy struts, not walls, that separate the cells. Inverse emulsion foams have distinguishable ring-shaped open cells with foam material concentrated where the pore-forming emulsion components came into contact during processing. TIPS usually results in the spinodal morphology, a morphology identified by its randomly bifurcated open-cells (5).

Ideal morphologies are rarely observed. In general, the microstructure of a foam will be a dispersion of the above-noted morphologies, present in varying amounts according to the process used for foam formation. Foams produced by phase separation often exhibit unique microstructures not associated with those mentioned above. Some microstructures have been described as lacy, flaky, spherulitic, and fibril-like. In many polymeric foam materials, several distinctly different morphologies can be observed (5).

## Methods of Characterization

Numerous methods exist for the testing of foam materials, facilitating the evaluation of the efficacy of different foam production techniques. ASTM test methods provide standardized guidelines for evaluating many foam characteristics including acoustical, electrical, thermal, and mechanical properties as well as cellular structure and flammability. The importance of mechanical properties depends on the end-use application. Key mechanical properties include response to compressive forces, abrasion resistance, creep, fatigue, indentation hardness, shock-absorbing qualities, and tensile strength (6). With the exception of cellular structure determination, these standardized test methods were developed through modifications of existing tests for solid materials.

Several other test methods are consistently used to evaluate the unique morphologies of polymeric microcellular foams. Scanning electron microscopy (SEM), light microscopy, permeability, mercury porosimetry, capillary flow porometry, and Brunauer-Emmett-Teller surface area measurement (BET) are commonly-used methods for characterizing the cells of a microcellular foam. SEM readily provides a qualitative assessment of cellular structure, producing a two-dimensional representation of a three-dimensional structure (2). Only a small area of a sample can be studied at one time, though, limiting the effectiveness of this method for foam evaluation. Light microscopy can only be used to examine relatively large cellular structures, limiting this evaluation method. Permeability measurements aid in determining the open- or closed-cell nature of a foam  but do not necessarily indicate cell size.

Mercury porosimetry is one method used to determine pore size distribution in a cellular solid *(7)*. Non-wetting liquid mercury is forced into an evacuated cellular material at a controlled pressure to determine the range of cell sizes present *(8)*. In general, this method is an effective research tool but it does have several inherent limitations. Using mercury porosimetry, it is difficult to differentiate between cells that do and do not penetrate entirely through the sample. Also, foams which are relatively fragile may be damaged by the associated high pressures. Distortion of the cell walls induced by high pressures can lead to erroneous results. Finally, this method is incapable of providing good resolution for cells in the 0.1-100 μm, a range of great importance to those investigating polymeric microcellular foams.

Capillary flow porometry is a related technique which can be used to provide information about the range of cell diameters for only those pores which extend all the way through the sample. The wetting liquid, typically a fluorocarbon, has low surface tension, low volatility, and low viscosity *(8)*. After the sample has been imbibed with the wetting solution, gas is forced through the sample at a controlled pressure. The pore size distribution is determined by recording changes in air flow across the sample with respect to the applied pressure. Choice of an appropriate wetting agent is critical as its properties influence the results of the test. Capillary flow porometry is valuable for evaluating polymeric porous materials, such as microcellular foams, while mercury porosimetry is typically used on more sturdy porous samples such as wood, sintered metal powders, clays, and carbon *(8)*.

The surface area of polymeric microcellular foams, as well as pore size distribution, pore structure, and pore volume, can also be determined using BET (Porous Materials Inc., Ithaca, NY, personal communication, 1992). Total surface area is determined by monitoring the physical adsorption of an adsorbate gas at a controlled temperature and pressure onto a porous material. Adsorption is realized through the van der Waal's type interaction of gas molecules with the cellular material. The static method of BET analysis is relatively simple, monitoring pressure changes of the single, adsorbate gas across the sample material. On the other hand, the dynamic method utilizes continuous flow of an adsorbate gas/carrier gas mixture and monitors thermal conductivity to determine the volume of gas adsorbed (Porous Materials Inc., Ithaca, NY, personal communication, 1992). The appropriateness of the method used is determined by the sample material and its applications. One limiting assumption of BET theory is that each sample has a surface with adsorption sites of equal energetic character *(9)*. Despite this theoretical limitation, BET is a valuable method for determining the surface area of a polymeric microcellular foam.

### Applications for Polymeric Microcellular Foams

Microcellular foams have found applications in many diverse fields, ranging from high energy physics to medicine. Each application necessitates the control of microstructure, the tailoring of foam morphology to enhance the end-use effectiveness of the specific foam. For instance, polyacrylonitrile (PAN) microcellular foams were developed for the collection of comet dust in outer space, engineered with very small cell sizes to facilitate particle deceleration. Carbon foams have been produced, from polymeric precursors, for use as electrodes, insulation, filters, composite components, and catalyst supports *(4)*. In the preparation of morphologically-tailored microcellular carbon foams, PAN foams were produced using TIPS  and then carbonized in an inert atmosphere to form a carbon microcellular foam. Foam density and morphology are readily controlled using this method. One application for these carbon foams is their use in composite constructs. Polymers or polymer-precursors can be drawn into the foam to produce composite materials. The electrical conductivity of these composites can be controlled, facilitating their use as shielding or absorbing electrical devices or device components *(4)*. The mechanical characteristics of these composites are also controlled by varying their composition. Although these foams are made for non-medical applications, their possible future use in biomedical devices should not be overlooked.

Numerous biomedical applications have been suggested for polymeric microcellular foams including their use as "timed-release drug delivery systems, vascular replacements, neural regeneration pathways, and artificial bone templates" *(4)*. The kinetics of drug release can be manipulated through control of foam morphology. Availability of absorbable polymeric foams will broaden the scope of potential biomedical foam applications. Such foams could enhance the effectiveness of drug delivery systems, providing suitable substrates for the modulated release of traditional drugs and/or proteins. Additionally, different types of microcellular matrices may find use in other applications such as gas and fluid filtration, constructs containing immobilized chemical and biological reagents for continuous chemical and biochemical processing, and three-dimensional constructs for tissue regeneration.

**Current Topics Relevant to Potential Biomedical Applications**

Advances in microcellular foam production range from refinements of existing technology to the development of new methods facilitating the formation of foams with tailored physical and chemical characteristics. One of the most promising areas of new foam technology is that concerned with the production of polymeric aerogels. Aerogels are traditionally used in high-energy physics applications and are members of a class of materials that LeMay et al. (1990) term "low-density microcellular materials." In their work, these investigators emphasized the unique physical nature of such materials. Unlike other microcellular materials, the cells of aerogels are extremely small in diameter (ranging from 0.1 to 0.3 μm) and the density of the material is quite low (0.01 to 0.30 g/cm$^3$) *(5)*. The kinetics of polymerization are responsible for these unique properties. The unusual physical characteristics of aerogels are associated with their uncommon mechanical, thermal, and optical properties. For instance, because of their minimal cell size, transparent aerogels can easily be produced.

Aerogels may have applications beyond high-energy physics. The ability to control the morphologies of aerogels could be exploited to create a new class of bioactive materials. Although the aerogel cell size is too small to encourage tissue ingrowth, the high surface area of these materials may be ideal for efficient filtration, fractionation, or biochemical processing.

Extensive research into control of the TIPS process is ongoing at the University of Texas at Austin where Lloyd and his colleagues have learned to control the TIPS process to produce polymeric microcellular membranes with unique and varied microstructures. Primarily, isotactic polypropylene (iPP) has been used in the experimental work. Additionally, microcellular membranes have been produced from high-density polyethylene, poly(chlorotrifluoroethylene), poly(4-methyl-1-pentene), and poly(vinylidene fluoride) *(10-11)*. Thus, mixtures of polymers and diluents are subjected to thermally-induced phase separation to produce microcellular membranes. The morphology that results is dependent on processing parameters, including the relative concentration of polymer and diluent in the mixture. Extraction is used to remove the diluent, revealing the polymeric microstructure produced by the induced phase separation. The resulting morphologies are varied and difficult to describe but can be characterized by the observance of pores, beads, flakes, stacked layers, struts, and myofibrils. In each of the microcellular membranes, one of these structures predominates, giving the membrane unusual structural characteristics.

A new foam formation process has been developed in this laboratory which has been used to produce polymeric microcellular foams with unique morphologies. Homochain and heterochain polymers have been converted into microcellular foams by this general, reproducible process. Foams of heterochain polymers could be made without discernible changes in polymer molecular weight. Micrographs of two representative foams are shown in Figures 1 and 2. Figure 1 illustrates a Nylon 12 open-cell foam with cells ranging from 5 to 30 μm in diameter. This foam exhibits a relatively high degree of uniformity in its cell size, shape, and dispersion. The distinct microstructure of a polypropylene open-cell foam is shown in Figure 2. SEM

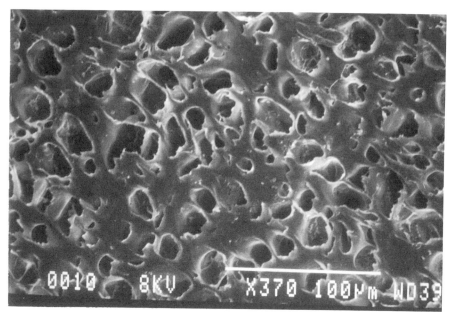

Figure 1.  Nylon 12 microcellular foam.

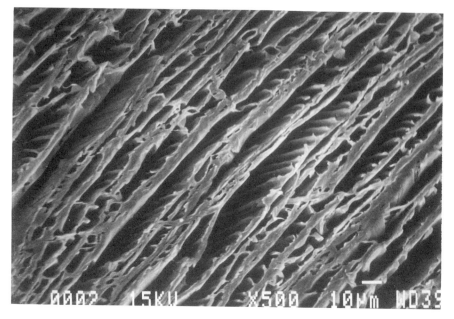

Figure 2.  Polypropylene microcellular foam.

analysis of both longitudinal and transverse sections of this foam have revealed that the foam is composed of stacked layers, spaced about 10 μm apart. Ridges, spaced approximately 5 μm apart, were observed to run longitudinally through the foam. At times, these ridges were observed to form pseudo-channels through the sample.

The successful production of unique foams from these and other polymeric materials indicates that the new process, developed in this laboratory, has potential application for the design of foam biomaterials with tailored morphologies. For many biomedical applications, there is a need for polymeric biomaterials which have a high surface area. Polymeric microcellular foams are potentially valuable for vascular prostheses, hard tissue regeneration, and time-modulated drug delivery systems. The ability of polymeric microcellular foams to be engineered morphologically underscores the potential importance of these materials as biomaterials.

References
1. Gibson, L. J.; Ashby, M. F. *Cellular Solids: Structure and Properties*; Pergamon Press: Oxford, England, 1988.
2. Aubert, J. H. *Journal of Cellular Plastics,* **1988**, *24*.
3. Williams, J. M.; Wrobleski, D. A. *Journal of Materials Science Letters*, **1989**, *24*, pp 4062-4067.
4. Aubert, J. H.; Sylwester, A. P. *Chemtech*, **1991**, *21*, pp 234-238, 290-295.
5. LeMay, J. D.; Hopper, R. W.; Hrubesh, L. W.; Pekala, R. W. *MRS Bulletin*, **1990**, *15*, pp 19-45.
6. Stengard, R. A. In *Plastic Foams:*; Frisch, K. C. and Saunders, J. H., Eds.; Marcel Dekker, Inc.: New York, NY, 1972, Vol. 1; pp 397-449.
7. Rootare, H. M. *Advanced Experimental Techniques in Powder Metallurgy*; Plenum Press: 1970; pp 225-252.
8. Lines, R. W. *Laboratory Practice*, **1987**.
9. Lowell, S.; Shields, J. E.; *Powder Surface Area and Porosity*; Chapman and Hall: New York, NY, 1984; pp 17-28.
10. Lloyd, D. R.; Kinzer, K. E.; Tseng, H. S. *Journal of Membrane Science*, **1990**, *52*, pp 239-261.
11. Lloyd, D. R. et al. *Journal of Membrane Science*, **1991**, *64*, pp 1-11, 13-29, 31-53, 55-68.

RECEIVED June 2, 1993

# SYNTHESIS, SURFACE ACTIVATION, AND CHARACTERIZATION OF BIOMATERIALS

Chapter 6

# Surface Biolization by Grafting Polymerizable Bioactive Chemicals

Y. Ito, K. Suzuki, and Y. Imanishi

Department of Polymer Chemistry, Kyoto University, Kyoto 606–01, Japan

A general method to surface design of materials for biocompatibility was provided. This is based on the surface-grafting of polymerizable biological chemicals on the materials. For the purpose the plasma-pretreated polymerization method was employed and the polymerizable biological chemicals was used or were synthesized. To design blood-compatible materials poly(vinyl sulfonate) and polymerizable thrombin-inhibitor were used. In addition to design cell-adhesive materials the cell-adhesion-peptide, Arg-Gly-Asp-Ser was surface-grafted after connecting with a vinyl group.

Biocompatibility and medico-functionality (elasticity, permeability, *etc.*) are qualities required in materials for artificial organs (*1,2*). The former usually depends on the surface properties, and the latter on the bulk properties. Surface modifications to enhance biocompatibility is a useful method in designs which avoid interferring with bulk properties as shown in Figure 1 (*1,2*). In addition to immobilization of biological macromolecules or cells on the surface (*3-5*), the surface modification has been mainly considered to be the physico-chemical modifications such as the wettability control. So far very few examples of biomimetic approach using low molecular weight chemicals containing biological activity were carried out (*6*). Such an approach has some advantages, *i.e.*, the designed materials can be more biospecifically active than that designed by the physico-chemical approach and more stable than that containing biological chemicals which are, for example, thermo-sensitive.

We recently developed a new general method to biolize surfaces for biocompatibility, such as blood-compatibility and tissue-compatibility. The basic concept of surface biolization is surface-grafting molecules which is made polymerizable in advance. In the case of polymerizable biological molecules, direct surface-polymerization can be used. Otherwise, a vinyl group can be conjugated to the site which is not related with biological activity of the biological molecules to polymerize. Vinyl sulfonate was used in this study as a polymerizable bioactive molecule because of its heparinoid activity. Other examples such as a thrombin inhibitor and a cell adhesive active peptide, Arg-Gly-Asp-Ser (RGDS), were also employed after conjugating with polymerizable vinyl groups. The chemicals and surface modification method are illustrated in Figure 2.

0097–6156/94/0540–0066$06.00/0
© 1994 American Chemical Society

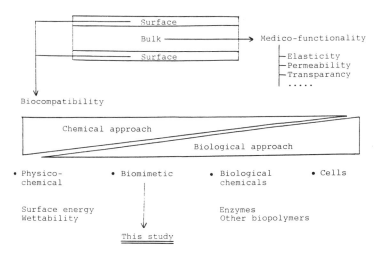

Figure 1.   Classification of approaches for biocompatible materials.

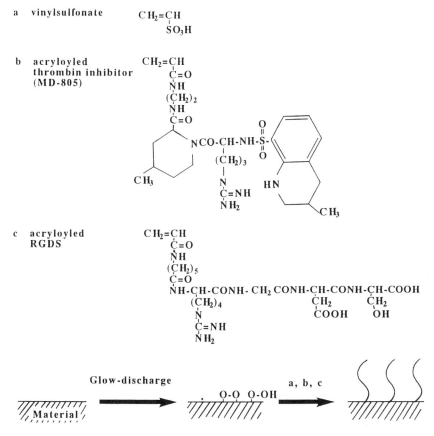

Figure 2.   Surface biolization by plasma-pretreated polymerization method.

### Surface Design for Blood-Compatibility

**Poly(vinyl sulfonate).** When an artificial material comes in contact with blood, thrombi are formed on the surface. The mechanism is illustrated in Figure 3. After the initial stage of blood protein adsorption, the platelet system and the intrinsic coagluation system are activated to form thrombus. Heparin has been the most widely used for inhibition of thrombus formation by inactivation of thrombin which is generated in the coagulation system and catalyzes fibrin formation. In addition to the immobilization of heparin onto materials (3), there have been a number of investigations of the syntheses of heparin-like (heparinoid) polymers (sulfonated polydienes, sulfated chitosan, sulfonated polystyrene, sulfonated dextran, and sulfonated polyurethanes) (7). However, in these earlier investigations, the synthetic methods were complicated and troublesome, and mechanical properties of the materials have not been taken into account. These properties are very important for practical use. Therefore, surface modified grafts of poly(vinyl sulfonate) by plasma-pretreated polymerization was performed on certain mechanically strong materials (8,9).
    Antithrombogenicity of water-soluble poly(vinyl sulfonate) (PVS) was nearly 7.5% that of heparin, though the activity depended on the molecular weight of PVS. Considering that the blood coagulation time in the presence of PVS was significantly prolonged by the addition of antithrombin III, the antithrombogenic effect was based on the interaction with antithrombin III as illustrated in Figure 4 (8,9). The higher molecualr weight of PVS should more siginificantly interact with antithrombin III to induce a confromational change of the protein because of the continuous sulfonate groups on PVS.
    PVS was grafted onto the surface of polyurethane, polystyrene, and poly(ethylene terephthalate) films by the plasma-pretreated method. Activated partial thromboplastin time (APTT) of PVS-grafted polyurethane films was greatly prolonged, and a fibrin network was not found at all on the film grafted with PVS in densities higher than 1.6 $\mu g/cm^2$, a level at which the surface is completely covered with PVS. *In vitro* thrombus formation on the film was suppressed with increasing amounts of grafted PVS, and no thrombus was formed during a 20-min contact with blood on the film with PVS grafted in densities higher than 1.6 $\mu g/cm^2$. The antithrombogenic mechanism is considered as shown in Figure 5a.

**Polymerizable Thrombin Inhibitor.** In order to directly deactivate thrombin without antithrombin III as shown in Figure 5b, a thrombin inhibitor was immobilized on polymer surfaces (10). Figure 6 shows a schematic drawing of the interaction of thrombin catalytic site with a thrombin inhibitor, which is named MD-805. Because the carboxylic group of MD-805 was not reported to be related to the inhibition activity, a vinyl group was connected to this group. No siginificant difference in the inhibition constant $K_i$ was found between MD-805 and its vinyl derivative. The polymerizable thrombin inhibitor was then grafted onto a polyurethane film. The film not only deactivated thrombin, but it also reduced platelet adhesion and activation, thus becoming antithrombogenic.

### Surface Design of Cell-Adhesive Materials

**Synthesis of RGDS-Immobilized Membrane.** Cell adhesion is a ubiquitous process that influences many aspects of cell behavior. For example, proliferation, migration, secretion, and differentiation of cells are triggered by adhesion to matrix. Since Pierschbacher and Ruoslathi found that the RGDS sequence in cell adheison proteins was an active site, a number of investigations into its applications have been carried

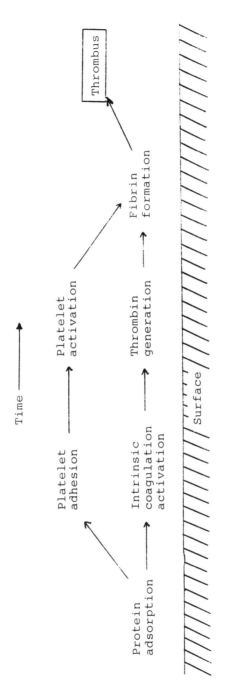

Figure 3. Sequence of events during blood-material interactions.

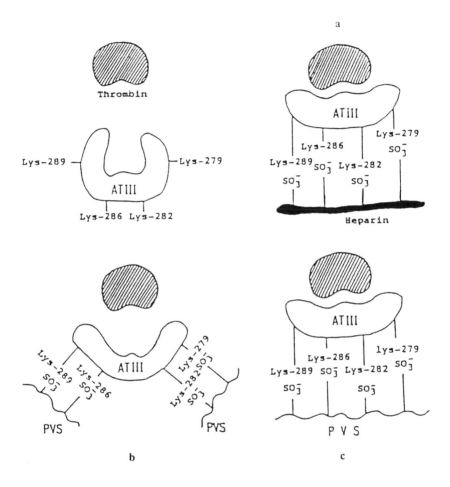

Figure 4.    Schematic representation of interactions among thrombin, antithrombiin III (ATIII) and heparin (a) or poly(vinyl sulfonate) (PVS) (b,c). The interaction is considered to depend on the steric position and continuity of sulfonate groups. The activity of PVS is less than that of heparin because of the difference of steric position of the sulfonate groups. On the other hand, the high molecular weight PVS (c) induces more conformational change of ATIII than the low molecular weight PVS (b) because of the continuity of the sulfonate groups.

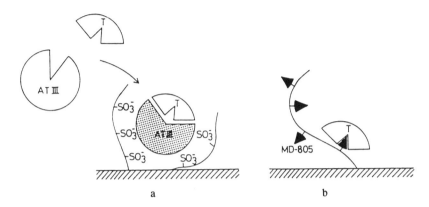

Figure 5.    Antithrombogenic mechanisms of (a) poly(vinyl sulfonate)-grafted and (b) polymerizable thrombin inhibitor-grafted materials.

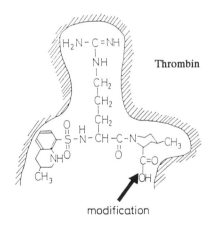

modification

Figure 6.    Schematic drawing of the interaction of thrombin with a thrombin-inhibitor, MD-805. The carboxylate group, which is coupled to a vinyl group, is indicated as the modification site.

out (11-13). The use of bioactive peptides instead of proteins has two principal advantages. One is that short-chain peptides are more resistant to the denaturation caused by ethanol, heat, and pH variations. Another is that peptides can be integrated in a large amount on the surface to compensate low unit activity when compared to the high molecular weight cell-adhesion proteins.

**Synthesis of $CH_2=CH-CONH-(CH_2)_5-COOSu$ (1).** 5-Aminhexanoic acid (5 g) and $Ca(OH)_2$ (5 g) were solubilized in water (75 ml). After cooling the solution in an ice bath, 3.5 ml of acryloyl chloride was added dropwise with vigorous stirring over 12 min. After the excess of $Ca(OH)_2$ was removed by filtration, the solution was acidified by concentrated HCl. The precipitate was washed with water and dried *in vacuo*. The product (1 g) was suspended in methylene chloride (20 ml), and *N*-hydroxysuccinimide (670 mg, 1 eq.) and 1-ethyl-3-(3-dimethylaminopropyl)carbodiimide (1.34 g, 1.2 eq.) were added to the suspension. After the suspension was stirred for 1 h at $0°C$, 15 ml of ethyl acetate was added and the mixture was stirred additionally for 15 min. The mixture was washed with $NaHCO_3$ aqueous solution and NaCl aqueous solution and dried on sodium sulfate. The product **1** was recrystallized from methylene chloride-diethyl ether. The yield was 768 mg (54 %). m.p., 122-124°C (Lit. 122-123°C).

**Coupling of RGDS to Product 1.** RGDS (150 mg), product **1** (5.0 eq.) and 48.5 ml (1.0 eq.) of triethylamine were dissolved in 10 ml of *N*,*N*-dimethylformamide and the mixture was stood overnight at room temperature. The coupling product (product **c** in Figure 1) was purified by LH20 and ODS (MeOH/H2O=3/7) columns and freeze-dried. Analysis calculated for $C_{24}H_{39}N_8O_{10}$: C, 47.99 %; H, 6.71 %; N, 18.99 %. Found: C, 48.23 %; H, 6.53 %; N, 18.52 %.

**Graft-Polymerization of Product c and Cell Adhesion Experiment.** A polystyrene (PST) film was glow-discharged (pressure; 0.2 Torr, electric current; 8 mA) over 1 min and immersed in aqueous solutions containing monomers (acrylamide, acrylic acid, *N*-(3-trimethylammoniumpropyl)acrylamide chloride, and product **c** of various concentrations for 8 h at 70°C. The grafted membrane was washed until no changes were observed in the washing solutions by means of pH and ultraviolet measurement.

The cell adhesion experiemnts were performed by using sub-cultured mouse fibroblast cells STO as reported previously (13).

**Cell-Adhesion Activity of the RGDS-Grafted Membrane.** Figure 7 summarizes the number of adhered mouse fibroblast cells STO onto various surface-grafted membranes. The glow-discharge treatment increased cell adhesivity. However, the acrylamide- or acrylic acid-grafting reduced the adhesivity markedly. On the other hand, the cationic polymer-grafted membrane had high cell adhesivity. These results indicate that hydrophilic polymer grafting reduced cell adhesion, however, that cationic polymers enhanced the adhesion. The membrane co-polymerized with product **c** promoted cell adhesion more than that polymerized with cationic monomer, and was comparable to a fibronectin-coated membrane in cell-adhesiveness. In addition, the co-polymerized film enhanced cell spreading more than cationic polymer-grafted film as shown in Figure 8. This study shows that the cell adhesion enhancement of the RGDS-immobilized materials was comparable to the fibronectin-coated materials as reported previously (11, 13)

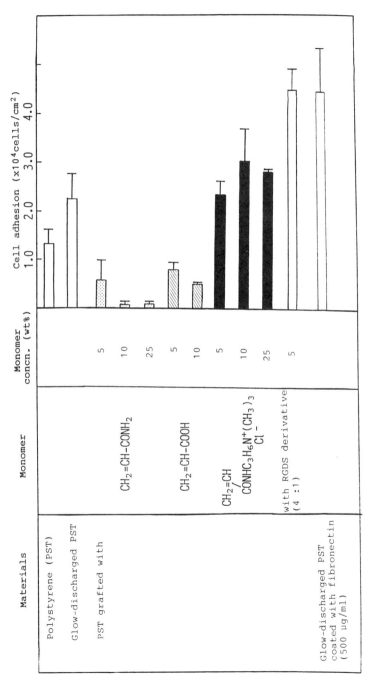

Figure 7.   Adhesion of mouse fibroblast cells STO onto various surface-grafted polystyrene films.

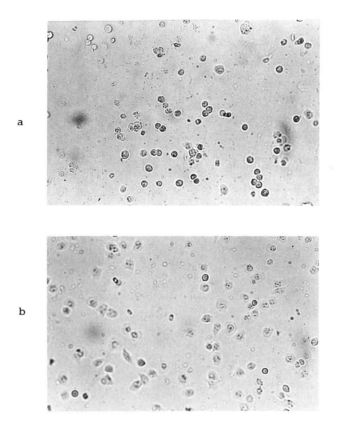

Figure 8.    Photograph of mouse fibroblast cells STO adhered onto (a) poly[N-(3-trimethylammoniumpropyl)acrylamide chloride]-grafted poly-styrene film and (b) poly[N-(3-trimethylammonium propyl chloride-co- RGDS)acrylamide]-grafted polystyrene film.

References
1. Ito, Y. In*Synthesis of Biocomposite Materials: Chemical and Biological Modifications of Natural Polymers;* Imanishi, Y., Ed.; CRC Press: Boca Raton, FL, 1992; pp15-84.
2. Ito, Y.; Imanishi, Y.*J. Biomat. Appl.* **1992**, *6,* 293-318.
3. Ito, Y.*J. Biomat. Appl.* **1987**, **2,** 235-265.
4. Ito, Y.; Imanishi, Y.*CRC Crit. Rev. Biocomp.* **1989**, *5,* 45-104.
5. Liu, S. Q.; Ito, Y.; Imanishi, Y.*J. Biomed. Mater. Res.* in press.
6. Ito, Y.; Liu, L.-S.; Imanishi, Y.*J. Biomat. Sci., Polym. Edn.* **1991**, *2,* 123-138.
7. Ito, Y.; Iguchi, Y.; Imanishi, Y.*Biomaterials* **1992**, *13,* 131-135.
8. Ito, Y.; Iguchi, Y.; Kashiwagi, T.; Imanishi, Y.*J. Biomed. Mater. Res.* **1991,** *25,* 1347-1361.
9. Ito, Y.; Liu, L. -S.; Imanishi, Y.*J. Biomed. Mater. Res.* **1991**, *25,* 99-115.
10. Ito, Y.; Liu, L. -S., Matsuo, R.; Imanishi, Y.*J. Biomed. Mater. Res.* **1992**, *26,* 1065-1080.
11. Ito, Y. In*Synthesis of Biocomposite Materials: Chemical and Biological Modifications of Natural Polymers;* Imanishi, Y., Ed.; CRC Press: Boca Raton, FL, 1992; pp245-284.
12. Imanishi, Y.; Ito, Y.; Liu, L. -S.; Kajihara, M.*J. Macromol. Sci. -Chem.* **1988,** *A25,* 555-570.
13. Ito, Y.; Kajihara, M.; Imanishi, Y.*J. Biomed. Mater. Res.* **1991,** *25,* 1325-1337.

RECEIVED May 13, 1993

# Chapter 7

# Hydrophilic, Lipid-Resistant Fluorosiloxanes

G. Friends, J. Künzler, R. Ozark, and M. Trokanski

Department of Chemistry and Polymer Development, Bausch and Lomb Inc., Rochester, NY 14692–0450

A contact lens material based on a methacrylate end capped poly (trifluoropropylmethylsiloxane), octafluoropentyl methacrylate and a wetting agent, 2-vinyl-4,4-dimethyl-2-oxazolin-5-one (VDMO), was developed. The methacrylate end capped fluoro siloxane was prepared by the co-ring opening polymerization of octamethylcyclotetrasiloxane with trifluoropropylmethylcyclotrisiloxane and a bis-methacrylate tetramethyldisiloxane. The copolymerization of the fluorosiloxane with varying concentrations of octafluoropentyl methacrylate and VDMO resulted in transparent, oxygen permeable, high strength, lipid resistant films. A copolymer composition of 80 parts of a methacrylate end capped 65 mole % trifluoropropyl siloxane, 20 parts of octafluoropentylmethacrylate and 5 parts VDMO gave a transparent, wettable film possessing a modulus of 120 g/mm$^2$, a tear strength of 5.8 g/mm, an oxygen permeability of 230 Barrers and an oleic acid uptake of only 1.4%.

The design of new polymeric materials for contact lens application requires an extensive knowledge of polymer chemistry, polymer properties and the physiology of the eye (1-4). The properties that must be optimized in designing a new contact lens material are: optical transparency, chemical and thermal stability, wettability to tears, mechanical properties, dimensional stability, biological compatibility and oxygen permeability.

Polydimethylsiloxane (PDMS) due to its low modulus of elasticity, optical transparency and high oxygen permeability has been extensively studied for soft contact lens application (5). PDMS possesses an oxygen permeability which is about 70 times higher than the oxygen permeability of the polymacon hydrogel poly(HEMA)(6). There are, however, two basic limitations of soft contact lenses based on PDMS. The first limitation is that PDMS based contact lenses have a high affinity for lipids (7). Lipid uptake can lead to swelling, deformation, cracking, loss of mechanical properties and discoloration. The second limitation is that PDMS based contact lenses are inherently non-wettable which leads to patient discomfort and poor visual acuity. The majority of research on PDMS based contact lenses has focused on methacrylate functionalized PDMS due to

0097–6156/94/0540–0076$06.00/0

the relative ease of methacrylate cure and property modification by copolymerization with other methacrylate monomers (8).

A variety of methods have been pursued to develop wettable PDMS based contact lens materials (9-10). The methods used to impart wettability include: plasma etching, plasma polymerization, copolymerization with a hydrophilic wetting agent and copolymerization with a protected wetting agent. The plasma based methods consist of either etching with an oxygen plasma or plasma polymerization of a hydrophilic monomer onto the lens surface. Acceptable, long term wettability is achieved using these methods, however, the plasma surface treatments are costly due to the additional manufacturing steps. The hydrophilic wetting agent method consists of copolymerization of the methacrylate functionalized PDMS with a hydrophilic wetting monomer such as hydroxyethyl methacrylate or methacrylic acid. This approach is also limited in that phase separation often occurs before acceptable wettability is achieved resulting in an opaque material. Alternatively, protected wetting agents have been used. This method consists of copolymerization of the methacrylate functionalized PDMS with a protected wetting agent, such as trimethylsilyl protected methacrylic acid. This method allows for the addition of large concentrations of wetting agent without phase separation. A wettable, clear film can be obtained by this route, however, a deprotection followed by an organic extraction is necessary to remove the deprotection by-products.

This paper describes a novel contact lens material based on a methacrylate end capped poly (trifluoropropylmethylsiloxane)(TFP), octafluoropentyl-methacrylate (OFPMA) and a wetting agent 2-vinyl-4,4-dimethyl-2-oxazolin-5-one (VDMO). The trifluoropropylmethyl siloxane, fluorinated methacrylate methacrylate and VDMO in the correct concentration provide a lipid resistant, oxygen permeable, hydrophilic contact lens material.

## Experimental
## Materials

Octafluoropentylmethacrylate (OFPMA-99% by GC) was purchased from PCR Inc. and was distilled before use (bp. 75-80°C). Isobornyl methacrylate (IBOMA-99+% by GC) was purchased from Rohm and Haas and was distilled before use (bp. 70°C/0.1mm Hg). VDMO was purchased from SNPE Inc. and was used as received (purity by GC 95%). T-butylperoctoate (TBO) was purchased from Attochem and was used as received. Octamethylcyclo-tetrasiloxane ($D_4$) and the trifluoropropylmethylcyclotrisiloxane ($D_3$-TFP) were purchased from Hüls America and were used as received. The 1,3-bis methacryloylbutyltetramethyl disiloxane ($M_2$) was prepared according to a procedure described in the patent literature (11). All other solvents and reagents were used as received.

## Synthesis of a DP 100 methacrylate end capped poly (65 mole % trifluoropropylmethylsiloxane)-co-(35 mole % dimethylsiloxane) (65-TFP) (Scheme 1)

Octamethylcyclotetrasiloxane (39.4g, 0.133 mole) trifluoropropylcyclotrisiloxane (154.3g, 0.33mole) and methacryloylbutyltetramethyldisiloxane (6.3g, 0.015mole) were added at room temperature to a round bottom flask under dry nitrogen. Trifluoromethanensulfonic acid (0.54g, 3.6mmole) was added and the reaction mixture was stirred for 24 hours. Sodium bicarbonate was then added to the viscous reaction product and the stirring continued for 16 hours. Following the neutralization procedure, chloroform (500mls) was added and the solution was dried over magnesium sulfate and filtered using a 5$\mu$ millipore Teflon filter. The filtrate was placed on a rotary evaporator and the chloroform was removed. The resultant prepolymer was added dropwise with rapid stirring to 500ml of methanol to remove the unreacted cyclics. The polymer layer was collected and the procedure was repeated twice. Following the third fractionation, the polymer was collected, dissolved in diethylether, dried over magnesium sulfate and again filtered through a $5\mu$ filter. The resultant solution was placed on the rotary evaporator and the diethylether was removed. The molecular structure of the purified 65-TFP (150g, 75%) was confirmed by NMR spectroscopy. $^1$H-NMR (CDCl$_3$, TMS, ppm): 0.25 (s, 357H, Si-CH$_3$), 0.8 (m, 134H, Si-CH$_2$-), 1.5, (m, 4H, -CH$_2$-CH$_2$-O-), 1.9 (s, 6H, -CH$_3$), 1.95 (t, 134H, -CH$_2$-CF$_3$ and -Si-CH$_2$-CH$_2$-CH$_2$-), 4.1 (t, 4H, -CH$_2$-O), 5.6 (s, 1H, =C-H), 6.2 (s, 1H, =C-H).

**Techniques**

Monomer purity was determined on a Hewlett-Packard HP5890A GC using a 6.1m X 0.32cm column of 10% SP-1000 80/100 Supelcoport. Monomer structure was confirmed by 60 MHz $^1$H-NMR spectroscopy using a Varian EM 360 spectrometer. Films were cast between silanized glass plates using a 0.3mm teflon spacer. The glass silanization procedure consists of dipping the glass plates (8.3cm X 10.2cm) in a 20% v/v solution of trimethylchlorosilane in heptane followed by rinsing the glass plates in distilled water and air drying. The cure conditions were established through a thorough investigation of initiators and temperature to optimize vinyl conversion and percent extractables. Vinyl conversion was monitored by NIR spectroscopy using a Perkin-Elmer Lambda 9 spectrophotometer. The optimum cure conditions consisted of one hour at 60°C, one hour at 80°C and two hours at 100°C using 0.5% TBO as initiator. The resultant films were extracted overnight at 25°C in toluene. The toluene was removed by subjecting the films to a vacuum of 30mm Hg at 50°C for four hours. The resultant films were then boiled in buffered saline for two hours to open the oxazoladinone ring (Scheme 2). The hydrolytic test procedure consists of heating the films in buffered saline at 80°C and monitoring the weight loss and percent water at 3, 5, 7, and 14 days. Following the hydrolytic test procedure, the films are again extracted in toluene at 25°C to determine the final weight loss. The oleic acid uptake procedure consisted of soaking films in oleic acid (NF grade) at 50°C and monitoring the weight gain. The oleic acid uptake test is a simple screening procedure used to determine the fat solubility of the test films and lenses. The mechanical properties of films were determined on an Instron Model 4500 using ASTM methods 1708 and 1938. Oxygen permeability (DK) was determined using the polarographic probe method (*12*). Contact angle measurements were performed on a Rame-Hart Model A-100 Goniometer.

**Results and Discussion**

The goal of this study was to design a high oxygen permeable, low modulus, lipid resistant, wettable contact lens material based on a methacrylate end capped trifluoropropyl side chain (TFP) siloxane and a copolymerizable wetting agent 2-vinyl-4,4-dimethyl-2-oxazolin-5-one (VDMO). It was our hope that the fluorinated siloxane would generate a polymer possessing a low affinity for lipids, acceptable mechanical properties and high oxygen permeability. The VDMO wetting agent, due to its moderately polar ring structure, would be compatible with the trifluoroside chain siloxanes and, on ring opening to a highly hydrophilic methyl alanine, would provide for acceptable surface wettability without the formation of by-products.

The physical and mechanical properties which we hoped to achieve in this study included: a Youngs modulus between 20 g/mm$^2$ and 200g/mm$^2$, a tear strength greater than 3 g/mm, a DK greater than 50 barrers and water contents

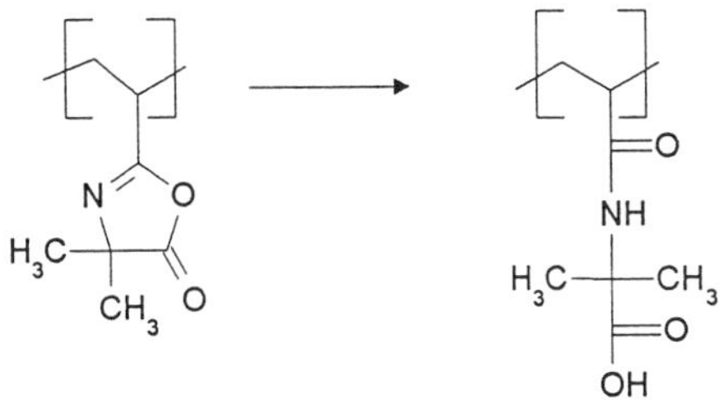

SCHEME 1.

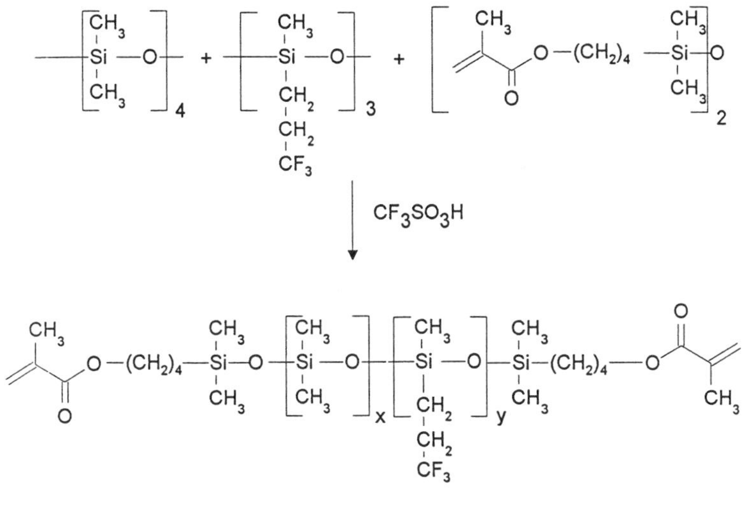

SCHEME 2

between 0 and 5%. These physical and mechanical property objectives were chosen based on clinical experience from a variety of commercial and experimental silicone lens materials.

Scheme 1 outlines the synthetic procedure used to prepare the methacrylate end capped trifluoropropyl side chain siloxanes. This consisted of the acid catalyzed ring opening polymerization of octamethylcyclotetrasiloxane ($D_4$), trifluoropropylmethylcyclotrisiloxane ($D_3$-TFP) and bis-methacryloylbutyltetra-methyldisiloxane ($M_2$). In this study a constant chain length of 100 was used and the mole substitution of trifluoropropyl side chain was varied. The 24 hour equilibration time was determined by monitoring the reaction using liquid chromatography techniques.

Films were first cast from the methacrylate end capped siloxanes with 12.5 mole%, 25 mole %, 40 mole % and 65 mole% of the trifluoropropyl side chain with varying levels of VDMO. The films were transparent and fully cured as determined by NIR spectroscopy. A significant decrease in oleic acid uptake occurred with an increase in the concentration of trifluoropropyl side chain. A 95/5 (65-TFP/VDMO) formulation absorbed 0.2% oleic acid, a 95/5 (40-TFP/VDMO) formulation absorbed 1.0% oleic acid and a 95/5 (DP 100 methacrylate end capped polydimethylsiloxane (no fluoro substitution)/VDMO) formulation absorbed 4% oleic acid following an overnight soak. The mechanical properties of these films, however, could not be evaluated due to their poor strength.

The next phase of this work consisted of attempting to improve the mechanical properties of the fluoro side chain siloxane films. Past research has shown that bulky monomers, such as isobornylmethacrylate or cyclohexylmethacrylate, improve the mechanical properties of siloxane based formulations by imparting rigidity to the polymer backbone (13). Films were cast from the 12.5 mole %, 25 mole %, 40 mole % and 65 mole % trifluoropropyl side chain siloxanes with varying concentrations of isobornymethacrylate (IBOMA) and VDMO. Table I summarizes the mechanical and physical property results from the TFP/IBOMA/VDMO films. An increase in modulus, tensile and tear strength occurred with increasing concentration of IBOMA. In addition, with an increase in the concentration of IBOMA a significant increase in oleic acid uptake occurs! The majority of films listed in Table I, in fact, result in oleic acid uptake values higher than pure polydimethylsiloxane (2-3%). Figure 1 shows the dependence of oleic acid absorption versus IBOMA concentration for a series of copolymers based on the 12.5 mole %, 25 mole %, 40 mole % and 65 mole % trifluoropropyl side chain siloxanes.

The final phase of this work consisted of attempting to improve the mechanical properties while maintaining the low oleic acid uptake of the trifluoropropyl side chain siloxanes. In these experiments, octafluoropentyl methacrylate was used as the strengthening agent. Films were cast from the 12.5 mole %, 25 mole %, 40 mole % and 65 mole % trifluoropropyl side chain siloxanes with varying concentrations of octafluoropentyl methacrylate (OFPMA) and VDMO. Table II summarizes the mechanical and physical property results from the TFP/OFPMA/VDMO films. A wide range in mechanical and physical properties were observed for the TFP/OFPMA/VDMO films. A dramatic increase in modulus, tensile and tear strength and a decrease in oleic acid uptake occurred with increasing concentration of OFPMA. Figure 2 shows the dependence of oleic acid absorption for a series of copolymers based on the 12.5 mole %, 25 mole %, 40 mole % and 65 mole % trifluoropropyl side chain siloxanes with varying amounts of OFPMA. A decrease in oleic acid uptake occurs with an increasing concentration of trifluoropropyl side chain. The 80/20/5 [65-

Table I. Mechanical and Physical Property Results for the TFP/IBOMA/VDMO Based Formulations

| Formulation (wt. %) | Modulus $(g/mm^2)$ | Tensile $(g/mm)$ | Elong. (%) | Tear $(g/mm)$ | Oleic % Uptake |
|---|---|---|---|---|---|
| 65mole % | | | | | |
| 95/5/5 | 81 | 45 | 80 | 2.8 | 3.1 |
| 90/10/5 | 122 | 104 | 106 | 5.6 | 5.8 |
| 85/15/5 | 273 | 197 | 99 | 11.0 | 10.1 |
| 40 mole % | | | | | |
| 95/5/5 | 90 | 52 | 89 | 2.1 | 5.1 |
| 90/10/5 | 116 | 87 | 107 | 3.2 | 8.5 |
| 85/15/5 | 245 | 201 | 130 | 7.3 | 12.0 |
| 25 mole % | | | | | |
| 95/5/5 | 118 | 56 | 66 | 2.2 | 7.2 |
| 90/10/5 | 153 | 83 | 80 | 3.5 | 10.2 |
| 85/15/5 | 390 | 192 | 104 | 8.8 | 14.1 |
| 12.5 mole % | | | | | |
| 95/5/5 | 139 | 60 | 57 | 1.7 | 10.9 |
| 90/10/5 | 179 | 76 | 62 | 2.8 | 14.4 |
| 85/15/5 | 333 | 122 | 70 | 5.7 | 11.5 |

DK in units of $(cm^3 \cdot O_2(STP) \cdot cm)/(sec. \cdot cm^2 \cdot mmHg) \times 10^{-11}$
All formulations consist of 0.5 percent TBO as initiator

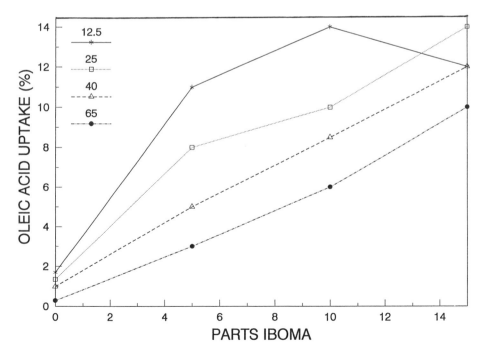

Figure 1. Dependence of oleic acid absorption versus parts IBOMA for a series of copolymers based on the 12.5 TFP, 25 TFP, 40 TFP and 65 TFP siloxanes.

Table II. Mechanical and Physical Property Results for the TFP/OFPMA/VDMO Based
Formulations

| Formulation (wt. %) | Modulus (g/mm²) | Tensile (g/mm) | Elong. (%) | Tear (g/mm) | Oleic % Uptake | DK |
|---|---|---|---|---|---|---|
| 65mole % | | | | | | |
| 80/20/5 | 120 | 136 | 135 | 5.8 | 1.4 | 234 |
| 60/40/5 | 409 | 427 | 145 | 17.2 | – | 206 |
| 55/45/5 | 712 | 400 | 136 | 32.0 | 0.7 | 240 |
| 50/50/5 | 1571 | 563 | 153 | 60.0 | 0.7 | 186 |
| 40/60/5 | 7630 | 730 | 135 | 158.0 | 0.7 | 102 |
| 20/80/5 | 27200 | 983 | 91 | 150.0 | 0.9 | 42 |
| 40 mole % | | | | | | |
| 80/20/5 | 127 | 89 | 102 | 5.6 | 1.5 | 246 |
| 60/40/5 | 224 | 248 | 120 | 12.2 | 1.1 | – |
| 55/45/5 | 709 | 535 | 125 | 25.0 | 1.0 | 267 |
| 50/50/5 | 1593 | 698 | 150 | 67.0 | 0.9 | 193 |
| 40/60/5 | 15000 | 725 | 177 | 218.0 | 0.9 | 67 |
| 20/80/5 | 34400 | 1030 | 63 | – | – | 30 |
| 25 mole % | | | | | | |
| 80/20/5 | 126 | 65 | 81 | 5.2 | 2.7 | – |
| 60/40/5 | 475 | 380 | 109 | 24.0 | 1.8 | – |
| 12.5 mole % | | | | | | |
| 80/20/5 | 148 | 104 | 97 | 4.6 | 3.4 | – |
| 60/40/5 | 544 | 453 | 106 | 22.0 | 2.3 | – |
| 40/60/5 | 2509 | 742 | 133 | 53.0 | 1.8 | – |
| 20/80/5 | 10760 | 887 | 142 | 241.0 | 1.3 | – |

DK in units of $(cm^3 \cdot O_2(STP) \cdot cm)/(sec. \cdot cm^2 \cdot mmHg) \times 10^{-11}$
All formulations consist of 0.5 percent TBO as initiator

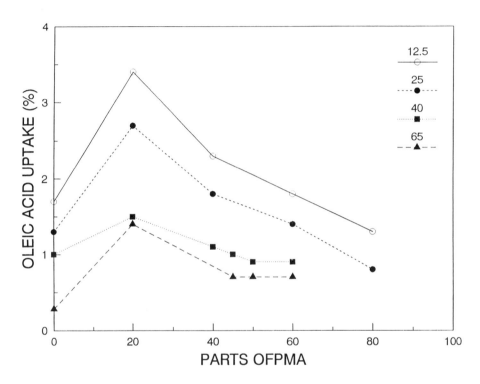

Figure 2.   Dependence of Oleic Acid Uptake versus parts OFPMA for formulations based on copolymers of 12.5 TFP, 25 TFP, 40 TFP and 65 TFP siloxanes with OFPMA and 5 parts VDMO.

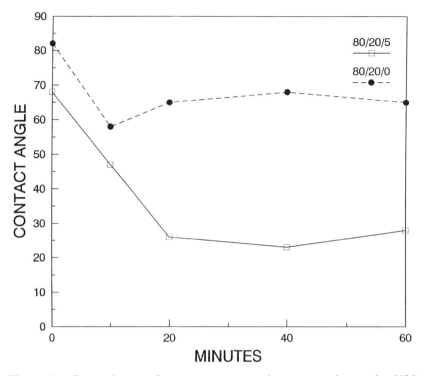

Figure 3.   Dependence of water contact angle versus minutes in 80°C buffered     saline     for     formulations     based     on     80/20/5     [65-TFP/OFPMA/VDMO and 80/20 [65-TFP/OFPMA].

TFP/OFPMA/VDMO] formulation had an oleic acid absorption of only 1.4%. An oleic acid absorption maxima is observed at 20 parts OFPMA with at present no explanation.  In addition, an increase in OFPMA concentration resulted in a decrease in % elongation and DK, however, the DK levels for all formulations are well above the level required for extended wear.    The 80/20/5 [65-TFP/OFPMA/VDMO] formulation resulted in a material possessing a tear strength of 5.8 g/mm, a modulus of 120 g/mm² and a DK of 234.  In contrast, the 60/40/5 [65-TFP/OFPMA/VDMO] formulation possessed a modulus of 7630 g/mm², a tear strength of 158 g/mm and a DK of 102.  All of the formulations had water contents of less than 1%.  Figure 3 shows the dependence of the sessile drop water contact angle versus time in 80°C buffered saline for a film containing 80 parts of 65-TFP, 20 parts of OFPMA and 5 parts of VDMO and for a film containing 80 parts of the 65-TFP, 20 parts of OFPMA and no VDMO. This plot clearly shows a significant drop in contact angle with 5 parts of VDMO following twenty minutes in 80°C buffered saline demonstrating the effectiveness of VDMO as a wetting agent.  Excellent on-eye wettability was achieved for the 80/20/5 [65-TFP/OFPMA/VDMO] formulation during short term clinical evaluation.  Finally, the hydrolytic stability of the TFP/OFPMA/VDMO films were measured.  Figure 4 shows the dependence of total weight loss (buffered saline and toluene extractable totals) versus heating time in buffered saline.  A significant increase in the total weight loss is observed with an increase in the concentration of VDMO, presumably due to an acid catalyzed degradation of the silicone backbone.

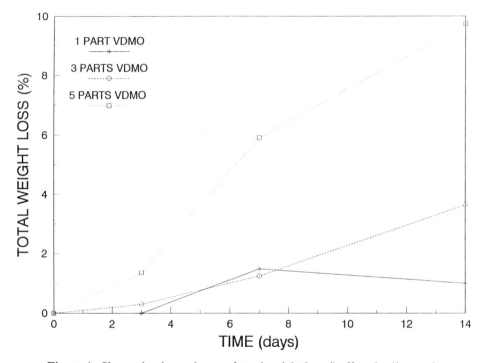

Figure 4. Shows the dependence of total weight loss (buffered saline and toluene extractable totals) versus heating time in 80°C buffered saline.

## Conclusion

A new contact lens material based on copolymers of methacrylate end capped trifluoropropyl side chain siloxanes, octafluoropentyl methacrylate and a wetting agent 2-vinyl-4,4-dimethyl-2-oxazolin-5-one (VDMO) was developed. These materials possess excellent mechanical properties, high oxygen permeability, wettability and a low affinity for oleic acid. Extended clinical evaluation needs to be performed on these materials to evaluate the long term wettability and deposit resistance.

## Literature Cited

1) N.A. Peppas and W.M. Yang, *Contact Lens*, **1981**, **7**, p 300.
2) P.L. Keogh, *Biomat., Med. Dev., Art. Org.*, **1979**, **7**,p 307.
3) B.J. Tighe and D. Pedley, *Proceed. Intern. Macro. Symp. Intern. Union Pure and Appl. Chem.*, **1980**, **27**, p 24.
4) M.J. Refojo, in *Encyclopedia of Polymer Science and Technology, Suppl. 1*, N.M. Bikales Ed., Wiley-Interscience, New York, 1976, pp 195-205.
5) B. Ratner, in *Comprehensive Polymer Science*, S.L. Aggarwal, Ed., Pergamon Press, New York, 1988, Volume 7, p 222.
6) *Polymer Handbook*, J. Brandrup and E.H. Immergut, Eds., 3rd ed., Wiley-Interscience, New York, 1989, p. VI 435.
7) D.E.Hart, R.R. Tidsale and R.A. Sack, *Ophthalmology*, **1986**, **93**, p 495.
8) J.W. Boretos, D.E. Detmer and J.H. Donachy, *J. Biomed. Mater. Res.*, **1971**, **5**, p 373.
9) C.P. Ho and H. Yasuda, *Polym. Mater. Sci. Eng.*, **1987**, **56**, p 705.
10) D.W. Fakes, J.M. Newton, J.F. Watts and M.J. Edgell, *Surf. Interface Anal.*, **1987**, **10**, p 416.
11) G. Friends and J. Künzler, U.S. Patent, 4,810,764, 1989.
12) I. Fatt, J.E. Rasson and J.B. Melpolder, *International Contact Lens Clinic*, **1987**, **14**, p. 38.
13) G. Friends, J. Melpolder, J. Künzler and J. Park, U.S. Patent, 4,495,361, 1985.

RECEIVED April 22, 1993

# Chapter 8

# Synthesis of Poly(ether urethane amide) Segmented Elastomers

A. Penhasi, M. Aronhime, and D. Cohn

Casali Institute of Applied Chemistry, The Hebrew University
of Jerusalem, 91904 Jerusalem, Israel

A new series of polyurethane block copolymers was synthesized, based
on the use of dicarboxylic acids as novel chain extenders.
Poly(tetramethylene glycol) of molecular weight 2000 was used as the
soft segment, which was reacted in a 1:2 ratio with hexamethylene
diisocyanate. Amide groups were formed in the reaction of isocyanate
groups with carboxylic groups. The effect of the type of diacid on the
kinetics of the polymerization was determined by monitoring both the
disappearance of isocyanate groups and the increase in intrinsic viscosity.
The type of diacid had a pronounced effect on the kinetics, with the rate
of reaction of diacid with isocyanate in the following order: oxalic acid >
fumaric acid > sebacic acid. The kinetics of a model system comprising
two moles of isocyanate and one mole of diacid were found to follow
second order kinetics modified by the addition of an autocatalytic term.
The polymer was characterized by elemental analysis, infrared
spectroscopy, nuclear magnetic resonance spectroscopy, viscometry and
thermal analysis. The time of the reaction was found to be important in
determining the structure and properties of the segmented polymer, as
observed by thermal analysis and viscometry. In addition, polymers
prepared with oxalic and fumaric acids were of lower molecular weight
when synthesized in THF as opposed to DMA, probably due to the
lower compatibility of the polymers formed in THF.

Thermoplastic polyurethane elastomers are one of the most important groups in the
class of thermoplastic elastomers (1-5). A typical polyurethane elastomer is a block
(or segmented) copolymer of the type (AB)$_n$, which is composed of alternating hard
and soft segments which are generally not compatible chemically. This incompatibility
leads to phase separation and the creation of micro-domains, in which hard block
domains are dispersed in the matrix of the soft segment. The hard segments serve as
physical crosslinking sites due to their local order at normal use temperatures and
impart strength and prevent flow. On the other hand, the soft segment glass transition
temperature ($T_g$) is normally well below the use temperature, thus imparting flexibility
to the system. The combination of physical crosslinks and flexibility allows for the
very high elongations typical of these thermoplastic elastomers (6). One additional

0097–6156/94/0540–0087$06.00/0

advantage of these kinds of materials is the fact that above the dissociation point of the hard segments, the material flows and can be processed like a normal thermoplastic.

In general, polyurethanes are chain extended by diols or diamines of low molecular weight, resulting in polyurethanes or poly(urethane urea)s, respectively. The hard block molecular weight usually varies between 300-3000 (7). The soft segment is often a poly(glycol ether) of molecular weight on the order of several thousand.

In the last few years our laboratory has developed a new group of polyurethanes based on the use of carboxylic acids as chain extenders (8-10). A carboxylic acid group, in the presence of a suitable catalyst, is capable of reacting with an isocyanate (11). An aliphatic isocyanate and an aliphatic carboxylic acid react to give an intermediate anhydride compound, which dissociates at normal temperatures to yield carbon dioxide and a stable amide (12):

$$
\underset{}{RNCO + R'COOH} \;\rightarrow\; RNHC \overset{O}{\underset{\parallel}{-}} O - CR' \overset{O}{\underset{\parallel}{}} \;\rightarrow\; RNHCOR' + CO_2\uparrow
$$

On the other hand, aromatic isocyanates react with aliphatic acids or weak aromatic acids at intermediate temperatures to yield acid anhydride, urea and carbon dioxide; the anhydride and urea can react at higher tempertures (160 °C) to form an amide and carbon dioxide (12).

The advantage in using diacids as chain extenders stems from the ability of the resulting amide groups to form strong and stable hydrogen bonds, which enhances phase separation and the formation of strong, rigid and often crystalline hard segment domains. The amide group is also more stable than the urethane linkage formed in the reaction between an isocyanate and a hydroxyl group, in that the urethane group can dissociate to the original reactants whereas the amide group cannot (13). In addition, unsaturated diacids (e.g., in fumaric, maleic and itaconic acids) can be used, which can further enhance phase separation due to the planar and rigid double bond; thus, the hard segments can be even stronger and more rigid. More importantly, the active double bonds  can also serve as sites for grafting of biologically active molecules, which can allow tailoring of the polymer in terms of its biological properties, in accordance with its specific clinical use (9,14).

In this work we investigated the kinetics of the reaction between diisocyanate and dicarboxylic acid, in the formation of the poly(ether urethane amide)s (PEUAms), by monitoring the disappearance of NCO groups and the increase in intrinsic viscosity of the polymer at different stages of the reaction. In addition, model systems were studied, consisting only of the diisocyanate and the diacid, in order to gain further insight into the kinetic mechanism of the reaction. Three diacids were studied: (1) oxalic acid, HOOC-COOH, the shortest and most active dicarboxylic acid; (2) fumaric acid, HOOCCH=HCCOOH, an unsaturated diacid of moderate activity; and (3) sebacic acid, HOOC-$(CH_2)_8$-COOH, a longer diacid of lower activity. We also investigated the importance of reaction time on overall conversion, and its effect on the phase separation and domain formation. Finally, the effect of reaction solvent was studied by comparing the intrinsic viscosity obtained with DMA as the reaction solvent as opposed to THF.

## EXPERIMENTAL

**Materials.** Hexamethylene diisocyanate (HDI) and dibutyl tin dilaurate (DBTDL) were purchased from Aldrich and were used as received. Sebacic acid (Aldrich), fumaric acid (BDH), and oxalic acid (Sigma) were dried overnight in vacuum at 60 °C prior to use. Poly(tetramethylene glycol) of molecular weight 2000 (PTMG, Polyscience) was dried overnight in vacuum at 80 °C before use. N,N-

Dimethylacetamide (DMA, Aldrich) was dried over 4 Å molecular sieves 48 hr prior to its use in the synthesis. Tetrahydrofuran (THF, Frutarom) was dried over KOH during one night and afterwards in a Na reflux for eight hours, followed by distillation with KOH pellets. N-butylamine (Aldrich) was dried over 4 Å molecular sieves for 48 hr, and 0.1 N $H_2SO_4$ (BDH) was used as received; these materials were used in the titration for determining isocyanate concentration. The quantity of water in the various materials was determined by Karl Fischer titration using a Mettler titrator. In all the materials which were used with the isocyanate the amount of water did not exceed 0.005%.

**Synthesis of model system.** A model hard block system was prepared by reacting two moles of diisocyanate to one mole of diacid, according to the following procedure. Into the reactor were introduced 10 mmole (1.68 gm) of HDI and 0.2 gm of the catalyst DBTDL, which were dissolved in 20 ml of DMA. The reactor was maintained at 67-70 °C in an oil bath, and the reaction was conducted in an atmosphere of dry nitrogen. A solution of 5 mmole of diacid in 20 ml DMA was added to the reactor contents. The reaction was continued for specific periods of time, and then stopped in order to take a sample and determine by titration the quantity of NCO remaining.

**Synthesis of poly(ether urethane amide).** The synthesis of these materials was conducted in two stages, consisting of a prepolymer stage followed by a chain extension step. In the first stage, 5 mmole of HDI dissolved in 12 ml DMA, with 0.2 gm DBTDL, were introduced into the reaction vessel. The reaction vessel was maintained at 67-70 °C and the reaction was conducted in an atmosphere of dry nitrogen. A solution of 2.5 mmole PTMG-2000 was dissolved in 25 ml DMA, which was added dropwise to the reactor during the course of 60 min. At the end of the glycol addition, the reaction was continued for an additional 60 min at the same temperature. For the second stage, 2.5 mmole of diacid dissolved in 12 ml DMA were added to the prepolymer during the course of 60 min. After the completion of the diacid addition, the reaction was stopped at different times to determine the amount of remaining NCO ( time zero was taken to be the time at the end of the diacid addition). After a sample was taken, excess methanol was added to the reactor to block the isocyanate groups. The polymer was then precipitated in deionized water at room temperature, washed several times with water, and the resulting white flakes were dried in vacuum overnight at 30-35 °C.

**Polymer analysis.** Several techniques were used to evaluate the polymers synthesized in this work, including intrinsic viscosity, NCO titration, calorimetry, Fourier Transform infrared spectroscopy (FTIR), and nuclear magnetic resonance spectroscopy (NMR).
 The intrinsic viscosity of the various polymers was determined in DMA at 25 °C. The amount of NCO remaining as a function of reaction time was determined by titration with butylamine using a standard procedure (*15*). The thermal transitions were determined in a differential scanning calorimeter (DSC) (Mettler TA 4000 Thermoanalyzer), in temperature scans from -120 to 350 °C at a heating rate of 10 °C/min, under dry $N_2$. Sample weights of 10-15 mg were used. Infrared spectra were recorded with an Analect FX-6260 Fourier Transform Infrared Spectrophotometer (FTIR) in the range of 400-4000 $cm^{-1}$. Spectra were recorded on samples cast from DMA onto NaCl plates, which were dried in vacuum overnight at 40 °C. Proton NMR spectra were determined on a Bruker AMX 400 H-NMR spectrometer at room temperature using deuterated DMF (dimethyl formamide) as the solvent. Finally, the polymers were subjected to elemental analysis to compare the theoretical with the actual composition.

## RESULTS AND DISCUSSION

**Kinetic Studies.** The model hard segments were prepared by reacting HDI with a diacid in the ratio of 2:1. For the model hard segments, 10 mmole HDI were reacted with 5 mmole diacid in DMA, as described above in the experimental section. The disappearance of NCO groups was followed by titration, and the results for the three diacids are shown in Figure 1. The strong effect of the type of diacid on the disappearance of NCO groups is evident from this figure. Oxalic acid resulted in the most rapid reaction, followed by fumaric acid and then by sebacic acid. These results are consistent with the reactivities of the carboxylic groups of the various diacids, in that there is a rapid decrease in reactivity for the first few members of a homologous series (*16*).

The kinetics of polyurethane reactions are usually characterized by second order kinetics (*17-20*), as follows:

$$\frac{dn}{dt} = -k_1 \, n \, c \tag{1}$$

where n = concentration of NCO groups at time t and c = concentration at time t of the second reactant, which in our case is carboxylic acid. The concentrations of NCO and COOH are related by stoichiometry:

$$n_0 - n = \text{moles NCO reacted} = \text{moles COOH reacted} = c_0 - c$$

where $n_0$ = moles of NCO at time 0
      $c_0$ = moles of COOH at time 0

Thus, $c = c_0 - n_0 + n = a + n$ where $a = c_0 - n_0$. Substituting into equation (1) yields

$$\frac{dn}{dt} = -k_1 \, n \, (\alpha+n) \tag{2}$$

Integrating results in

$$\frac{1}{\alpha} \ln \left(\frac{n+\alpha}{n}\right) = k_1 t + \text{constant} \tag{3}$$

A plot of the left-hand-side (L.H.S.) of equation (3) vs time should result in a straight line of slope $k_1$ if the reaction follows second-order kinetics. The results for the three acids are shown in Figure 2. The values of the second-order rate constant $k_1$, taken from the initial linear portions of the curves, are shown in Table I. As expected, oxalic acid has the highest value, followed by fumaric and then sebacic. Even thought the general trend is clear, the values obtained for oxalic acid in this and the following analyses should be treated somewhat cautiously due to the limited number of data points. All three figures indicate that the reactions are apparently initially second-order, but that at higher conversions large positive deviations are observed, indicating autocatalysis or some other change in the reaction mechanism (*18,19*). Autocatalysis by the urethane groups in the isocyanate-alcohol reaction is well known (*17,18,21*), whereas the autocatalytic effect of the amide groups formed by the isocyanate-carboxylic acid reaction has not been reported, to the best of the authors' knowledge. However, a catalytic effect of added amides and ureas in the amine-isocyanate reaction has been observed (*21*). It can also be seen in Figure 2 that the deviations from second-order kinetics become stronger in the order sebacic<fumaric<oxalic, consistent with a decrease in chain length and increase in reactivity. In other words, the

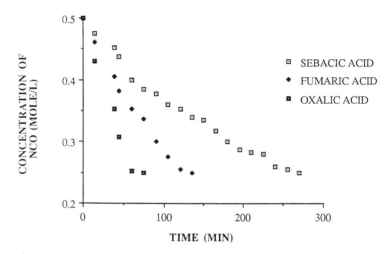

Figure 1: Isocyanate concentration (mole/liter) vs time (min) for the three diacids, for the model system.

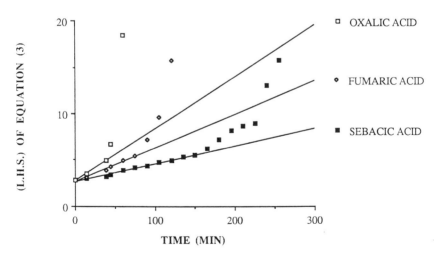

Figure 2: The left-hand-side of Equation (3) vs time (min) for the three diacids, indicating the apparently second-order nature of the reactions for short times.

autocatalytic effect becomes more pronounced as chain length decreases (see later). Also note from Table I that the second-order model is appropriate for all three acids to about the same %NCO conversion, about 60%.

**Table I: Summary of Kinetic Data for Model Systems**

| Acid | $k_1$ 2nd order | Until % Conversion | K | $k_2$ | Until % Conversion |
|------|-----------------|--------------------|------|-------|---------------------|
| Oxalic | 9.35 | 59 | 3.5 | 32.7 | 77 |
| Fumaric | 6.04 | 65 | 4.3 | 26.0 | 90 |
| Sebacic | 3.19 | 66 | 3.0 | 9.6 | 88 |

Units of rate constants: $k_1$ : liter/mole•sec x $10^4$
$k_2$ : $liter^2/mole^2$•sec x $10^4$

In light of the above statements, a more appropriate kinetic model, accounting for the autocatalytic effect of the formed amide group, is given by the following equation:

$$\frac{dn}{dt} = -k_1 \ n \ c - k_2 \ n \ c \ m \tag{4}$$

where m = moles of the amide group at time t and $k_2$ is the rate constant associated with the autocatalytic mechanism. By stoichiometry

moles NCO reacted = moles COOH reacted = moles amide formed

or

$$n_0 - n = c_0 - c = m$$

Thus, equation (4) can be written as

$$\frac{dn}{dt} = -k_1 \ n \ (\alpha+n) - k_2 \ n \ (\alpha+n) \ (n_0-n)$$

or

$$\frac{dn}{dt} = -k_1 \ n \ (\alpha+n) \ [1 + \frac{k_2}{k_1} \ (n_0 - n)] \tag{5}$$

If we let $K = k_2/k_1$ and $\beta = 1 + Kn_0$ then

$$\frac{dn}{dt} = -k_1 \ n \ (\alpha+n) \ [\beta - Kn] \tag{6}$$

Equation (6) can be rearranged to yield

$$\frac{dn}{n \ (\alpha + n) \ (\beta - Kn)} = -k_1 \ dt$$

which is integrated to

$$-\frac{\ln n}{\alpha\beta} + \frac{\ln (n + \alpha)}{\alpha (\beta + \alpha K)} + \frac{K \ln (\beta - Kn)}{\beta (\beta + \alpha K)} = k_1 \ t + \text{Constant} \qquad (7)$$

Thus, a plot of the L.H.S. of equation (7) vs. time should yield a straight line of slope $k_1$, which should be equivalent to the $k_1$ obtained from equation (3). A trial-and-error approach was used, with different values of K, to find the appropriate values which yield $k_1$ from equation (7) equivalent to that from equation (3). The results of this analysis are shown in Figure 3 for the three acids. The values of K obtained for the three acids are reported in Table I along with the other kinetic data.

The values of $k_2$ decrease in the order oxalic>fumaric>sebacic, which is consistent with the increasing distance between the two ends of the diacid. The value of K is greater for fumaric acid than that for sebacic, which could also be explained by the shorter chain length, thus resulting in a stronger autocatalytic effect for fumaric acid as opposed to sebacic acid. The value of K for oxalic acid is perhaps somewhat suspect due to the limited number of data points avaliable for analysis.

Table I also shows that the modified kinetic analysis is applicable for conversions up to about 90%. Beyond 90% conversion, more complicated kinetic mehcanisms, as well as limitations in the experimental procedure, could limit the agreement between data and model.

**Polymerization Reaction.** In order to demonstrate the success of the carboxylic acid-isocyanate reaction in the course of the polymerization, as opposed to the model systems, the polymers were examined by FTIR, NMR and elemental analysis. Figure 4a shows the infrared spectrum of a typical PEUAm at the end of the reaction, with fumaric acid as the chain extender (PEUFA). First of all, it can be seen that the characteristic peak associated with NCO, at about 2270 cm[-1], does not appear, indicating the completion of the isocyanate reaction. The peak at about 3325-3330 cm[-1] can be assigned to N-H groups which participate in hydrogen bonds (*22,23*).

A detailed spectrum of the carbonyl region is shown in Figure 4b. The peak associated with H-bound carbonyl amide appears at 1675-1680 cm[-1] (Figure 4b); the carbonyl amide results from the reaction between carboxylic acid and isocyanate. There is no peak (although perhaps a shoulder is evident) at 1690 cm[-1], which would indicate free carbonyl amide; thus almost all the amide groups apparently participate in hydrogen bonding. In contrast to the H-bound amide groups, the peak at 1722 cm[-1] can be associated with a carbonyl urethane group (the result of the HDI-PTMG reaction) which is not H-bound (*22,23*). Finally, there is a shoulder at 1705 cm[-1] which indicates a small quantity of hydrogen-bound urethane groups (*22,23*).

The H-NMR spectrum of PEUAm with fumaric acid as the chain extender (PEUFA) is shown in Figure 5. The peak assignments are listed in the figure. The NMR spectrum of PEUOXA is similar to that of PEUFA, except for the peak associated with the double bond of fumaric acid, which appears at 6.6 ppm.

Finally, the samples were analyzed for their elemental compositions, and the results for PEUFA are listed in Table II. It can be seen that the experimentally found values correspond with the theoretical ones, lending additional support to the proposed reaction mechanism shown in Figure 6 for the synthesis of PEUAms.

The effect of the different diacids on the polymerization kinetics (%NCO conversion) is summarized in Figure 7. In this figure, time zero refers to the end of the addition of the diacid. It can be seen that the same order of reactivity as observed in the model systems is preserved in the polymerizations. At time zero, for oxalic acid, 69% of the NCO which was in the system at the end of the prepolymer reacted, compared to 48% for fumaric acid and 35% for sebacic acid. The chain extension

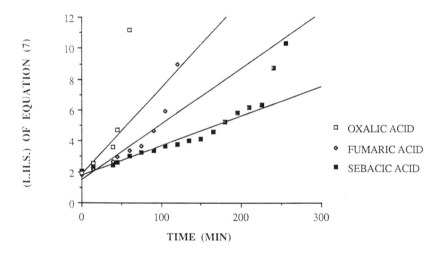

Figure 3: The left-hand-side of Equation (7) vs time (min) for the three diacids, indicating good agreement between the model incorporating an autocatalytic term and the data.

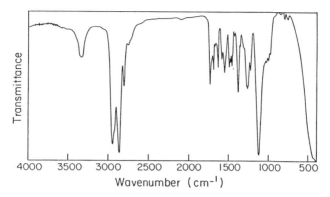

Figure 4a: Infrared spectrum of a typical PEUAm, in this case using fumaric acid as the chain extender.

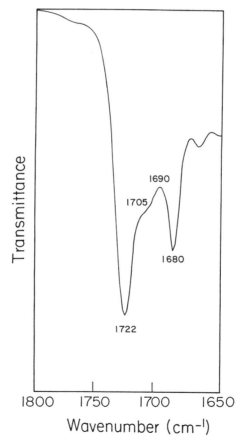

Figure 4b: Infrared spectrum of the carbonyl region of a fumaric acid-chain extended PEUAm.

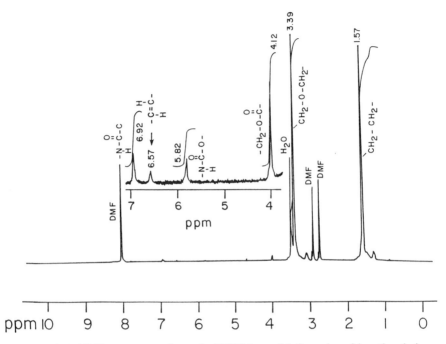

Figure 5: H-NMR spectrum of a typical PEUAm, with fumaric acid as the chain extender.

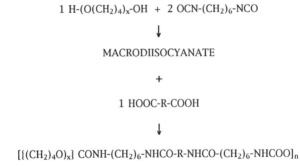

Figure 6: Proposed reaction mechanism for PEUAm synthesis.

## Table II: Elemental Analysis for PEUFA

| | Carbon (%) | Hydrogen (%) | Nitrogen (%) | Oxygen (%)* |
|---|---|---|---|---|
| Calculated | 65.0 | 10.6 | 2.3 | 22.1 |
| Found | 64.8 | 10.8 | 2.3 | 22.1 |

*% Oxygen calculated by difference

stage was completed after 45 min with oxalic acid, as opposed to 120 min for fumaric acid.

It can be seen clearly that the reactions with oxalic and fumaric acids go essentially to completion, whereas the reaction with sebacic reaches only about 93% conversion of NCO. The reaction with the latter was continued for 16 hr, and no further increase in conversion of NCO beyond 93% was evident, apparently due to the lower reactivity or amount of dissociation of this acid. It must be noted that the conversion was calculated according to the disappearance of NCO groups, as determined by titration. At conversions very close to 100% (or complete disappearance of NCO) the procedure cannot discriminate between different levels of reaction, so 100% conversion was assumed, even though NCO groups could still have been in the reaction system.

As another measure of conversion of the polymerizations, the intrinsic viscosity of the various polymers as a function of time was measured. The results (see later) indicated that the reactions had reached completion or ceased (in the case of sebacic acid), consistent with the results obtained by titration of NCO groups.

The intrinsic viscosity was also plotted against %NCO conversion for the three acids (Figure 8). The three polymers seem to have the same intrinsic viscosity-%NCO conversion relationship, since all the data apparently fall on one curve. Figure 8 indicates the importance of very high degrees of conversion in obtaining high intrinsic viscosities and therefore polymers of high molecular weight. Polymers of high molecular weight are necessary to obtain materials with acceptable mechanical, physical and chemical properties. Other investigators have indicated the importance of the relationship between polymer molecular weight and properties of the polymer (*24-27*), whereas other reports indicate that beyond a certain molecular weight, the molecular weight no longer influences the mechanical properties (*28*). In some of these references (*24-26*) the increase in molecular weight increased the efficiency of phase separation and therefore the creation of more distinct and ordered hard segments. The effect of reaction time on the development of mechanical properties of PEUAms will be the subject of a future publication.

Similar molecular weight-property effects were observed with the PEUAms of the present work. Figure 9 shows DSC thermograms of PEUOXA at different stages of the polymerization, from the prepolymer until the final polymer, one hour after the completion of the diacid addition. The low temperature transition at 2-12 °C can be ascribed to the soft segment (PTMG-2000) melting point whereas the high temperature transitions can be assigned to the hard segments. As expected, high temperature transitions do not exist for the prepolymer, as seen in curve A. For the early stages of the polymerization (time 0 and 15 min), a rather broad transition appears at about 200 °C, which can be associated with disordered hard segments. Very small peaks appear at about 250-270 °C, which can be related to the small number of hard segments, due to the low molecular weight of the polymer. Only at the later stages of the polymerization (30-60 min) does the high temperature peak grow in size and move to higher temperatures, around 280 °C in the final polymer. The maximum value of the

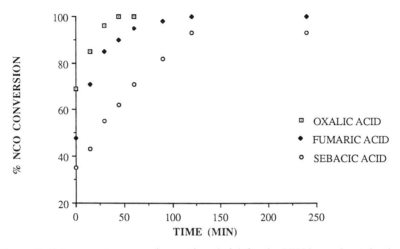

Figure 7: % Isocyanate conversion vs time (min) for the PEUAm polymerizations in DMA for the three diacids.

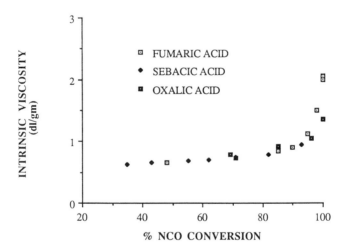

Figure 8: Intrinsic viscosity (dl/gm) vs % isocyanate conversion for the PEUAm polymerizations in DMA for the three diacids.

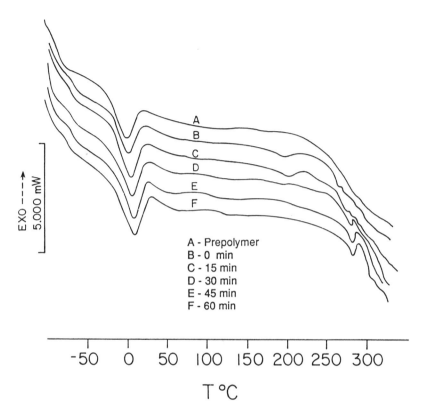

A - Prepolymer
B - 0 min
C - 15 min
D - 30 min
E - 45 min
F - 60 min

-50   0   50   100   150  200  250  300

T °C

Figure 9: DSC thermograms for a typical PEUAm, using oxalic acid as the chain extender, for different stages of the polymerization process. The times indicate the time from the end of the addition of oxalic acid.

enthalpy of fusion occurs for the polymer after 60 min (curve F), which indicates improved phase separation and ordering of the hard segments.

**Effect of solvent on the synthesis.**  The effect of the solvent on the synthesis and molecular weight of polyurethanes has been reported previously (*25,29*), in that certain solvents were observed to result in materials with lower molecular weights. This same effect was observed in this work, when comparing synthesis in DMA vs. synthesis in THF.  The effect of THF on the intrinsic viscosities of the polymers is shown in Figures 10a-c, where comparison with the results from the DMA synthesis is presented.  It can be seen that in general polymers of lower molecular weight were received when THF was used.  The differences were quite significant for PEUOXA and PEUFA, and less so for PEUSEBA.  The effect of the solvent can perhaps be understood by realizing that PEUOXA and PEUFA are more polar than PEUSEBA, and thus their solubilities are lower in THF than in DMA.  This was manifested by the noticeable cloudiness and phase separation of polymer from THF during the syntheses of PEUOXA and PEUFA.  Thus the incompatibility of these materials with the solvent led to their precipitation at a certain molecular weight, preventing the attainment of high molecular weight polymer.  PEUSEBA, on the other hand, is more compatible with THF, and thus it remained in solution, allowing for the production of high molecular weight material, similar to that received when DMA was used as a solvent.

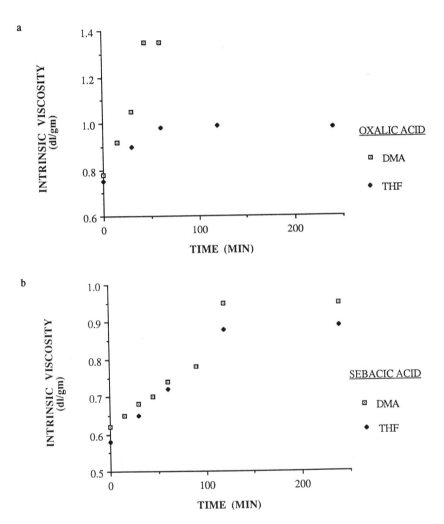

Figure 10: Intrinsic viscosity (dl/gm) vs time (min) for PEUAms synthesized in DMA as opposed to THF: (a) oxalic acid; (b) fumic acid.

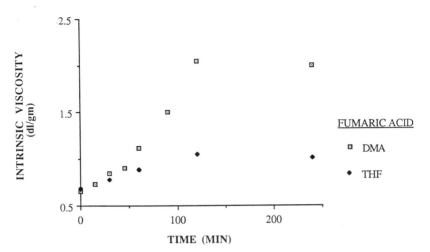

Figure 10—*Continued:* Intrinsic viscosity (dl/gm) vs time (min) for PEUAms synthesized in DMA as opposed to THF: (c) sebacic acid.

## CONCLUSIONS

The synthesis and kinetics of a new series of polyurethanes, with a dicarboxylic acid as the chain extender, were investigated. The use of dicarboxylic acids as chain extenders led to the introduction of amide groups into the polyurethane, resulting in very strong, stable and rigid hard segment domains, as evidenced by the very high hard segment transition temperatures observed in the DSC. The kinetics of the isocyanate-carboxylic acid reaction were studied by using a model system consisting of two moles of isocyanate to one mole of acid. The kinetics could be modelled by the addition of an autocatalytic term, arising from the amide groups formed in the reaction, to a normal second-order kinetic term. Of the three diacids studied, oxalic acid was observed to react most rapidly with isocyanate, followed by fumaric and then sebacic acids. This order was seen both in the model system and polymerization reaction, and is consistent with a decreasing reactivity of functional groups as the chain length increased. For the polymerization reactions, high extents of conversion are necessary to obtain polymers of high molecular weight with good development of hard segment domains. DSC was used to confirm the importance of high extents of conversion in obtaining polymers with well developed morphologies. Finally, a change in reaction medium from DMA to THF was observed to result in materials of lower molecular weight, presumably as a result of the reduced compatibility of the polymers in THF.

## REFERENCES

1. Estes, G.M.; Cooper, S.L.; Tobolsky, A.V. *J. Macromol. Sci., Rev. Macromol. Chem.*, **4**, 313 (1970).
2. Schneider, N.S.; Desper, C.R.; Illinger, J.L.; King, A.O. *J. Macromol. Sci.-Phys.*, **B11**, 527 (1975).
3. *Block Copolymers*; Noshay, A.; McGrath, J.E., Eds.; Wiley: NY, 1973.
4. *Multiphase Polymers*, Cooper, S.L.; Estes, G.M., Eds.; Adv. Chem. Ser. 176, ACS: Washington, D.C., 1979.
5. Gibson, P.E.; Vallance, M.A.; Cooper, S.L. In *Properties of Polyurethane Block Polymers*, Goodman, I., Ed.; Developments in Block Copolymers; Elsevier Applied Sci. Publ.: London, 1982.

6. Bonart, R. *Polymer*, **20**, 1389 (1979).
7. Lelah, M.D.; Cooper, S.L. *Polyurethanes in Medicine*; CRC Press, Inc.: Boca Raton, Fla., 1986.
8. Cohn, D.; Yitzchaik, S.; Bilenkis, S. U.S. Patent 5,100,992.
9. Cohn, D.; Bilenkis, S.; Penhasi, A.; Yitzchaik, S. In *Clinical Implant Materials*, Heimke,G.; Soltesz, U.; Lee, A.J.C., Eds.; Adv. in Biomaterals, 9, 473 (1990).
10. Cohn, D.; Penhasi, A. *Clinical Materials*, **8**, 105 (1991).
11. Dieckman, W.; Breest, F.; *Ber.* **39**, 3052 (1906).
12. Saunders, J.H.; Frisch, K.C.; *Polyurethane Chemistry and Technology*, Part I, Chemistry, Interscience: NY, 1962.
13. Zapp, R.L.; Serniuk, G.E.; Minckler, L.S. *Rubber Chem. Technol.*, **43**, 1154 (1970).
14. Tanzi, M.C.; Barzaghi, B.; Anouchinsky, R.; Bilenkis, S.; Penhasi, A.; Cohn, D. *Biomaterials*, **13**(7), 425 (1992).
15. Siggia, S.; Hanna, J.G. *Anal. Chem.*, **20**, 1084 (1948).
16. Flory, P.J. *Principles of Polymer Chemistry*, Cornell University Press: Ithaca, NY, 1953; p.69.
17. Saunders, J.H.; Dobinson, F. In Chemical Kinetics, Vol. 15; Bamford, C.H.; Tipper, C.F.H., Eds.; Elsevier Scientific Publishing Co.: Amsterdam, 1976; Ch. 7.
18. Sato, M. *J. Am. Chem. Soc.*, **82**, 3893 (1960).
19. Ephraim, S.; Woodward, A.E.; Mesrobian, R.B. *ibid.*, **80**, 1326 (1958).
20. Kogon, I.C. *J. Org. Chem.*, **24**, 438 (1959).
21. Briody, J.M.; Narinesingh, D. *Tetrahedron Letters*, **44**, 4143 (1971).
22. Grasel, T.G.; Cooper, S.L. *Biomaterials*, **7**, 315 (1986).
23. Marchant, R.E.; Zhao, Q.; Anderson, J.M.; Hiltner, A. *Polymer*, **28**, 2032 (1987).
24. Kenn, W.; Davidovits, J.; Rauterkus, K.J.; Schmidt, G.F. *Makromol. Chem.*, **43**, 106 (1961).
25. Lyman, D.J. *J. Polym. Sci.*, **XLV**, 49 (1960).
26. Yu Xue-Hai, M.R.; Nagarajan, T.G.; Gibson, P.E.; Cooper, S.L. *J. Polym. Sci., Polym. Phys. Ed.*, **23**, 2319 (1985).
27. Hespe, H.F.; Zembrod, A.; Cama, F.J.; Lantman, C.W.; Seneker, S.D. *J. Appl. Polym. Sci.*, **44**, 2029 (1992).
28. Schollenberger, C.S.; Dinbergs, K. *J. Elast. Plast.* **5**, 222 (1973).
29. Yilgor, I.; Sha'aban, A.K.; Steckle, Jr., W.P.; Tyagi, D.; Wilkes, G.L.; McGrath, J.E. *Polymer*, **25**, 1800 (1984).

RECEIVED April 22, 1993

# Chapter 9

# Interaction of Water with Polyurethanes Containing Hydrophilic Block Copolymer Soft Segments

N. S. Schneider[1], J. L. Illinger[2], and F. E. Karasz[3]

[1]Geo-Centers, Inc., 7 Wells Avenue, Newton, MA 02159
[2]Army Research Laboratory, Materials Directorate, AMSRL-MA-PB, Watertown, MA 02172
[3]Polymer Science and Engineering Department, University of Massachusetts, Amherst, MA 01003

The saturation water uptake and the nonfreezing bound water, determined as a function of temperature in a polyurethane (I) containing the pure polyethylene oxide soft segment are only weakly dependent on temperature from 288 to 333 K. In polyurethanes (II) containing block copolymer polyethylene oxide/polypropylene oxide soft segments with various ratios of the hydrophilic and hydrophobic segments, there is a strong decrease with temperature and a steep drop above 303K. This behavior is interpreted in terms of the temperature dependent phase compatibility of the polyethylene oxide and polypropylene oxide segments of II. The depression of $T_g$ appears to be governed solely by the nonfreezing bound water and is much larger in sample I than in samples of set II. The more limited effect in the samples of set II is attributed to restricted mobility arising from coupling of the short terminal polyethylene oxide segments to rigid hard segment units. The simple Fox mixing equation was judged to provide a more reliable fit to the data from sample I than the available free volume approach, which required unrealistically high values of the thermal expansion coefficient for water.

Polyurethanes based on the incorporation of a polyethylene oxide soft segment exhibit substantial water uptake and, related high moisture vapor transmission rates (MVT). The variable and controlled degree of swelling with the related water and solute permeability could be useful in certain biomedical applications, including controlled release of medications from the water swollen polymer. In earlier work Tobolsky and coworkers (*1*) described the

0097–6156/94/0540–0103$06.00/0

properties of a series of hydrophilic polyurethanes prepared for possible reverse osmosis separations. The water solubility was controlled by varying the proportion of polyethylene oxide and polypropylene oxide used to form the mixed soft segment. It was shown that the saturation water concentration was directly proportional to the polyethylene oxide concentration. A variation on this approach was undertaken in studies by Illinger (2,3). In this work the soft segment consisted of a block copolymer containing a central segment of polypropylene oxide and terminal segments of polyethylene oxide. Samples were prepared in which the proportions of these two components in the soft segment were varied. However, in a recent review of this study it was realized that the dependence of the saturation water concentration and the state of the sorbed water on temperature were unusual and deserved more detailed consideration. In addition, it appeared worthwhile to attempt a more detailed analysis of the effect of water on the glass transition temperature. One paper on this subjects (4) has been published and a second is in press.

**Experimental**

Polyurethane samples were formed from diphenyl methane diisocyanate (MDI), butanediol (BD), and a macroglycol (PE) in several mole ratios. The soft segment consisted of pure polyethylene oxide, molecular weight 1450 (Union Carbide), pure polypropylene oxide or of one of three block copolymers of polypropylene oxide-polyethylene oxide, molecular weight 2000 (Wyandotte Corp.), each with a different ratio of the two components. The notation 5PE33 designates a sample with 50% by weight of polyethylene oxide in the soft segment and 33% MDI in the hard segment, which corresponds to a 4/3/1 mole ratio of MDI, BD and PE. Details of the synthesis, sample composition and other properties have been reported earlier (2). The composition and certain properties of the polyurethanes are summarized in Table I. In calculating the molecular weight of the block copolymer soft polyether from the size of the segments, given as the number of ethylene oxide and propylene oxide units, EO/PPO, in the second column it is necessary to double the ethylene oxide contribution since the polymer consists of a central polypropylene oxide segment terminated by polyethylene oxide segments of equal length.

　　　Equilibrium sorption measurements were performed on a preweighed sample of polymer immersed in distilled water which was maintained at the proper temperature, as required. The sample was removed from the water, blot dried and placed in a tared weighing bottle to determine the weight gain. DSC runs were made using a Perkin-Elmer DSC-2 with subambient accessory. Samples were prepared from films cast from DMF soution and dried under vacuum for 48 hours at 50°C. The polymer discs of known dry weight were equilibrated with water and transferred to custom fashioned, gold foil pans. The excess water was allowed to evaporate on the microbalance to the desired water content before the pans were hermetically sealed. The sealed samples were equilibrated at 10° intervals from 273 to 323 K (Teq) in a constant temperature bath or in the DSC. The time for equilibrium was longer at the lower temperatures, varying from 4 hours at 323 K to 24 hours at 273 K. The

Table I. Composition and Properties of Polyurethane Samples

| Sample | [a]EO/PPO Mole Ratio MDI/BD/SS | [b]Wt% HS | Tg (K) | [c]Wt% Water | [d]Wt% Water | [e]$H_2O$/EO |
|--------|------------------|-------|--------|--------------|--------------|-------------|
| 10PE33 | 36/0   | 4.20/3/1 | 46.2 | 242 | 58 | 107 | 2.62 |
| 5PE33  | 11/17  | 4.20/3/1 | 39.8 | 236 | 25 | 84  | 2.05 |
| 3PE33  | 7/23   | 4.20/3/1 | 39.8 | 230 | 8  | 43  | 1.04 |
| 1PE33  | 2.5/33 | 4.20/3/1 | 39.8 | n.d. | 3 | 48  | 1.16 |
| 0PE33  | 0/35   | 4.20/3/1 | 39.8 | 232 | 2 | (3.3) | (0.11) |
| 5PE28  | 11/17  | 3.15/2/1 | 32.6 | 236 | 40 | 121 | 2.96 |
| 5PE33  | 11/17  | 4.20/3/1 | 39.8 | 236 | 25 | 84  | 2.05 |
| 5PE40  | 11/17  | 6.30/5/1 | 50.3 | 236 | 15 | 61  | 1.49 |

n.d. not determined. a. Number of ethylene oxide and propylene oxide units in soft segment blocks. b. Weight % hard segment. c. Grams of water per 100 grams of polymer at 30°C. d. Grams of water per 100 grams of PEO at 30°C. e. Ratio of moles of water to moles of ethylene oxide units.

sample was transferred from the constant temperature bath to the DSC that was held at the equilibration temperature. The DSC was quenched at a setting of 320° per minute to 150 K, then scanned at 20° per minute to Teq.

## Results

**Immersion Measurements of Water Uptake.** The water uptake in the polyurethanes of varying composition determined at 30°C by immersion is summarized in Table I. The first column of water data lists the values as grams of water per 100 grams of polymer. These results are consistent with the general expectation that the water uptake will reflect the amount of the hydrophilic component, polyethylene oxide, and, therefore, decrease with increasing polypropylene oxide in the soft segment or with increasing hard segment content. However, the results in the next column indicate that the amount of sorbed water is not simply proportional to the polyethylene oxide content. The ratio, grams of water per 100 grams of polyethylene oxide, decreases with decreasing polyethylene oxide content and with increasing hard segment content. In the last column the results are presented as moles of water per ethylene oxide repeat unit. The ratio in 10PE33 is close to three, the value suggested as the water of hydration per ethylene oxide unit (5). Therefore, in this case it is expected that the water molecules would be bound to ethylene oxide units. This expectation can be tested by examining the DSC trace to determine the fraction of nonfreezing water and will be discussed later.

Measurements of water uptake made at several temperatures over the range 276 to 333 K are instructive about the nature of the interactions determining the sorption levels. The results are summarized in Table II as the ratio of moles of water to moles of the ethylene oxide repeat unit. In almost

Table II. Immersion Water Uptake as a Function of Temperature
(moles of water per mole of EO)

| Sample | 276 K | 299 K | 303 K | 323 K | 333 K | Ratio |
|--------|-------|-------|-------|-------|-------|-------|
| 10PE33 | 3.08 | 2.93 | 2.62 | 2.00 | 1.63 | 1.89 |
| 5PE33 | 4.01 | 3.12 | 2.05 | 0.76 | 0.68 | 5.89 |
| 3PE33 | 4.01 | 2.32 | 1.04 | 0.71 | 0.67 | 5.89 |
| 1PE33 | 2.37 | 1.59 | 1.16 | 1.42 | 1.42 | 1.67 |
| 0PE33 | (0.15) | (0.13) | (0.11) | (0.15) | (0.15) | 1.0 |
| 5PE28 | 5.92 | 4.40 | 2.96 | 0.88 | 0.73 | 8.10 |
| 5PE33 | 4.01 | 3.12 | 2.05 | 0.76 | 0.67 | 5.89 |
| 5PE40 | 2.98 | 2.30 | 1.49 | 0.67 | 0.62 | 4.81 |

SOURCE: Reproduced with permission from reference 4. Copyright 1993.

all cases the values are highest at the lowest temperature, 276 K and decrease with increasing temperature. The exception is 1PE33, where the value at 303 K, seems to be lower than at the other temperatures but otherwise follows the same trend. It is noteworthy that the number of moles of water to ethylene oxide at 276 K is about 30% higher for 5PE33 and 3PE33 than for the pure PEO containing polymer, 10PE33. This is due to the fact that the latter sample has a higher hard segment content, because of the lower soft segment molecular weight (see Table I). The comparison of the results for the three samples with different hard segment content shows that reducing the hard segment content from 50.3% in 5PE40 to 39.8% in 5PE33 also increases the saturation water content by 38%.

A measure of the change in the water uptake with temperature is given by the values in the last column, which represent the ratio of the water uptake at 276 K to that at 333 K. The values are close to 6 for 5PE33 and 3PE33, compared with 1.9 for 10PE33. The trend in the data is illustrated in Figure 1, where the solubility, S, is plotted as moles of $H_2O$ per EO against the reciprocal of absolute temperature, T, in the manner appropriate for determining the heat of solution, $\Delta H$, according to the usual relation:

$$\Delta H = -R(d\ln S/d(1/T)) \tag{1}$$

The heats of solution are negative throughout the temperature range with the possible exception of 0PE33 at 303 K. The contrast between the behavior of 10PE33 and that of the other polymers is clearly in the comparison with the three samples with different hard segment contents, shown in Figure 1.

**DSC Measurements of the State of Sorbed Water.** An example of DSC results obtained for 10PE33 with 48% added water is given in Figure 2 of reference 5. For this condition, there is a broad endotherm at all Teq with an onset temperature of approximately 260 K. A second sharp endotherm appears in the samples equilibrated at 303 K and higher. It is customary to identify the broad endotherm as bound freezing water, implying that the behavior is

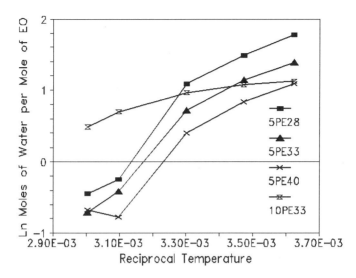

Figure 1. Water Uptake as a Function of Temperature.

influenced by strong interactions with the matrix. The water contributing to
the sharp endotherm has been labeled free water. NMR measurements have
shown that the mobility of dissolved free water is much lower than that of
bulk water (6). However, this endotherm could also be due to water that
exists in the free space within the sample cell. In the present case, the added
water exceeds the solubility at 323 K and is borderline at 313 K so the sharp
endotherm must be due to the excess water.

DSC measurements were made on the various samples in this study at
two levels of added water at various equilibration temperatures from 273 to
323 K. The higher amount of added water was well below saturation at 273 K
but in all cases exceeded saturation at 303 K and above whereas, the lower
amount of added water was below saturation at all temperatures. The amount
of bound nonfreezing water was calculated by subtracting the total amount of
water represented by the endotherms from the added water, assuming the
applicability of the heat of fusion for bulk water, 79.8 calories per gram.
Only the results at the higher water content are recorded in Table III. The
additional data that appears in this table on bound water and depression of
the Tg will be discussed later.

The behavior of the nonfreezing water in 10PE33 is very different from
that of the samples with a block copolymer polyether. In 10PE33, the amount
of nonfreezing water decreases slowly with increasing equilibration
temperature, as evidenced in Table III, and did not change significantly with
the added water level. Thus, at 283 K the amount of nonfreezing water was
was 60% of the lower amount of added water, 78% and 42% of the higher
amount of added water, 124% (g water per 100 g polyethylene oxide). In

5PE33, in contrast, the amount of nonfreezing bound water at the lower added water level, 31%, was essentially equal to the amount of added water up to 303 K and then decreased sharply between 303 and 313 K. At the higher amount of added water, 111%, the amount of nonfreezing has increased significantly (see Table III), shows a somewhat stronger dependence on equilibration temperature in the range 273 to 303 K and decreases sharply above 303 K. This behavior is similar to that seen for the saturation water content in the immersion experiments. Also, although not illustrated here, in 5PE33 the nonfreezing water represents an appreciable fraction of the saturation concentration at the three highest temperatures where the higher added water level exceeds saturation.

**Table III. Effect of Changes in the Amount of Bound Water on the Depression of $T_g$**

| Sample | Added Water | Equil. Temp. | Nonfreez. Water[a] | Freez. Water[a] | Nonfreez. Water[b] | $\Delta T_g$ |
|--------|-------------|--------------|---------------------|------------------|---------------------|--------------|
| 10PE33 | 124% | 273 | 62 | 62 | 1.52 | 49 |
|        |      | 283 | 52 | 72 | 1.27 | 46 |
|        |      | 293 | 54 | 63 | 1.32 | 48 |
|        |      | 303 | 57 | 50 | 1.39 | 50 |
|        |      | 313 | 50 | 44 | 1.22 | 48 |
|        |      | 323 | 19 | 63 | 0.46 | 34* |
| 5PE28  | 157% | 273 | 120 | 37 | 2.93 | 31 |
|        |      | 283 | 116 | 41 | 2.83 | 31 |
|        |      | 293 | 42 | 115 | 1.02 | 9* |
|        |      | 303 | 102 | 21 | 2.49 | 21 |
|        |      | 313 | 85 | 0 | 2.07 | 26 |
|        |      | 323 | 28 | 12 | 0.68 | 7* |
| 5PE33  | 111% | 273 | 105 | 6 | 2.56 | 24 |
|        |      | 283 | 98 | 13 | 2.39 | 25 |
|        |      | 293 | 66 | 45 | 1.61 | 19* |
|        |      | 303 | 71 | 13 | 1.74 | 18 |
|        |      | 313 | 19 | 38 | 0.46 | 2* |
|        |      | 323 | 16 | 15 | 0.39 | 1 |
| 5PE40  | 79% | 273 | 79 | 0 | 1.93 | 27 |
|        |     | 283 | 79 | 0 | 1.93 | 27 |
|        |     | 293 | 68 | 11 | 1.66 | 24 |
|        |     | 303 | 23 | 38 | 0.56 | 8* |
|        |     | 313 | 37 | 6 | 0.90 | 12 |
|        |     | 323 | 22 | 5 | 0.54 | 6* |

a. Amount of water as grams of water per 100 grams of PEO
b. Amount of nonfreezing water as moles water/mole EO
* Values of $\Delta T_g = T_g$ (dry) - $T_g$ (wet) that illustrate the correlation with sudden changes in the amount of nonfreezing bound water
SOURCE: Reproduced with permission from reference 4. Copyright 1993.

Data are also presented in Table III on 5PE40 which, in most respects, is similarly to 5PE33. At the lower added water level, 29%, the nonfreezing water content is equal to the added water up to 323 K. At the higher water content, 79%, the nonfreezing water has increased by nearly threefold, essentially in proportion to the increase in added water, and equals the added water at 273 and 283 K. As recorded in Table III, above 293 K there is a sharp drop in nonfreezing water. Since the nonfreezing water at 303 K is equivalent to 5 g/100 g polymer, compared to a saturation concentration of 15 g/100 g polymer at this temperature, the change is not due to the marked decrease in saturation concentration that occurs in this temperature range although it probably reflects the same cause.

**Effect of Water on the Glass Transition Temperature.** The Tg of the isolated soft segments, pure PEO, pure PPO, and block PEO/PPO/PEO polymers all showed a single Tg at approximately 210 K, indicating that the Tg of PPO and PEO are very close. As the data in Table I show, the Tg of the soft segment in the polyurethanes is increased by 20 to 30 degrees. This is due to some mixing of urethane hard segments with the soft segment phase. A set of DSC traces for 10PE33 equilibrated at 323 K, with various amounts of added water up to 88% (g water per 100 g PEO), appears in Figure 4 of reference 5. These results illustrate the change in the state of water with increasing added water, from nonfreezing bound water at low amounts of added water to the onset of a broad endotherm, representing bound freezing water, and finally a sharp peak at 273 K which represents the melting of free dissolved water. It is also apparent that the glass transition temperature decreases continuously with the amount of added water. It is this latter aspect of the behavior that is the focus of this section.

Table III provides a record of the state of the added water and the resulting depression of the glass transition temperature. In addition to the amount of bound nonfreezing water discussed earlier, the table includes results for bound freezing water. At the three lower equilibration temperatures, where the amount of added water is below saturation, the freezing bound water was estimated as the difference between the amount of nonfreezing bound water and the amount of added water. The saturation water content decreases with increasing temperature with the result that the added water exceeds the saturation content at 303 K and the two higher equilibration temperatures in all cases. For these conditions the amount of freezing bound water was calculated as the difference between the nonfreezing bound water and the saturation concentration. It is possible to have water that melts at 273 K, even at an added water content that is below saturation. As mentioned earlier, this represents free water that is dissolved in the polymer and has a mobility much lower than bulk water (5). The amount of freezing bound water will be overestimated to the extent that this has occurred. However, it will develop that the analysis of interest for the glass transition behavior is not dependent on accurate values of the freezing bound water.

The first question to be addressed is whether the glass transition temperature is dependent on the total amount of dissolved water or only on the amount of nonfreezing bound water. The last column of Table III records the depression of the Tg which results following equilibration of the sample at the various temperatures. The Tg for 10PE33 is essentially independent of temperature. The Tg for the other samples changes in a marked way only at 303 K and above. Some insight on the relative importance of the state of water is indicated by results in Table III that represent instances of a marked decrease in Tg relative to neighboring values. In almost all these cases there is a marked decrease in the amount of nonfreezing bound water and a corresponding increase in the amount of freezing bound water. This observation suggests that the controlling factor is the amount of nonfreezing bound water rather than the total amount of dissolved water. This conclusion is consistent with the assumption that Illinger made in analyzing her results (3). The Tg depression appears comparable for the different samples in Table I. However, in comparison to the amount of nonfreezing bound water, the Tg depression is much larger in 10PE33 than the other samples.

One method of calculating values of the glass transition temperature involves the use of the Fox equation. In the following equation $w_1$ represents the weight fraction of nonfreezing bound water in the polyethylene oxide component of the soft segment, $w_2$ represents the weight fraction of polyethylene oxide in the water polyethylene oxide mixture, $T_{g1}$ denotes an appropriate value for the glass transition temperature of water and $T_{g2}$ is the glass transition temperature of polyethylene oxide in the dry polymer:

$$\frac{1}{T_g} = \frac{w_1}{T_{g1}} + \frac{w_2}{T_{g2}} \qquad (2)$$

There is uncertainty concerning the correct value of $T_{g1}$. Earlier work suggested a value of 137 K but a later study (7), utilizing DSC measurements, indicated that no glass transition is observed for water at that temperature. It was concluded that the transition occurs at a much higher temperature but is obscured by the rapid crystallization of the quenched glass in the range of 150 to 162 K. For the purpose of the present calculations 165 K has been used for the alternative value. It is apparent from the examples in Table IV that the value for $T_{g1}$ has a marked effect on the calculated values of $T_g$. Use of $T_{g1}$ = 137 K provides a close fit to the data for 10PE33 with 124% (g of water per 100 g PEO) added water but a poor fit to the data for all other samples, exemplified by 5PE33. The calculations with $T_{g1}$ = 165 K provide a close fit to the data for 10PE33 with 78% added water. The fit to the data for the other samples is somewhat improved over the previous calculations, as indicated again by results for 5PE33, but is still poor.

It is also possible to carry out calculations of the depression of $T_g$ in terms of free volume concepts, using the following expression derived by Bueche and summarized by Meares (8) who gave examples of the application:

$$T_g = \frac{v_2 T_{g2}(a_l - a_g) + v_1 T_{g1} a_1}{v_2(a_l - a_g) + v_1 a_1} \qquad (3)$$

Table IV. Calculated and Experimental Values for $T_g$
Using the Fox Mixing Relation

| Sample | Added Water | Equil. Temp. | [a]Water | [b]Calc. $T_g$ | Exptl. $T_g$ | [c]$\Delta T_g$ | [d]$\Delta T_g$ |
|--------|-------------|--------------|----------|----------------|--------------|-----------------|-----------------|
| 5PE33  | 111%        | 273          | 0.510    | 171.4          | 210.3        | -38.9           | -18.1           |
|        |             | 293          | 0.294    | 182.2          | 214.1        | -33.6           | -13.0           |
|        |             | 313          | 0.160    | 208.9          | 230.5        | -21.6           | 12.6            |
|        |             | 323          | 0.138    | 211.7          | 231.8        | -20.1           | -12.1           |
| 10PE33 | 78%         | 273          | 0.355    | 190.3          | 204.6        | -14.3           | 3.0             |
|        |             | 293          | 0.338    | 192.2          | 206.0        | -13.8           | 3.1             |
|        |             | 313          | 0.324    | 193.8          | 203.9        | -0.9            | 6.3             |
|        |             | 323          | 0.281    | 199.2          | 222.0        | -10.1           | -8.0            |
| 10PE33 | 124%        | 273          | 0.383    | 187.1          | 193.8        | -6.7            | 11.5            |
|        |             | 293          | 0.351    | 190.7          | 194.4        | -3.7            | 13.6            |
|        |             | 313          | 0.333    | 192.8          | 194.7        | -1.9            | 14.7            |
|        |             | 323          | 0.160    | 215.6          | 208.2        | 7.4             | 17.0            |

a. Nonfreezing bound water expressed as the weight fraction in PEO. b. $T_g$ calculated using a water $T_g$ = 137 K. c. $\Delta T_g$ using results calculated with a water $T_g$ = 137 K. d. $\Delta T_g$ using results calculated with a water $T_g$ = 165 K. Note: in the above $\Delta T_g$ = calculated $T_g$ - experimental $T_g$. Adapted from reference 4.

Here $T_g$ is the depressed glass transition temperature; $T_{g2}$ and $T_{g1}$ are the glass transition temperatures of polymer and diluent, respectively; $v_2$ and $v_1$ are the corresponding volume fractions; $(a_l-a_g)$ is the difference in expansion coefficients of polymer liquid and glass; and $a_1$ is the thermal expansion coefficient of diluent. Because the thermal expansion coefficient of water at the temperature of the polymer glass transition is not known, it was used as a fitting parameter.

The results in the first column of data in Table V reproduce some of calculations carried out by Illinger using the following values; $a_1$ = $2.07*10^{-4}$, $(a_l-a_g)$ = $4.8*10^{-4}$ and $T_g$ = 137 K. The results are again recorded as the difference between the calculated and measured glass transitions for the three samples chosen to illustrate the trend. The predicted $T_g$ on average is about 6 degrees lower than the measured value for the block copolymer sample but more than 15 degrees higher than results for 10PE33 with 78% water and almost 30 degrees too high with 124% water. With the alternative water value,

$T_{g1}$ = 165 K, it is necessary to increase $a_1$ to 3.0 * $10^{-4}$ to match the previous results for 5PE33, as shown in the next column. These results are esseniallty identical to the previous set of values. To bring the predicted values for 10PE33 into line with the measured values, it is necessary to increase $a_1$ to 8 * $10^{-4}$. The results in the last column show that a match is achieved for 10PE33 with 124% water, but the depression of $T_g$ for 5PE33 is seriously overestimated. Furthermore, this value of the thermal expansion coefficient is unrealistically high. The thermal expansion coefficient of water ranges from about 2.6 to 5.0 x $10^{-4}$ at temperatures between 30 and 60°C.

**Table V. Difference Between Calculated and Experimental Values of $T_g$ Using the Free Volume Relation**

| Sample | Added Water | Equil. Temp.(K) | $^a \Delta T_g$ | $^b \Delta T_g$ | $^c \Delta T_g$ |
|--------|-------------|-----------------|------------------|------------------|------------------|
| 5PE33  | 111%        | 273             | -7.7             | -4.7             | -38.5            |
|        |             | 293             | -1.8             | -0.4             | -30.4            |
|        |             | 313             | -5.7             | -5.6             | -21.3            |
|        |             | 323             | -5.9             | -5.9             | -19.8            |
| 10PE3  | 78%         | 273             | 17.3             | 17.7             | -12.8            |
|        |             | 293             | 17.1             | 17.4             | -12.2            |
|        |             | 313             | 20.1             | 20.3             | -8.6             |
|        |             | 323             | 4.9              | 4.9              | -21.4            |
| 10PE3  | 124%        | 273             | 26.0             | 26.7             | -5.2             |
|        |             | 293             | 27.8             | 28.2             | -2.1             |
|        |             | 313             | 28.7             | 29.0             | -0.4             |
|        |             | 323             | 25.8             | 25.6             | 8.5              |

a. $\Delta T_g$ calculated with water $T_g$ = 137 K and $a_1$ = 2.07*$10^{-4}$.
b. $\Delta T_g$ calculated with water $T_g$ = 165 K and $a_1$ = 3.00*$10^{-4}$.
a. $\Delta T_g$ calculated with water $T_g$ = 137 K and $a_1$ = 8.00*$10^{-4}$.
Note: in the above $\Delta T_g$ = calculated $T_g$ - experimental $T_g$.
Adapted from reference 4.

The glass transition temperature of polyethylene oxide is not known with certainty. However, when incorporated into the polyurethane, an increase in the glass transition temperature is expected. In part, this is due to mixing of short urethane segments with the soft segment phase and the resulting hydrogen bonding between the miscible urethane groups and the ether oxygen of the soft segment. Thus, in addition to the free volume contribution, the effectiveness of water in lowering the glass transition temperature might be due to a reduction in the urethane-to-ether hydrogen bonding. If this effect could be taken into account, it should permit fitting of the 10PE33 data by the free volume expression with somewhat lower values of the thermal expansion coefficient for water. However, this value is so large that it would probably still lie outside the physically acceptable range.

## Discussion and Conclusions

This study has shown that there are marked differences in the interaction of water with the polyurethane based on the pure polyethylene oxide soft segment compared with the samples containing the block copolymers consisting of polyethylene oxide and polypropylene oxide. In 10PE33 the saturation water content is only weakly dependent on temperature. This finding is consistent with the results of NMR studies of aqueous polyethylene oxide solutions showing that the state of hydration is stable to temperatures as high as 80 °C (*4*). In the polymers containing the block copolymer soft segment the saturation concentration decreases rapidly with increasing temperature and drops abruptly above 303 K. When plotted in the Arrhenius form, the behavior of the saturation concentration suggests the occurrence of a transition above 303 K. The two types of polymer show equally marked differences in the temperature dependence of the nonfreezing water determined by DSC.

An explanation for the composition and the temperature dependence of the saturation and nonfreezing water content is suggested by measurements on the compatibility of mixtures polyethylenoxide and polypropylene oxide (*9*). Polymers of sufficiently high molecular weight were found to be incompatible. The upper critical consolute temperature was strongly dependent on the molecular weight of the two polymers. For a polypropylene oxide molecular weight of 2000 and polyethylene oxide molecular weights of 550 and 750 the critical consolute temperatures were 67 °C and about 112 °C, respectively. On the basis of these results, it can be assumed that the segments of the block copolymer corresponding to 5PE55 and 3PE33 would also exhibit incompatibilty. The phase separation temperature would be lowered as a result of the linkage between the segments and the lower molecular weight of the segments; 986 and 484 for the molecular weights of the PPO and PEO segments in 5PEO, and 1334 and 308 in 3PEO. Despite the observation of only a single soft segment Tg, these results suggest that the two soft segment components are incompatible at low temperature. Therefore, the transition in water solubility above 303 K in these two polymers could be a result of a transition from incompatibility to compatibility in the soft segment phase.

The problem is to explain why this type of organization leads to normalized saturation water concentrations comparable to that in the pure polyethylene oxide at 273 K but shows a much stronger reduction with increasing temperature and a sudden drop with the onset of the suggested phase compatibility. Based on the preceding model, the drop in water solubility at the temperature of soft segment phase compatibility must be due to the more hydrophobic nature of the polypropylene oxide and to a reduction in the possibility of water molecules bridging ethylene oxide segments. There might also be secondary effects arising from propylene oxide/ethylene oxide interactions. The strong temperature dependence of the saturation water content in the block copolymer samples would be similarly related to the progressive homogenization of regions of different stability in the soft segment phase with increasing temperature.

It was noted at the outset that the saturation water concentration in polyurethanes with various ratios of polypropylene oxide and polyethylene oxide was directly proportional to the polyethylene oxide content (1). It is now clear that the departure from this behavior in the current samples, documented in Table I, is due to the partial compatability of the two components of the soft segment at 30 °C. However, as shown in Table II, at 276 K where the two components of the soft segment can be expected to be completely phase segregated, the solubility is essentially proportional to the polyethylene oxide content.

The DSC examination of the state of water in the various samples as a function of equilibration temperature indicates that the depression of the glass transition temperature is controlled by the amount of nonfreezing bound water, with very little contribution from the other forms of water. It is somewhat surprising that the bound freezing water, which is present in considerable amounts under certain conditions, has so little effect. Moreover, there is a striking difference in the effect that the nonfreezing bound water has in  lowering the glass transition temperature of the two types of samples. The effect is much larger in the sample with the pure polyethylene oxide soft segment than the samples with block copolymer soft segments. In comparing the effect in the two types of sample it is assumed that the Tg which is measured is that of the water plasticized polyethylene oxide component. This assumption is supported by the observation that water has an antiplasticizing effect in 0PE33, which has a pure polypropylene oxide soft segment (3). However, it is not clear why only a single soft segment Tg is observed under these conditions.

It is suggested that the smaller effect of bound nonfreezing water on the Tg depression in the block copolymer soft segment samples is due to the short length of the polyethylene oxide segments, which are the terminal segments in the block copolymer soft segment, and from the restrictions on mobility due to coupling to the rigid hard segment units as well as to the unswollen polypopylene oxide segments. It is not expected that these restrictions to mobility would be offset by the free volume contributed by the added water. As a consequence, it cannot be expected that any single relation will provide an equally close fit to the results for both types of samples. In view of the restrictions to the mobility of the block copolymer soft segment, it appears that the results for 10PE33 are to be preferred in testing predictive relations for the diluent depression of the glass transition temperature. On this basis, the simple Fox mixing equation is more useful than the free volume relation used here. Although a superior fit to the experimental results is obtained with a water $T_g$ of 165 K , it cannot be assumed that this is the correct value due to uncertainties in the present data. The difficulty is represented by the observation of a somewhat greater $T_g$ depression with 124% versus 78% water when the comparison is made at the equal amounts of nonfreezing water.

It was mentioned that a value close to three has been reported for the water to EO molar stochiometry (5). This result was obtained from chemical shift NMR measurements as a function of water content. More recently, deuterium NMR studies by Hey and coworkers (*10*) have shown that the concentration at which mobile water is first observed coresponds closely to the first appearance of a water endotherm by DSC. This occurs at about one mole of water per mole of EO. The conflict between these two conclusions might be due to the fact that the chemical shift data is measuring less specific interactions between water and ethylene oxide. However, the results of Table III, when converted to moles of nonfreezing water per mole EO, indicate that the nonfreezing water is temperature dependent and close to 1.3 in 10PEO33 and between 2 and 3 in the block copolymer samples. Like the saturation water, the nonfreezing water decreases with increasing hard segment content, suggesting that the nonfreezing water is sensitive to constraints of the matrix. This could be a factor in the result obtained by Hey and coworkers, since the measurements were carried out on a highly crystalline polyethylene oxide sample. In any case the various unresolved questions indicate a need for more reliable data combining DSC with solid-state NMR measurements on the state of water and its effect on Tg and on polymer mobility.

**Literature Cited**

1. Chen, C. T.; Eaton, R. F.; Chang, Y. S.; Tobolsky, A. V.; *J. Appl. Polym. Sci.* **1962**, *16*, 2195 .
2. Illinger, J. L.; Schneider, N. S.; Karasz, F. E. In *Permeability of Plastic Films and Coatings to Gases, Vapors and Liquids*; Hopfenberg, H. D., Ed.; Plenum Publishing Corp.: New York, New York, 1975; 183-196.
3. Illinger, J. L. In *Polymer Alloys*; Klempner, D.; Frisch, K. C., Eds.; Plenum Publishing Corp.: New York, New York, 1977; 313-325.
4. Schneider, N. S.; Illinger, J. L.; Karasz, F. E.; *J. Appl. Polym. Sci.* **1993**, *47*, 1419.
5. Liu, K. J.; Parsons, J. L.; *Macromolecules* **1969**, *2*, 529.
6. Froix, M. F.; Nelson, R.; *Macromolecules* **1975**, *8*, 726.
7. McFarlane, D.R.; Angell, C.A.; *J. Phys. Chem.* **1984**, *88*, 759.
8. Meares, P.; *Polymer Structure and Bulk Properties*; Van Nostrand: London, 1965.
9. Takahashi, T.; Kyu, T.; Tran-Cong, Q.; Yano, O.; Soen, J.; *J. Polym. Sci., Polym. Phys.* **1991**, *29*, 1419.
10. Hey, M.J.; Ilett, S.M.; Mortimer, M; Oates, G.; *J. Chem. Soc., Faraday Trans.* **1990**, *86*, 2673.

RECEIVED May 13, 1993

# Chapter 10

# Formation and Reactivity of Surface Phosphonylated Thermoplastic Polymers

K. R. Rogers and Shalaby W. Shalaby

Department of Bioengineering and Materials Science and Engineering Program, Clemson University, Clemson, SC 29634–0905

Two types of components are currently used in orthopaedic prostheses, metallic intramedullary stems and polymeric articulationg surfaces. Metallic components have experienced problems with stable fixation in bone with both cemented and cementless approaches, with the most stable interface with hard tissue having been achieved with the use of hydroxyapatite (HA) coatings on metal substrates. Questions concerning the biocompatibility and mechanical performance of metallic materials have also arisen. It has been suggested that light-weight, high performance polymeric composite materials would eliminate problems associated with the biocompatibility and mechanical properties of metal components. However, the problem of bone fixation would still be an issue as in the current use of polymeric articulating surface components. Therefore, the current study was initiated to develop processes which would create highly reactive surfaces on biomedically relevant polymers that are capable of integrating, molecularly, with hydroxyapatite and thus, facilitate direct bonding to bone. Representative homochain and heterochain aliphatic polymers such as polyethylene and nylon-12 were surface modified by creating phosphonyl dichloride moieties which were then subsequently converted to other biomedically relevent functionalities, such as potassium and calcium phosphonates. All surfaces were extensively characterized using SEM, EDX, ATIR, contact angle measurements, and ESCA. Exposure of the modified surfaces to fibroblasts revealed a delayed attachment of the cells but no discernible gross toxicity. Finally, it was shown that the treated polymer surfaces were capable of interacting with a hydroxyapatite salt solution, immobilizing Ca ions on the surface in a crystalline order similar to that of calcium phosphate.

The current technology in orthopaedic prostheses involves implants which consist of two general types of components, an intramedullary stem and an articulating surface. Intramedullary stems are composed of metals and metal alloys, utilizing titanium, alumina, cobalt, chromium, and stainless steel. Articulating surfaces require certain tribological properties of the surface material. It has been found that polymeric materials, primarily ultra-high molecular weight polyethylene (UHMWPE), possess both the mechanical and frictional properties to perform in this capacity. Both types of components have been successful, but have encountered certain problems.

0097–6156/94/0540–0116$06.00/0
© 1994 American Chemical Society

The traditional method of fixation for the metal stems is to employ the use of a grouting material such as poly (methyl methacrylate) (PMMA) bone cement. However, the issue of loosening of these cemented components has resulted in the investigation of the cementless intramedullary stem which utilizes alternate methods of fixation (*1*). Both porous metal coatings and "bioactive" material coatings, such as hydroxyapatite, have been attempted in order to alleviate the problems associated with the use of bone cement. Porous metal coatings can provide a mechanical fixation through the direct growth of bone into the porous structure of the coating, while "bioactive" coatings feature direct chemical bonding of the coating to the bone. Both types of coatings have also encountered problems with adequate fixation, although the hydroxyapatite coatings have been shown to adhere irreversibly to bone (*2,3,4*). Other problems associated with the use of metals in orthopaedic prostheses include the possibility of bone resorption in the tissue surrounding the metal implant, and concerns regarding the possible carcinogenic effect of metal ion release products (*5,6*).

It has been suggested that a possible solution would be to replace these metal components with light-weight, high performance composite materials consisting of polymeric materials. Such a composite material is believed to provide a relatively low elastic modulus as compared to metals, the absence of metal ion release products, and the ability to tailor the strength of the material to suit individual design requirements. However, the problem of fixation in hard tissue of a polymeric component would still exist, as in the case of the current use of UHMWPE components.

UHMWPE articulating components are used in hip, knee, and elbow prostheses. Their fixation sometimes involves the use of a metal backing in conjunction with PMMA bone cement to interface with the bone. This method of fixation has realized limitations, however. The use of a metal backing decreases the thickness of the polymeric component, which has been suggested to contribute to the wear problems associated with the use of UHMWPE components. Also, the polymer/metal interface is sometimes unstable, resulting in migration of the polymeric component. The application of PMMA has also led to problems associated with its use in intramedullary stems, including loosening.

Clearly, a need exists to develop an optimum polymer/bone interface which will provide a direct, stable, permanent fixation in hard tissue for both present and future polymeric components in orthopaedic prostheses. This need provided a strong incentive to pursue the present study on surface activation and to investigate the development of methods of creating hydroxyapatite-like surfaces on polyethylene, the currently used orthopaedic polymer. The surface activation entailed the selective surface phosphonylation of polymeric films made of polyethylene and nylon-12. Nylon-12 was chosen as a representative heterochain polymer whose chain structure closely resembles polyethylene and yet contains hydrolyzable functionalities similar to those of nylon-6, a widely used biomaterial. To study the biocompatibility of a typical new surface, the current study also involved the interaction of a modified polyethylene film with fibroblasts and hydroxyapatite salt solution.

## Experimental

**Surface Modification Processes .** The process of surface phosphonylation for polyethylene and nylon-12 has been adequately described elsewhere (*7*). Following phosphonylation, the phosphonyl dichloride moieties were then subsequently converted to potassium and calcium phosphonate moieties as also described elsewhere (*7*).

**Surface Characterization.** Several analytical methods were used in characterizing the modified polymer surfaces. Energy Dispersive X-ray (EDX) spectra and electron micrographs of the surfaces were obtained using a JEOL JSM-IC848 scanning electron microscope in order to confirm the presence of phosphorus and chlorine, potassium, or calcium on the polymer surface. The concentration of different elements present were quantified relatively using the area under the peaks. Attenuated Internal Reflection (ATIR) spectra were obtained for the untreated and treated polymer surfaces using a BIO-RAD FTS-4R infrared spectrometer and then the two spectra were compared to identify characteristic peaks which are associated with the bonds formed by the addition of the phosphonate side groups on the polymer backbone.

Contact angle measurements were made on untreated and modified polyethylene films using a contact angle goniometer from Garner Scientific, Inc. Ten drops of approximately 1 milliliter of deionized water were placed on each film using a sterile pipette and the static contact angle was measured for each drop. An average was obtained for each film and the contact angle was then compared for the untreated vs. treated material in order to evaluate a difference in surface energy for the treated films.

Electron Spectroscopy for Chemical Analysis (ESCA) analysis was performed on treated polyethylene films at the University of Florida using a KRATOS XSAM 800 photo-electron spectrometer with DS800 software. The ESCA data was used to determine the chemical composition of the surface at about 50 A depth.

An Instron Universal test machine (model 1321) was used to perform tensile strength testing of both untreated and surface treated polyethylene films. The films were phosphonylated according to the process described elsewhere (7) and cut to a dogbone shape of the following dimensions: 5 cm length, 1.5 cm grip width, 0.75 cm testing width, and 45 micron thickness. Twenty samples of both untreated and treated films were pulled until failure using the following parameters: gage length of 1 cm, load cell of 500 pounds at 5% range, and a ramp rate of 9.9 mm/minute. The cross section of each sample was determined using a micrometer and the ultimate tensile strength (psi) was calculated.

Cytotoxicity Study . The gross cytotoxicity of phosphonylated polyethylene and polyethylene with potassium phosphonates present on the surface was evaluated using a modified ASTM standard F813-83, Direct Contact Cell Culture Evaluation of Materials for Medical Devices. Fifteen polyethylene films were phosphonylated and five of these films were subsequently converted to potassium phosphonate according to the methods described elsewhere (7). After thorough drying under vacuum, the films were cleaned using a Harrick Plasma Cleaner and subsequently placed in a sterile saline solution until testing. Immediately before testing, the films were rinsed thoroughly in sterile, ultrapure water.

A strain of L929 mouse fibroblasts was obtained from Sigma Chemical Company and maintained according to the ASTM standard. Twenty-five 35x10 mm style Corning polystyrene petri dishes were prepared with the test materials in the following manner. Ten untreated polyethylene films as well as the fifteen phosphonylated films were cut to fit tightly on the bottom of the small petri dishes and used to line the bottom surface of the petri dishes. Ten unlined petri dishes were used as the negative control and five plates lined with polyethylene films coated with methyl cyanoacrylate were used as the positive control.

After preparation of the petri dishes, each dish was seeded with two milliliters of cell medium containing fibroblasts and allowed to incubate at 37° C for a total of one week. Photographs of each plate were taken every twenty-four hours.

Hydroxyapatite Formation Study. The interaction of synthetic hydroxyapatite with treated polyethylene films was evaluated using a modification of a procedure published by Golomb and Wagner (8). Polyethylene films were phosphonylated and converted to potassium and calcium phosphonates.

Solutions used for incubation were prepared with a concentration product of calcium ($CaCl_2.2H_2O$) and phosphate ($K_2HPO_4$) of 9 $mM^2$. For this, calcium chloride ($CaCl_2.2H_2O$) and potassium phosphate ($K_2HPO_4$) were used as solutions of 3.87 mM and 2.32 mM concentrations, respectively. Each salt solution was prepared in 0.05 tris(Hydroxymethyl) aminomethane-hydrochloric acid buffer, pH=7.4. Equal volumes of each salt solution were mixed in glass tubes containing the test material mounted by surgical silk suture thread. The final incubation solution yielded a Ca/PO4=1.67, as in the hydroxyapatite component of bone. Two tubes contained phosphonylated polyethylene and two contained polyethylene with potassium phosphonate surfaces. Two tubes containing untreated polyethylene films were used as a control. The glass tubes were allowed to incubate for one week at 60° C.

After incubation, the films were sonicated for one hour in deionized water, rinsed three times and oven dried for twenty-four hours. The amount of calcium and phosphate present and surface microtexture were evaluated using EDX analysis and SEM, respectively.

## Results and Discussion

**Surface Modification .** The surface phosphonylation of polyethylene and nylon-12 resulted in the presence of phosphorus and chlorine on the polymer surfaces as reflected in the EDX spectra for each material as shown in Figures 1 and 2. These phosphonyl dichloride moieties were then successfully converted to potassium and calcium phosphonate groups as also reflected in the EDX spectra shown in Figures 1 and 2.

ATIR spectra of the treated polymer surfaces were obtained and compared with the spectra for untreated polymers in order to determine the presence of group frequencies due to the modified polymer surface. The characteristic peaks identified for treated polyethylene and nylon-12 appeared at wave numbers 1129, 994, 935, and 783 cm$^{-1}$ as shown elsewhere (9). In many cases, there was the appearance of a shoulder at 2956 cm$^{-1}$ which has been assigned to a -P-C-H- group.

The contact angle measurements were conducted for untreated and phosphonylated polyethylene. The angle determined for the untreated material was 90° and this angle was decreased to 65° for the treated polymer. The lowering of the contact angle reinforces the suggestion that hydrophilic phosphonate moieties were created on the surface of the hydrophobic substrate.

ESCA analysis was performed on samples which were phosphonylated and subsequently converted to the potassium phosphonate functionality bearing surfaces. Table I lists the elements present on the sample surfaces, their position, and the atomic concentration of each element from the high resolution scan for specific elements.

### Table I. ESCA data for modified PE surfaces

| Element | Position | Atomic Concentration % | |
| | | Sample 1 | Sample 2 |
|---|---|---|---|
| P | 133.20 | 2.13 | 2.48 |
| C | 285.10 | 63.54 | 62.52 |
| O | 531.45 | 18.87 | 20.21 |
| K | 293.25 | 8.82 | 11.49 |
| Si | 102.60 | 4.24 | 1.90 |
| Cl | 200.10 | 2.41 | 1.40 |

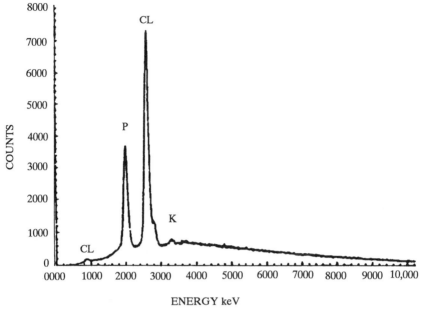

a

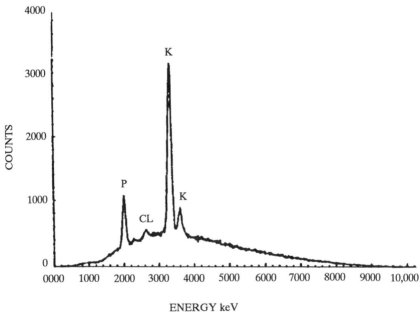

b

Figure 1.  Typical EDX spectra for modified PE surfaces containing (a) phosphonyl dichloride and (b) potassium phosphonate moieties.

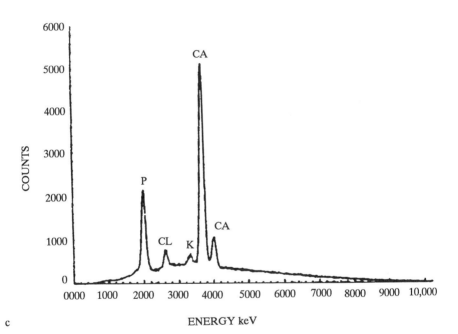

c

Figure 1—*Continued*.   Typical EDX spectra for modified PE
surfaces containing (c) calcium phosphonate moieties.

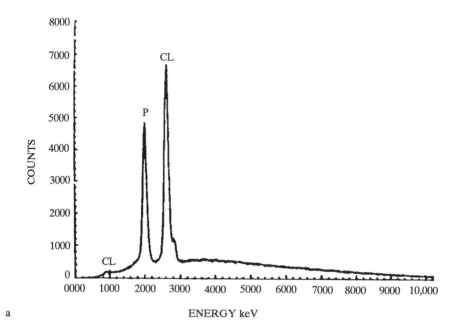

a                    ENERGY keV

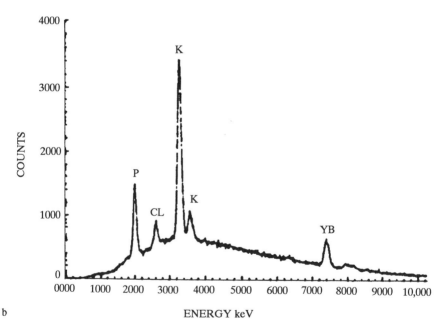

b                    ENERGY keV

Figure 2.  Typical EDX spectra for modified nylon-12 surfaces containing (a) phosphonyl dichloride and (b) potassium phosphonate moieties.

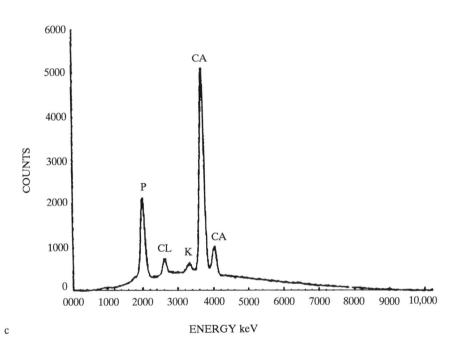

c                                    ENERGY keV

Figure 2—*Continued*.  Typical EDX spectra for modified nylon-
12 surfaces containing (c) calcium phosphonate moieties.

The ESCA analysis revealed the presence of chlorine on both samples in trace amounts, which concurred with the EDX data for the same samples. The EDX data showed only one type of chlorine present while there were two types of potassium present. This leads to the conclusion that the chlorine ions are present as potassium chloride while the prevalent funtionality present on the polyethylene surface is potassium phosphonate. The ESCA data also revealed that there is one phosphorus per 27 chain carbons, which means only about 4% of the methylene groups had their hydrogen replaced by a potassium phosphonate functionality. The calculated O/K ratio is 2.5, where theoretically the ratio should be 1.5 for completely converted potassium phosphonate moieties and 3.0 for a phosphonic acid bearing surface. The difference between the theoretical and calculated ratios is probably indicative of a partially neutralized surface.

Tensile strength measurements were made for untreated and phosphonylated polyethylene films in order to ensure that the bulk tensile properties of the treated polymer were not affected by the surface modification. Twenty samples of both treated and untreated films were pulled until failure. The average ultimate tensile strength for untreated polyethylene was 2,437 psi with a standard deviation of 308 and the average for the treated films was 2,339 psi with a standart deviation of 729. A student's t-test revealed no statistically significant difference in the tensile strength between the two polymers. This tensile data, coupled with the percentage of the material affected by the surface treatment as calculated from ESCA data confirms that the material modification was limited to the material surface and did not penetrate the bulk of the material.

Cytotoxicity Study. The cytoxicity study was carried out in order to evaluate the gross toxicity of polyethylene films with phosphonyl dichloride and potassium phosphonate moieties present on the surface. The behaviour of fibroblasts seeded on these surfaces was compared with their behavior on materials surfaces whose toxicity is known. Polyethylene films coated with methyl cyanoacrylate were used as a positive control because it is well known that methyl cyanoacrylate has an extremely toxic effect on cells (10). The fibroblasts were killed within 24 hours of being placed on the positive control films. Photographs of the fibroblasts seeded on these positive control films are shown in Figure 3, exhibiting total cell death and lysis.

Polystyrene petri dishes were used as the negative control because these dishes are used in cell cultures; cells readily attach, spread, and proliferate on this material surface. By 24 hours, the fibroblasts had attached and begun spreading. At 3 days, the cells had formed a confluent monolayer, and at 1 week the cells had become overcrowded in the dishes. The fibroblasts' behavior mirrored this behavior on the untreated polyethylene film surfaces. Photographs displaying this cell behavior are shown in Figure 4. Figure 5 provides photographs displaying the fibroblast reaction to the phosphonylated polyethylene, which paralleled that of the cells on the potassium phosphonate bearing surfaces. At 24 hours, all fibroblasts remain suspended in the medium, but show no signs of death or lysis. At 3 days, a few cells have attached while the majority remain in suspension, but still display no signs of lost vitality. At one week, more cells have attached to the film surface and have begun to spread, displaying behavior similar to that of the cells at one day on the negative control surface. This condition of deferred attachment reflects no toxic effects on the cells by the treated films, but suggests that the treated surfaces reduce the cells' propensity for immediate attachment. This reluctance to attach is probably due to the fact that the modified surfaces possess a negative charge which has been well documented to repel cell attachment to a material surface.

Hydroxyapatite Formation Study. The directed growth of hydroxyapatite crystals on the modified polyethylene surfaces was attempted in vitro in order to assess, indirectly, the propensity of the materials for integration with natural HA in vivo. Untreated PE films as well as PE films bearing phosphonyl dichloride and potassium phosphonate moieties were incubated in the calcium and phosphate salt solution, used

Figure 3. Fibroblast death and lysis on a polymethyl cyanoacrylate coated surface.

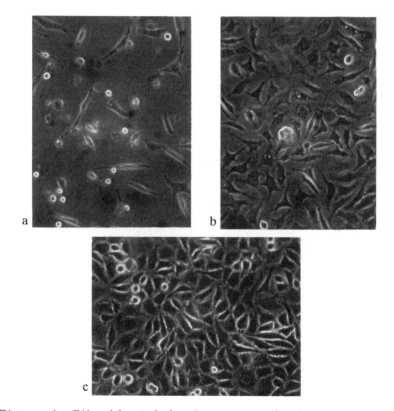

Figure 4. Fibroblast behavior on a polystyrene
surface at (a) one day, (b) three days, and (c) one
week.

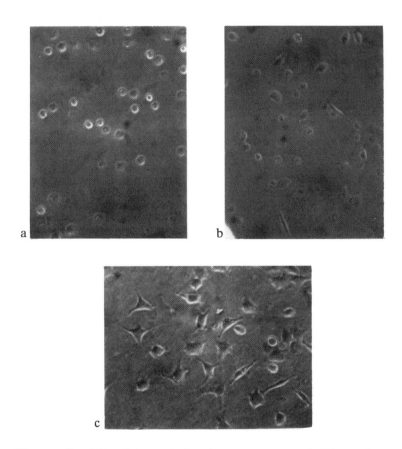

Figure 5. Fibroblast behavior on treated PE surfaces at (a) one day, (b) three days, and (c) one week.

earlier as precursors of HA (8). The films were incubated for one week at 60°C. EDX spectra were used in examining the presence and relative concentration of phosphorus and calcium on these surfaces. The EDX spectra are shown in Figures 6 and 7. Given in Table II are the values for the integration of the peaks from the spectra, which gives relative quantities of calcium and phosphorus present.

Table II. Relative Concentrations of Elements Present on Film Surfaces of the HA Formation Study Samples

| | -POCl_2 Bearing Surface | | | -PO(KO)_2 Bearing Surface | | |
|---|---|---|---|---|---|---|
| | | Day | | | Day | |
| Element | 0 | 1 | 3 | 0 | 1 | 3 |
| P | 43692 | 32835 | 10383 | 14587 | 15803 | 25403 |
| Cl | 29326 | 4123 | 1289 | 6988 | 10512 | 4905 |
| K | | | | 24719 | 55759 | 43124 |
| Ca | | 61850 | 10286 | | 12298 | 167554 |

The phosphonyl dichloride bearing surface lost a major fraction of its chlorine after one day, and virtually all the chlorine is removed after three days indicating that the majority of phosphorus is present as phosphonic acid and calcium phosphate and/or phosphonate. The Ca/P ration was 1.88 for day one, but decreases to 1.0 by day three. The Ca/P ratio for HA is 1.67, which compares to day one but is not exact.

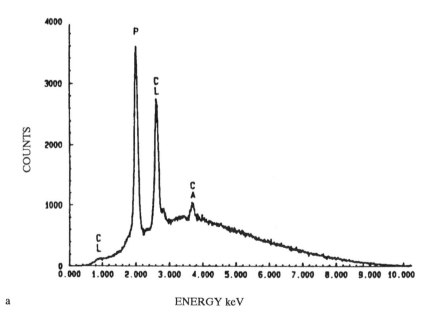

a                                    ENERGY keV

Figure 6.    EDX spectra for phosphonyl dichloride bearing PE surfaces in HA study at (a) day 0.

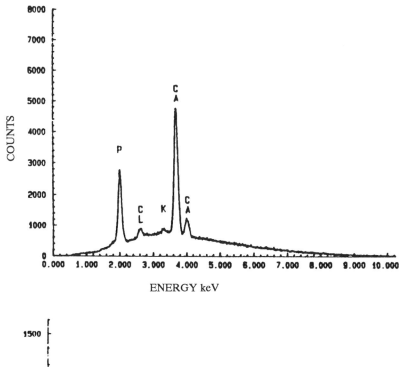

b    ENERGY keV

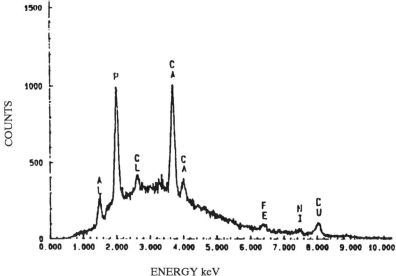

c    ENERGY keV

Figure 6—*Continued*.   EDX spectra for phosphonyl dichloride
bearing PE surfaces in HA study at (b) day 1 and (c) day 3.

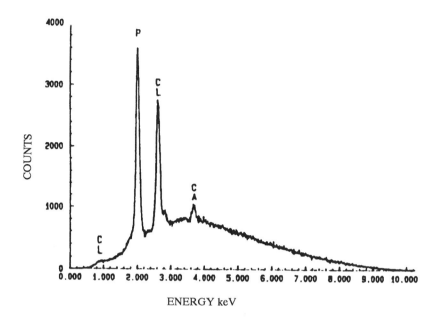

a          ENERGY keV

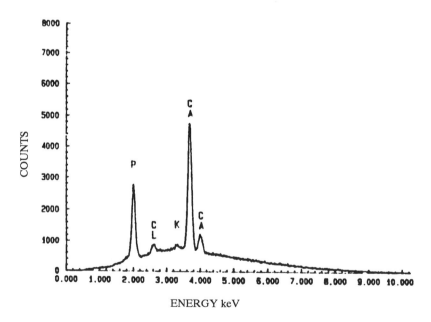

b          ENERGY keV

Figure 7. EDX spectra for potassium phosphonate bearing PE surfaces in HA study at (a) day 0 and (b) day 1.

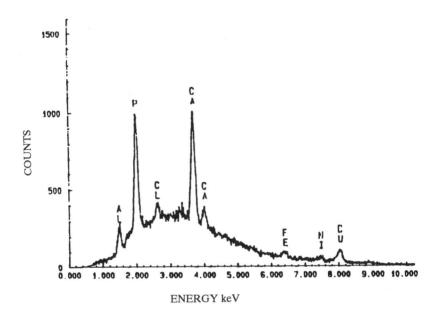

c                                                ENERGY keV

Figure 7—Continued.   EDX spectra for potassium phosphonate
bearing PE surfaces in HA study at (c) day 3.

The obtained Ca/P ratios and the distinct increase in calcium content over time
indicates that the immobilization of Ca is possible on the modifed surfaces, but
suggests that a stable calcium phosphate compound has not been formed in the given
time period. SEM photographs, shown in Figure 8, taken at day one and day three
show a distinct difference in the appearances of the surfaces. At day one, the texture
of the surface is foam-like with a porosity of <0.5 microns. After 3 days, the texture
of the surface is much finer with a dramatically decreased porosity. This may be due
to the recrystallization of the inorganic uppermost surface and/or the increased salt
formation and/or deposition on the film surface. An x-ray diffraction pattern, shown
in Figure 9, reveals a crystalline structure resembling that of calcium phosphate at
day 3. At this time, it is not fully understood how the calcium phosphate compounds
react with the treated polymer surfaces.

Conclusions

      Based on the results of the current study, the following conclusions have been
drawn. Processes for the selective surface phosphonylation of polyethylene and
nylon-12 and the subsequent conversion to other biologically relevant moieties have
been developed and basic characterization of prevailing functionalities was achieved.
Immobilization of calcium ions on a modified polyethylene surface is feasible and has
been achieved by incubation of the modified surface in a saturated calcium phosphate
solution. An exposure to fibroblasts in a typical culture medium to a phosphonate
bearing surface decreased cell attachment but elicited no discernible toxicity.

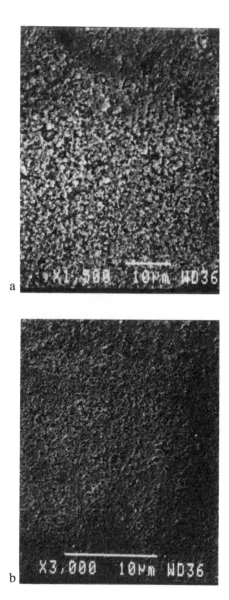

Figure 8. SEM photographs of HA study sample at (a) one day, and (b) three days.

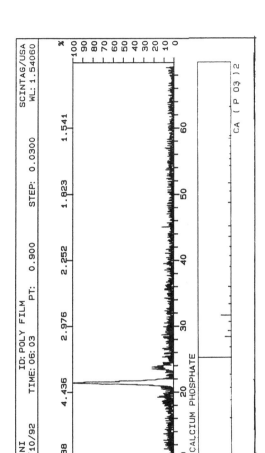

Figure 9. X-ray diffraction pattern of a HA study sample at three days compared to the pattern for calcium phosphate.

References
1. Stauffer, R.N. *J. Bone Joint Surg.* 1982, *64A*, 983.
2. Klein, C.P.; Abe, Y.; Hosono, H.;  de Groot, K. *Biomaterials* 1984, *5*, 362.
3. Holmes, R.E.; Hagler, H.K. *J. Oral Maxillofac. Surg* 1987, *45*, 421.
4. Geesink, R.; de Groot, K.; Klein, C.P. Clin. Orthop. 1987, 225, 147.
5. Burke, D.W.; Gates, E.I.; Harris, W.H. *J. Bone Joint Surg.* 1984, *66-A*, 1265.
6. Dumbleton, J.H.; Black, J. In *Some Long Term Complications*; Ling, R.S.M., Ed.; Complications of Total Hip Replacement; Churchill-Livingstone: New York, New York, 1984.
7. Shalaby, S.W.; Rogers, K.R., U.S.Patent Application #07/840,020, 1992
8. Golomb, G.; Wagner, D. *Biomaterials* 1991, *12*, 397.
9. Rogers, K.R. "The Formation and Reactivity of Surface Phosphonylated Thermoplastic Polymers." M.S. Thesis. Clemson University, Clemson, SC. 1992.
10. Linden, C.L. "Cyanoacrylate Systms as Absorbable Tissue Adhesives." M.S. Thesis. Clemson University, Clemson, SC. 1992.

RECEIVED June 28, 1993

# Chapter 11

# Surface Modification of Polymeric Biomaterials with Poly(ethylene oxide)

## A Steric Repulsion Approach

**Mansoor Amiji[1] and Kinam Park**

**School of Pharmacy, Purdue University, West Lafayette, IN 47907**

Surface modification with hydrophilic polymers, such as poly(ethylene oxide) (PEO), has been beneficial in improving the blood compatibility of polymeric biomaterials. Surface-bound PEO is expected to prevent plasma protein adsorption, platelet adhesion, and bacterial adhesion by the steric repulsion mechanism. PEO-rich surfaces have been prepared either by physical adsorption, or by covalent grafting to the surface. Physically adsorbed PEO homopolymers and copolymers are not very effective since they can be easily displaced from the surface by plasma proteins and cells. Covalent grafting, on the other hand, provides a permanent layer of PEO on the surface. Various methods of PEO grafting to the surface and their effect on plasma protein adsorption, platelet adhesion, and bacterial adhesion is discussed.

A wide variety of natural and synthetic materials have been used for biomedical applications. These include polymers, ceramics, metals, carbons, natural tissues, and composite materials (1). Of these materials, polymers remain the most widely used biomaterials. Polymeric materials have several advantages which make them very attractive as biomaterials (2). They include their versatility, physical properties, ability to be fabricated into various shapes and structures, and ease in surface modification. The long-term use of polymeric biomaterials in blood is limited by surface-induced thrombosis and biomaterial-associated infections (3,4). Thrombus formation on biomaterial surface is initiated by plasma protein adsorption followed by adhesion and activation of platelets (5,6). Biomaterial-associated infections occur as a result of the adhesion of bacteria onto the surface (7). The biomaterial surface provides a site for bacterial attachment and proliferation. Adherent bacteria are covered by a biofilm which supports bacterial growth while protecting them from antibodies, phagocytes, and antibiotics (8). Infections of vascular grafts, for instance, are usually associated with *Pseudomonas aeruginosa*, *Escherichia coli*, *Staphylococcus aureus*, and *Staphyloccocus epidermidis* (9).
    Since the interactions leading to surface-induced thrombosis and biomaterial-associated infections occur at the biomaterial-blood interface, appropriate surface modification is beneficial in improving the blood compatibility of biomaterials (10).

[1]Current address: Northeastern University, School of Pharmacy, Boston, MA 02115

Hoffman reviewed the methods of surface modification of biomaterials and grouped them into two general categories: physico-chemical and biological methods (11). The physico-chemical methods are surface coating, chemical modification, graft copolymerization, and plasma treatment. The biological methods include, pre-adsorption of proteins, drug or enzyme immobilization, cell seeding, and preclotting. Ikada (12) proposed that a biomaterial surface with diffuse hydrophilic surface layer would be blood compatible. Water-soluble polymers can be grafted onto the biomaterial surface to provide a diffuse hydrophilic layer (13). Surface-bound water-soluble polymers are thought to prevent protein adsorption, platelet and bacterial adhesion by the steric repulsion mechanism.

## Steric Repulsion with Surface-Bound Hydrophilic Polymers

Surface-bound hydrophilic polymers have been used in steric stabilization of many different colloidal dispersions (14). Steric stabilization occurs as a result of repulsion between the two overlapping polymer layers (15). Steric repulsion of plasma proteins, platelets, and bacteria by surface-bound hydrophilic polymer is illustrated in Figure 1. For effective steric repulsion, the hydrophilic polymer in the diffuse layer must satisfy the following three requirements. First, the polymer molecule should be tightly bound to the surface. Second, a segment of the polymer should also extend into the bulk aqueous environment. Extension and flexibility of the polymer segment determines the dominance of steric repulsion over the van der Waals attractive forces. Due to these opposing requirements, block copolymers with hydrophilic and hydrophobic segments are more effective in steric stabilization of colloidal particles than homopolymers (16). Steric repulsion with homopolymers is effective only when they are covalently grafted to the surface. Finally, the polymer molecules should cover the surface completely. Steric repulsion is not effective if a significant portion of the surface remains exposed.

In this article, we will discuss the steric repulsion of plasma proteins, platelets, and bacteria by surface-bound poly(ethylene oxide) (PEO). PEO, a neutral hydrophilic polymer, has been used most widely for surface modification of biomaterials.

## Conformation of PEO in Aqueous Environment

PEO is a polyether type of water-soluble synthetic polymer as shown below.

$$-(CH_2CH_2O)_n-$$

Low molecular weight polymers (i.e., less than 10,000 daltons) are referred to as poly(ethylene glycol) (PEG), while those with higher molecular weights are known as poly(ethylene oxide) or polyoxyethylene (17). Compared to other polyethers such as poly(propylene oxide) (PPO), PEO is highly water-soluble (18). Kjellander and Florin (19) explained the aqueous solubility of PEO by a good structural fit between water molecules and the PEO chains which results in hydrogen bonding between the ether oxygen of PEO and water molecules. The aqueous solubility decreases with increasing temperature. The decreased aqueous solubility at elevated temperature is due to decrease in hydrogen bonding and corresponding increase in hydrophobic interactions between polymer chains (20). Using $^{13}$C-NMR, Bjorling et al. (21) have recently shown that PEO adapts a gauche conformation in a polar solvent such as water and a trans conformation in a nonpolar medium. The gauche conformation is more suitable for hydrogen bonding.

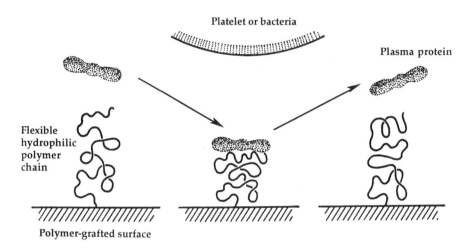

Figure 1.  Steric repulsion of plasma proteins, platelets, and bacteria by surface-grafted hydrophilic polymer chain.

## Steric Repulsion with Surface-Bound PEO

At the solid-water interface, terminally attached PEO will interact with water molecules and extend into the bulk aqueous medium. Using the surface force technique, Luckam (22) has observed that steric repulsion with PEO mainly occurs due to osmotic repulsion between interdigitated PEO chains. Jeon et al. (23,24) have theoretically modeled protein-surface interactions in the presence of PEO and found that steric repulsion by surface-bound PEO chains is mainly responsible for the prevention of protein adsorption on PEO-rich surfaces.

## Effect of PEO on Biocompatibility

PEO-rich surfaces have been prepared by physical adsorption of PEO homopolymer (25). Only high molecular weight PEO homopolymers (Mol. Wt. > 100,000 daltons) are effectively adsorbed on hydrophobic surfaces (26). Chromatographic supports for separation of proteins and cells have been treated by physical adsorption of high molecular weight PEO (25). Physically adsorbed PEO homopolymers, however, can be displaced by other macromolecules which have higher affinity to the surface. In blood, for instance, many proteins and cells can easily displace physically adsorbed PEO from the surface.

    Surface-adsorbed PEO-containing amphipathic block copolymers would be more stable than homopolymers. The water-insoluble segment of the copolymer can interact with a hydrophobic surface by hydrophobic interactions, while the water-soluble PEO chains can extend into the bulk aqueous medium. PEO/PPO/PEO triblock copolymers (Pluronics®) has been used for prevention of protein adsorption and cell adhesion on hydrophobic surfaces. When Pluronics with 30 propylene oxide (PO) residues were adsorbed on hydrophobic surfaces, decrease in albumin adsorption was not significant (27). Pluronics with 30 PO residues were weakly bound to the surface. Albumin could easily displace weakly bound Pluronics and interact directly with the surface. To improve the stability of copolymers on the surface, Lee et al. synthesized copolymers of PEO-methacrylates containing alkyl chains for tight binding to hydrophobic surface (28). Tight binding to hydrophobic surfaces can also be achieved with Pluronics containing longer hydrophobic PPO segments. Using ten different Pluronics with varying PO and ethylene oxide (EO) segments, we have found that on dimethyldichlorosilane-treated glass (DDS-glass), Pluronics having 56 or more PO residues were more effective in preventing plasma protein adsorption and platelet adhesion than Pluronics with 30 PO residues (29). Pluronic F-108, which has 56 PO and 129 EO residues, minimized the adsorption of plasma proteins onto polystyrene latex particles (30). The adhesion of cells and bacteria onto octadecyldimethylsilane-treated glass was also inhibited, when the surface was coated with Pluronic F-108 (31). Bridgett et al. (32) showed that the adhesion of *Staphylococcus epidermidis* onto Pluronic-treated polystyrene surfaces decreased by more than 97% compared to control. Blainey and Marshall (33) found a decrease in the adhesion of marine bacteria by 87-100% on F-108-treated surfaces for up to 4 days post-treatment. Upon long-term exposure to marine environment, however, Pluronics were not effective in preventing bacterial adhesion. The adsorption of Pluronics onto hydrophobic surfaces presents the simplest method for short-term modification for the prevention of protein adsorption and cell adhesion.

    Desai and Hubbell (34) have entrapped PEO chains to the surface by partially dissolving the base polymer in a suitable solvent. Poly(ethylene terephthalate) (PET) with entrapped PEO (Mol. Wt. 18,500) decreased albumin adsorption by 80% and platelet adhesion by more than 95% as compared to the control PET. In addition, the adhesion of *Staphylococcus epidermidis*, *Staphylococcus aureus*, and *Pseudomonas aeruginosa* was reduced by 70-95% on PEO-modified PET compared to control PET

(35). Ruckenstein and coworkers (36,37) have used a similar approach to entrap PEO-containing block copolymers to poly(methyl methacrylate), polystyrene, and poly(vinyl acetate) surfaces. The base polymer was swollen in an organic solution containing PEO block copolymer. The hydrophobic segment of the copolymer was thought to be entangled with the swollen surface. The swollen substrate was then collapsed by passing through a water phase. For entrapment of PEO homopolymers or block copolymers, one has to consider the issues of toxicity from the use of organic solvents, long-term stability of the entrapped PEO, and poor adaptability of the technique to other polymeric substrates.

Covalent grafting of PEO to the surface is the most effective way of creating a permanent PEO layer. Merrill et al. (38) initially reported that polyurethanes with PEO soft-segment are highly blood compatible. Ito and Imanishi (39) in their review article have presented the work of many investigators who have prepared polyurethanes with polyether soft-segment. The results showed that segmented polyurethanes containing high molecular weight PEO had improved blood compatibility compared to those containing other types of polyethers. Yu et al. (40) have developed hydrogels of PEO-containing polyurethaneurea as coatings for biomedical products. Polyurethanes with the PEO grafted side chains have been found to be highly blood compatible (41,42). Figure 2 shows the reaction scheme for synthesis of polyurethanes with methoxy-PEG side chains. Chaikof et al. (43) have recently developed an interpenetrating polymer network (IPN) of PEO and polyether substituted polysiloxane. A significant decrease in platelet adhesion was observed when the PEO (Mol. Wt. 8,000) content in the IPN was increased up to 65%. It is clear from these studies that PEO in polyurethane soft segments or in IPN's should extend into the bulk aqueous environment and remain flexible for improved blood compatibility.

Allmer et al. (44) have covalently coupled PEO chains to glycidyl methacrylate-bound polyethylene surfaces. Akizawa et al. (45) have coupled methoxy-PEG with terminal carboxyl group to form ester linkage with the hydroxyl groups of cellulose dialysis membranes. Plasma protein adsorption, platelet adhesion, and complement activation was significantly decreased when PEG-monoacid was grafted to cellulose membranes (46). Desai and Hubbell (47) grafted cyanuric chloride-activated PEO to amine-derivatized PET surfaces. Figure 3 shows the preparation of amine-derivatized PET surface and subsequent covalent grafting of cyanuric chloride-activated PEO. About 50% decrease in plasma protein adsorption and more than 90% decrease in platelet adhesion was observed on PEO-grafted surfaces when the molecular weight of PEO was either 18,500 or 100,000. Gombotz et al. (48) have reported coupling of *bis*-amino PEO to cyanuric chloride-derivatized PET films. The adsorption of albumin and fibrinogen was found to decrease with increasing molecular weight of immobilized PEO. Chemical coupling of PEO to polymeric surfaces is possible only if the surface has functional groups that can react with PEO derivatives. For inert polymers such as polyethylene, PEO coupling is possible only when the surface is pre-modified with reactive functional groups (44). Grafting by use of UV or gamma irradiation, however, may not require pre-modification of the polymer surface.

Mori and Nagaoka (49) prepared PEO-rich surfaces by photoinduced grafting of methoxy poly(ethylene glycol) methacrylates to poly(vinyl chloride) surface in the presence of dithiocarbamate. With increasing PEO chain length up to 100 EO residues, plasma protein adsorption and platelet adhesion was significantly decreased on PEO-grafted surfaces. Clinical application of the PEO-grafted PVC tubes showed reduced potential for thrombogenicity compared to control PVC tubes (50). Tseng and Park (51) have synthesized PEG-phenylazide for photoinduced grafting to various polymeric surfaces. Upon irradiation with UV light, azide groups are converted into highly reactive nitrene groups. The reaction scheme for synthesis of

HO - PEBD - OH   +   MDI $\longrightarrow$   Prepolymer $\longrightarrow$

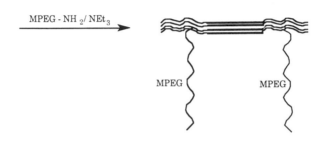

HO - PEBD -     Epoxidised polybutadiene-diol;    $M_n$ = 3020, Epoxy groups = 14

MPEG: Methoxy polyethylene glycol-amine     $CH_3-(OCH_2CH_2)_n-NH_2$;   $M_n$ = 100

DEG: Diethylene glycol     $OH-CH_2-CH_2-O-CH_2-CH_2-OH$

MDI: 4,4'-diphenylmethane di-isocyanate     NCO-⬡-$CH_2$-⬡-OCN

NEt$_3$ : Triethylamine     $CH_3CH_2-\underset{\underset{CH_2CH_3}{|}}{\overset{\overset{CH_2CH_3}{|}}{N}}$

Figure 2. Reaction scheme for coupling of methoxy-poly(ethylene glycol) (MPEG) to the polyurethane soft segment. (From reference 41).

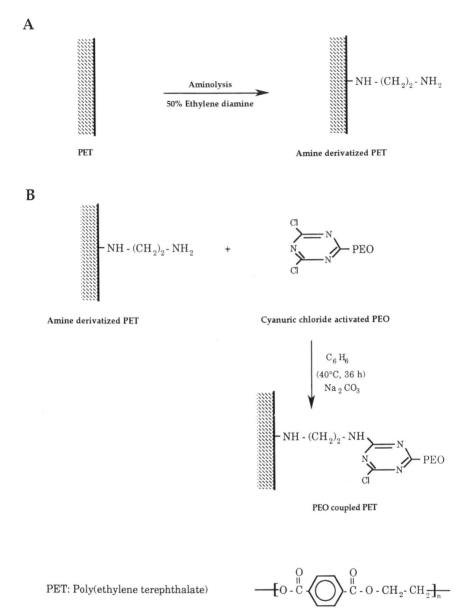

Figure 3. Reaction scheme for the preparation of amine-derivatized poly(ethylene terephthalate) (PET) surface (A) and chemical coupling of cyanuric chloride-activated PEO to amine-derivatized PET (B). (From reference 47).

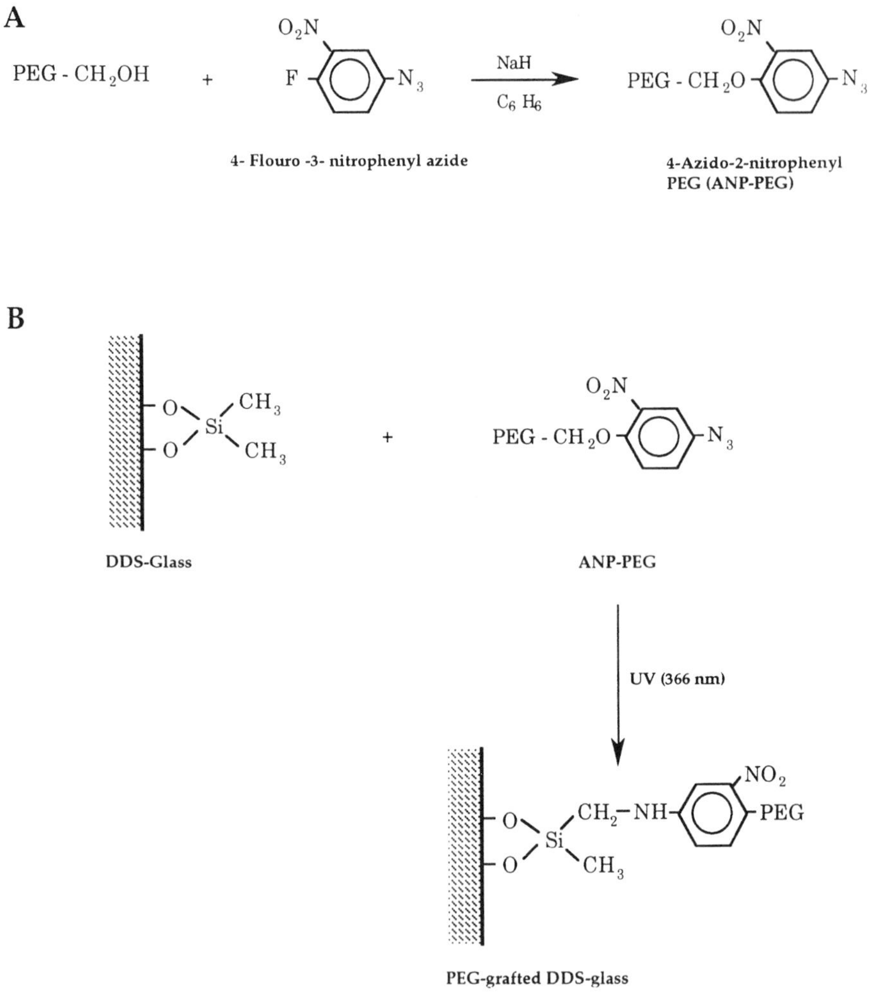

**A**

4- Flouro -3- nitrophenyl azide

4-Azido-2-nitrophenyl
PEG (ANP-PEG)

**B**

DDS-Glass                              ANP-PEG

UV (366 nm)

PEG-grafted DDS-glass

Figure 4. Reaction scheme for the preparation of (A) 4-azido-2-nitrophenyl PEG (ANP-PEG) and (B) photo-induced grafting of ANP-PEG to DDS-glass. (From reference 51).

PEG-phenylazide and photoinduced grafting onto DDS-glass is shown in Figure 4. Platelet adhesion and activation was prevented on PEG-grafted DDS-glass.

Sheu et al. (52) have recently introduced a method for grafting PEO-containing block copolymers (Brij ) by exposing the adsorbed copolymers to glow discharge treatment. High energy gamma or electron beam irradiation can also be used to graft PEO to various surfaces. Pluronic copolymer was grafted to DDS-glass by gamma irradiation (53). Platelets adhesion was prevented when Pluronic F-68-treated DDS-glass was irradiated in the presence of an aqueous buffer. Sun et al. (54) have grafted PEG-methacrylates to Silastic films by mutual irradiation in the presence of $Cu^{2+}$ ions to prevent homopolymer gellation. Approximately 72% decrease in fibrinogen adsorption was observed when the number of EO residues in grafted PEG were 100. Gamma irradiation at high doses may unfavorably alter the bulk properties of some polymers such as polypropylene.

## Concluding Remarks

Surface modification with PEO has been beneficial in improving the blood compatibility of polymeric biomaterials. Terminally-attached PEO layer is effective in preventing plasma protein adsorption, platelet adhesion, and bacterial adhesion by the steric repulsion mechanism. PEO-rich surfaces have been prepared by physical adsorption of PEO, entrapment of PEO, and by covalent grafting of PEO to the surface. For long-term applications involving blood contact, covalent grafting would be the best approach in providing a permanent layer of PEO on the biomaterial surface. Most grafting methods, however, suffer from many practical limitations. They require multiple steps and only a few types of polymers have functional groups available for modification. Advances in the grafting methods should be useful in simplifying the process and hopefully in the development of a biocompatible polymer.

## Acknowledgements

This study was supported by the National Heart, Lung, and Blood Institute of the National Institute of Health through grant HL 39081.

## References

1.  A.S. Hoffman. Synthetic polymeric biomaterials. In C.G. Gebelein (ed.) *Polymeric Materials and Artificial Organs*. ACS Symposium Series Vol 256. Am. Chem. Soc. Washington, DC. 1984 pp 13-29.
2.  E.P. Goldberg and A. Nakajima (eds.) *Biomedical Polymers, Polymeric Materials and Pharmaceuticals for Biomedical Use*. Academic Press. New York, NY. 1980.
3.  J.D. Andrade, S. Nagaoka, S.L. Cooper, T. Okano, and S.W. Kim. Surfaces and blood compatibility. Current hypothesis. *Trans. Am. Soc. Artif. Intern. Organs* **33**: 75-84 (1987).
4.  J.S. Hanker and B.L. Giammara. Biomaterials and biomedical devices. *Science* **242**: 885-892 (1988).
5.  J.D. Andrade and V. Hlady. Protein adsorption and material biocompatibility: A tutorial review and suggested hypothesis. *Adv. Polymer Sci.* **79**: 1-63 (1986).
6.  J.M. Anderson and K. Kottke-Marchant. Platelet interactions with biomaterials and artificial devices. *CRC Crit. Revs. Biocomp.* **1**: 111-204 (1985).
7.  A.G. Gristina. Biomaterial-centered infection: Microbial adhesion versus tissue integration. *Science* **237**: 1588-1595 (1987).

8. J.W. Costerton, T.J. Marrie, and K.J. Cheng. Phenomena of bacterial adhesion. In D.C. Sawage and M. Fletcher (eds.) *Bacterial Adhesion: Mechanisms and Physiological Significance.* Plenum Press, New York, NY. 1985 pp 3-43.
9. J. Dankert, A.H Hogt, and J. Feijen. Biomedical polymers: Bacterial adhesion, colonization, and infection. *CRC Crit. Revs. Biocomp.* **2:** 219-301 (1986).
10. S.W. Kim and J. Feijen. Surface modification of polymers for improved blood compatibility. *CRC Crit. Revs. Biocomp.* **1:** 229-260 (1980).
11. A.S. Hoffman. Modification of material surfaces to affect how they interact with blood. *Ann. NY. Acad. Sci.* **516:** 96-101 (1987).
12. Y. Ikada. Blood-compatible surfaces. *Adv. Polymer Sci.* **57:** 103-140 (1984).
13. S. Nagaoka, Y. Mori, T. Tanzawa, Y. Kikuchi, F. Inagaki, Y. Yokota, and Y. Nioshiki. Hydrated dynamic surfaces. *Trans. Am. Soc. Artif. Intern. Organs* **33:** 76-78 (1987).
14. D.H. Napper. Steric stabilization. *J. Colloid Interface Sci.* **58:** 390-407 (1977).
15. J. Klein. Surface interactions with adsorbed macromolecules. *J. Colloid Interface Sci.* **111:** 305-313 (1986).
16. D.H. Napper. Polymeric stabilization. In J.W. Goodwin (ed.) *Colloidal Dispersions.* The Royal Society of Chemistry, London, UK. 1981 pp 99-128.
17. F.E. Bailey and J.V. Koleske. *Poly(ethylene oxide).* Academic Press, New York, NY. 1976.
18. P. Molyneux. Nonionic polymers - polyoxides, polyethers, and poly(ethylene imine). In *Water-Soluble Synthetic Polymers: Properties and Behavior.* Volume I, Chapter 2. CRC Press, Boca Raton, FL. 1983 pp 19-73.
19. R. Kjellander and E. Florin. Water structure and changes in thermal stability of the system poly(ethylene oxide)-water. *J. Chem Soc., Faraday Trans.* **77:** 2053-2077 (1981).
20. F.E. Bailey and J.V. Koleske. Configuration and hydrodynamic properties of the polyoxyethylene chain in solution. In *Non-ionic Surfactants: Physical Chemistry.* By M.J. Schick. Marcel Dekker, Inc. New York, NY. 1987 pp 927-969.
21. M. Bjorling, G. Karlstrom, and P. Linse. Conformational adaptation of poly(ethylene oxide). A [13]C NMR study. *J. Phys. Chem.* **95:** 6706-6709 (1991).
22. P.F. Luckham. Measurement of the interaction between adsorbed polymer layers: The steric effect. *Adv. Colloid Interface Sci.* **34:** 191-215 (1991).
23. S.I. Jeon, J.H. Lee, J.D. Andrade, and P.G. DeGennes. Protein-surface interactions in the presence of polyethylene oxide I. Simplified theory. *J. Colloid Interface Sci.* **142:** 149-158 (1991).
24. S.I. Jeon and J.D. Andrade. Protein-surface interactions in the presence of polyethylene oxide II. Effect of protein size. *J. Colloid Interface Sci.* **142:** 159-166 (1991).
25. G.L. Hawk, J.A. Cameron, and L.B. Dufault. Chromatography of biological materials on polyethylene glycol treated controlled-pore glass. *Prep. Biochem.* **2:** 193-203 (1972).
26. T. Kato, K. Nakamura, M. Kawaguchi, and A. Takahashi. Quasielastic light scattering measurements of polystyrene latices and conformation of poly(oxyethylene) adsorbed on the latices. *Polymer J.* **13:** 1037-1043 (1981).
27. J.H. Lee, J. Kopecek, and J.D. Andrade. Protein-resistant surfaces prepared by PEO-containing block copolymer surfactants. *J. Biomed. Mater. Res.* **23:** 351-368 (1989).
28. J.H. Lee, P. Kopeckova, J. Kopecek, and J.D. Andrade. Surface properties of copolymers of alkyl methacrylates with methoxy (polyethylene oxide)-methacrylates and their application as protein-resistant coatings.*Biomaterials* **11:** 455-464 (1990).

29. M. Amiji and K. Park. Prevention of protein adsorption and platelet adhesion on surfaces by PEO/PPO/PEO triblock copolymers. *Biomaterials* **13:** 682-692 ( 1992).
30. J. Lee, P.A. Martic, and J.S. Tan. Protein adsorption on Pluronic copolymer-coated polystyrene particles. *J. Colloid Interface Sci.* **131:** 252-266 (1989).
31. N.F. Owens, D. Gingell, and P.R. Rutter. Inhibition of cell adhesion by synthetic polymer adsorbed to glass shown under defined hydrodynamic stress. *J. Cell Sci.* **87:** 667-675 (1987).
32. M.J. Bridgett, M.C. Davis, and S.P. Denyer. Control of staphylococcal adhesion to polystyrene surfaces by polymer surface modification with surfactants. *Biomaterials* **13:** 411-416 (1992).
33. B.L. Blainey and K.C. Marshall. The use of block copolymers to inhibit bacterial adhesion and biofilm formation on hydrophobic surfaces in marine habitat. *Biofouling* **4:** 309-318 (1990).
34. N.P. Desai and J.A. Hubbell. Solution technique to incorporate polyethylene oxide and other water soluble polymers into the surfaces of polymeric biomaterials. *Biomaterials* **12:** 144-153 (1991).
35. N.P. Desai, S.F.A. Hossainy, and J.A. Hubbell. Surface-immobilized polyethylene oxide for bacterial repellence. *Biomaterials* **13:** 417-420 (1992).
36. E. Ruckenstein and D.B. Chung. Surface modification by a two-liquid process deposition of A-B block copolymers. *J. Colloid Interface Sci.* **123:** 170-185 (1988).
37. J.H. Chen and E. Ruckenstein. Surface modification by a two-phase deposition of a surfactant. *J. Colloid Interface Sci.* **142:** 545-553 (1991).
38. E.W. Merrill, E.W. Salzman, S. Wan, N. Mahmud, L. Kushner, J.N. Lindon, and J. Curme. Platelets-compatible hydrophilic segmented polyurethanes from polyethylene glycols and cyclohexane diisocyanate. *Trans. Am. Soc. Artif. Intern. Organs.* **28:** 482- (1982).
39. Y. Ito and Y. Imanishi. Blood compatibility of polyurethanes. *CRC Crit. Revs. Biocomp.* **5:** 45-104 (1989).
40. J. Yu, S. Sundaram, D. Weng, J.M. Courtney, C.R. Moran, and N.B. Graham. Blood interactions with novel polyurethaneurea hydrogels. *Biomaterials* **12:** 119-120 (1991).
41. S.Q. Liu, Y. Ito, and Y. Imanishi. Synthesis and non-thrombogenicity of polyurethanes with poly(oxyethylene) side chains in soft segment. *J. Biomater. Sci., Polymer Edn.* **1:** 111-122 (1989).
42. E. Brinkman, A. Poot, L. Van der Does, and A. Bantjes. Platelet deposition on copolyether urethane modified with poly(ethylene oxide). *Biomaterials* **11:** 200-205 (1990).
43. E.L. Chaikof, E.W. Merrill, J.E. Coleman, K. Ramberg, R.J. Connolly, and A.D. Callow. Platelet interactions with poly(ethylene oxide) networks. *A AIChE J.* **36:** 994-1002 (1990).
44. K. Allmer, J. Hilborn, P.H. Larsson, A. Hult, and B. Ranby. Surface modification of polymers. VI. Biomaterial Applications. *J. Polym. Sci. Part A: Polym. Chem.* **28:** 173-183 (1990).
45. T. Akizawa, K. Kino, S. Koshikawa, Y. Ikada, M. Yamashita, and K. Imamura. Efficiency and biocompatibility of a polyethylene glycol grafted cellulosic membrane during hemodialysis. *Trans. Am. Soc. Artif. Intern. Organs* **35:** 333-335 (1989).
46. A. Kishida, K. Mishima, E. Corretge, H. Konishi, and Y. Ikada. Interactions of poly(ethylene glycol)-grafted cellulose membranes with proteins and platelets. *Biomaterials* **13:** 113-118 (1992).

47. N. Desai and J.A. Hubbell. Biological responses to polyethylene oxide modified polyethylene terephthalate surfaces. *J. Biomed. Mater. Res.* **25:** 829-843 (1991).
48. W.R. Gombotz, W. Guanghui, T.A. Horbett, and A.S. Hoffman. Protein adsorption to poly(ethylene oxide) surfaces. *J. Biomed. Mater. Res.* **25:** 1547-1562 (1991).
49. Y. Mori and S. Nagaoka. A new antithrombogenic material with long polyethylene oxide chains. *Trans. Am. Soc. Artif. Intern. Organs* **28:** 459-463 (1982).
50. S. Nagaoka and A. Nakao. Clinical application of antithrombogenic hydrogel with long poly(ethylene oxide) chains. *Biomaterials* **11:** 119-121 (1990).
51. Y.C. Tseng and K. Park. Synthesis of photo-reactive poly(ethylene glycol) and its application to the prevention of surface-induced platelet activation. *J. Biomed. Mater. Res.* **26:** 371-391 (1992).
52. M.S. Sheu and A.S. Hoffman. A new gas discharge process for preparation of n on-fouling surfaces on biomaterials. *Polymer Preprints* **32:** 239-240 (1991).
53. M. Amiji and K. Park. Surface modification by radiation-induced grafting of PEO/PPO/PEO triblock copolymers. *J. Colloid Interface Sci.* (in press).
54. Y.H. Sun, W.R. Gombotz, and A.S. Hoffman. Synthesis and characterization of non-fouling polymer surfaces. I. Radiation grafting of hydroxyethyl methacrylate and polyethylene glycol methacrylate onto Silastic films. *J. Bioactive Compatible Polym.* **1:** 316-334 (1986).References

RECEIVED May 13, 1993

# Chapter 12

# Ascorbic Acid as an Etchant–Conditioner for Resin Bonding to Dentin

James E. Code[1], Gary E. Schumacher[1], and Joseph M. Antonucci[2]

[1]Clinical Center/CODC, National Institutes of Health, Bethesda, MD 20892
[2]Dental and Medical Materials Group, Polymers Division, National Institute of Standards and Technology, Gaithersburg, MD 20899

L-ascorbic acid (AA) was evaluated as an etchant/conditioner for dentin bonding. A solution of AA (17.6% wt.%, in $H_2O$, pH 2.0) was applied to freshly cut dentin sections for time intervals of 15-120 s. The dentin sections were then rinsed with distilled $H_2O$, air dried, and evaluated for smear layer removal using scanning electron microscopy (SEM). Optimal time for smear layer removal was 30-60 s. Tensile bond strengths (TBS) were measured after dentin surfaces were treated sequentially with various solutions of AA (60 s), N-phenylglycine (NPG) in acetone (60 s), an acetone solution of a surface-active monomer, SAM, (60 s), and finally with a chemically cured composite. SEM results demonstrate significant smear layer removal from dentin using aqueous AA and TBS measurements demonstrate significant dentin bonding using a NPG/SAM-resin system with aqueous AA as the dentin/etchant conditioner.

Dentin, like bone, is a natural composite consisting essentially of apatitic mineral in a collagenous matrix. Because of its heterogeneous and nonuniform nature, the composition and structure of dentin is complex. Overall, the composition of dentin is 70% by weight mineral (45% by volume), and 20% by weight organic, mainly collagen, (33% by volume). A smaller but significant part of dentin is water, 10% by weight and 22% by volume. Some of this water is considered to be tightly bound, i.e., structural water. Dentinal tubule structure, both peritubular and intertubular, varies depending on the plane of the dentin surface and the distance of the tubule from the pulp. In addition, this heterogenous, vital tissue allows dentinal fluids to flow to its surface (1).

Because of its complexity bonding to dentin has presented more of a challenge than bonding to enamel. Adhesion of restorative resins to dental tissues has been attributed to chemical and/or mechanical factors (2). Theoretically, chemical bonding can occur with either the inorganic (apatite) or organic (collagen) constituents of dentin, with the formation of ionic or covalent bonds.    Mechanical adhesion is envisaged as occurring by the penetration and subsequent polymerization of monomers in intertubular dentin as well as in dentinal tubules.    In many bonding procedures smear layer removal is considered critical to good bonding.    Many adhesive systems use acidic treatments of dentin as the first step in bonding protocols to remove the smear layer.    To have effective surface wetting a bonding resin must have a lower surface free energy than the dentin surface to which it is applied. Ground dentin surfaces are covered with a smear layer which presents a low energy surface.    Etching the dentin surface removes the smear layer producing a cleaner, higher  surface energy substrate across which resin can wet and spread (3).

Effective bonding to dentin has been achieved by pretreatment with acid agents, e.g., phosphoric acid, citric acid plus ferric chloride, various forms of EDTA, etc., and acidic monomers, e.g., 2-methacryloxyethyl phenylphosphoric acid (phenyl-P), 4-methacryloxyethyl trimellitic anhydride (4-META), etc. (4). An acidic solution based on ferric oxalate (FO), which both removes the smear layer and mordants the cleansed surface to give improved bonding sites for coupling agents, has been developed (5).    It was anticipated that the ferric ions would form insoluble phosphates while the oxalate ions would form insoluble precipitates with calcium. The precipitated mineral would then solidify among the collagen strands and the restructured surface would be microporous, rigid and receptive to chemical agents. High bond strengths to dentin were obtained using a protocol of sequential applications of acidified FO, an N-aryl $\alpha$-amino acid and a surface-active monomer (SAM) (6).    Later it was found that the removal of the smear layer was primarily due to the presence of nitric acid as a contaminant in FO.    Subsequently, an effective etchant/conditioner consisting of an aqueous solution of 6.8 wt.% FO and 2.5 wt.% nitric acid was developed (7).

L-ascorbic acid (AA) or vitamin C (Figure 1) is a unique acid with chelating and antioxidant properties.    AA's moderate acidity and solubility in water and the fact that many of its salts, including its calcium salts, are water soluble give it potential as an etchant/conditioner for enamel and dentin.    Additionally, AA by virtue of its reducing properties can be used in a number of redox initiator systems for the polymerization of dental resins (8).    These properties, coupled with its chelation ability, give AA many desirable features that potentially could be useful in a resin bonding system.    The purpose of this study was to evaluate the use of AA as an etchant/conditioner for dentin bonding.

**Materials and Methods**

All of the materials used in this study were from commercial sources with the exception of N-phenylglycine (NPG) which was synthesized (9).    The SAMs used (Figure 2) were mono(2-methacryloyloxy)ethyl phthalate, MMEP, (Rhom Tech., Inc., Malden, MA), biphenyl dimethacrylate, BPDM, and diphenyl sulfone dimethacrylate, DSDM, (Bisco Inc., Itasca. IL), all derived by the reaction of 2-hydroxyethyl methacrylate with the appropriate anhydride.    Adaptic (Johnson and Johnson, East Windsor, NJ) was used as the chemically activated composite restorative material.    Extracted non-carious human molars, which were stored in

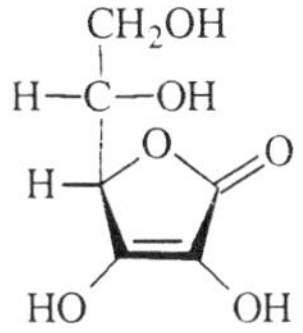

Figure 1. Chemical structure of ascorbic acid (AA).

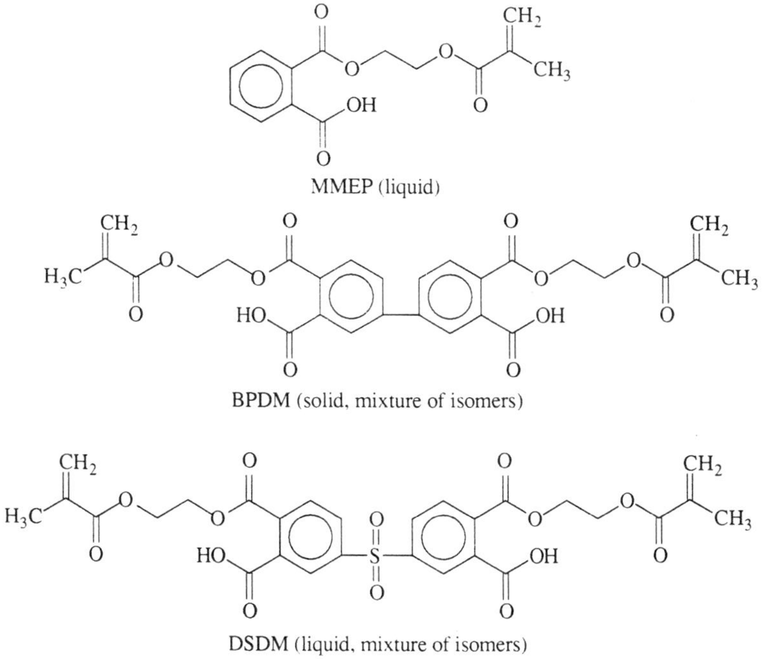

Figure 2. Chemical structures of surface-active monomers (SAM).

distilled water, were used to test the bonding of the composite resin to dentin. The dentin surface used for bonding was obtained by removing the occlusal surface perpendicular to the long axis of the tooth with a slow speed diamond sectioning blade (Isomet, Buehler Ltd., Lake Bluff, IL) running under water.

Tensile bond strength (TBS) were determined using a testing protocol and assembly previously described (6). To assess the efficacy of smear layer removal by aqueous AA the surface of 1 mm thick dentin cross sections were pretreated with one drop (0.05 mL) of AA (17.6 wt.% in distilled $H_2O$; pH = 2.0). The durations of AA treatment were: 15, 30, 45, 60, and 120 s. Each AA treated dentin surface was rinsed with distilled water for 10 s and then was air dried. The dentin specimens were then sputter coated with gold for evaluation by scanning electron microscopy (SEM).

A series of experiments were conducted using AA in a three step protocol to bond composite resin to dentin. Ten teeth were used for each bonding experiment. The bonding method employed was as follows: the dentin surface was first pretreated with one drop (0.05 mL) of AA in an aqueous solution for 60 s. The AA treated dentin surface was then rinsed with distilled water for 10 s and was blown dry with compressed air; a drop (0.05 mL) of either 10 wt.% or 5 wt.% NPG in acetone was placed on the dentin surface for 1 min. The acetone evaporated leaving a dry surface; one drop (0.05 mL) of the SAM in acetone was placed on the NPG treated surface for 1 min and then gently air blown to remove any excess acetone. The mixed composite paste was then applied. After 24 h of storage in distilled water at 24°C the bonded specimens were fractured in tension on a universal testing instrument (Model 1130, Instron Corp., Canton, MA) at a crosshead speed of 0.5 cm/min.

## Results

Compared to the SEM photomicrographs of an untreated dentin surface, photomicrographs of the AA treated dentin cross sections for times of 15, 30, 45, 60, and 120 s all showed significant surface changes(Figures 3-8). Optimal smear layer removal and opening of dentinal tubules occurred after 30-60 s of AA etching.

The mean TBS of the composite resin to dentin using the three step protocol of applying sequential solutions of AA, NPG and MMEP are shown in Table I. ANOVA calculations found no differences for groups A-E (p< 0.05). The mean TBS of the composite resin to dentin using a three step protocol of applying sequential solutions of AA, NPG and BPDM or DSDM are shown in Table II. The use of difunctional monomers in this experiment was to demonstrate the general applicability of using AA as a dentin etchant/conditioner for dentin bonding.

## Discussion

SEM results clearly demonstrate removal of the smear layer from dentin using AA as a dentin etchant/conditioner. Various water soluble salts are probably formed, i.e., calcium hydrogen phosphates or calcium ascorbates, during interaction of AA

Figure 3.   SEM photomicrograph of dentin surface with smear layer untreated (3000 X).

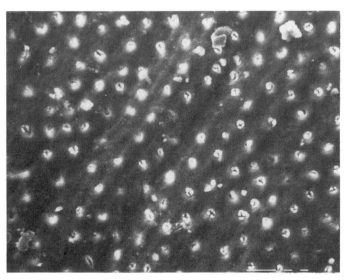

Figure 4.   SEM photomicrograph of dentin surface treated with 17.6 wt.% aqueous AA for 15 s (2000 X).

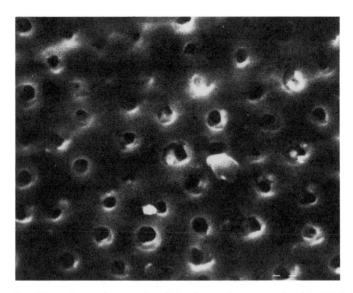

Figure 5.   SEM photomicrograph of dentin surface treated with 17.6 wt.% aqueous AA for 30 s (5000 X).

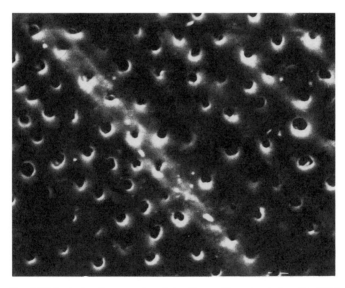

Figure 6.   SEM photomicrograph of dentin surface treated with 17.6 wt.% aqueous AA for 45 s (3000 X).

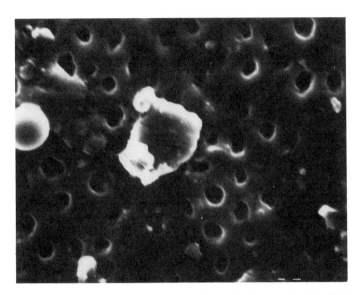

Figure 7.   SEM photomicrograph of dentin surface treated with 17.6 wt.% aqueous AA for 60 s (5000 X).

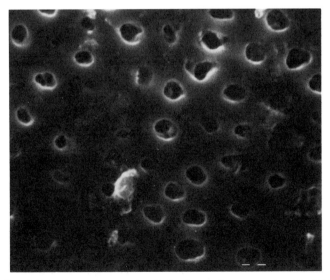

Figure 8.   SEM photomicrograph of dentin surface treated with 17.6 wt.% aqueous AA for 120 s (5000 X).

Table I.  Tensile Bond Strengths of MMEP to Dentin

| Bonding Experiment | Composition in wt.% of solution used in 3 step protocol | Tensile Bond Strength in MPa (Std Dev) | Number of Measurements, n |
|---|---|---|---|
| A | AA  32.0 %<br>NPG  10.0 %<br>MMEP 17.3 % | 8.9 (1.9) | 10 |
| B | AA  25.0 %<br>NPG  10.0 %<br>MMEP 20.7 % | 7.7 (2.5) | 10 |
| C | AA  17.1 %<br>NPG  10.0 %<br>MMEP 20.7 % | 7.6 (2.5) | 10 |
| D | AA  30.0 %<br>NPG   5.0 %<br>MMEP 23.2 % | 6.2 (1.9) | 10 |
| E | AA  25.0 %<br>NPG   5.0 %<br>MMEP 20.7 % | 6.8 (3.3) | 10 |

Table II.  Tensile Bond Strength of Difunctional
Aromatic Carboxylic Acid Methacrylates to Dentin

| Bonding Experiment | Composition in wt.% of solution used in 3 step protocol | Tensile Bond Strength MPa (Std Dev) | Number of Measurements, n |
|---|---|---|---|
| F | AA    20 %<br>NPG   10 %<br>BPDM  21.5 % | 10.3 (3.2) | 9 |
| G | AA    20 %<br>NPG   10%<br>DSDM  15 % | 5.06 (2.1) | 10 |
| H | AA    20 %<br>NPG   10 %<br>DSDM  15 %  * | 7.8 (3.9) | 9 |

* 2 Coats of DSDM applied to dentin

with the smear layer and intact dentin. Rinsing with distilled water removes these soluble salts and leaves a clean dentin surface suitable for resin-mediated bonding. Theoretically, AA can chelate calcium mineral in the smear layer or even the calcium phosphate mineral in intact dentin. The vicinal enolic OH groups of AA (or as a partially oxidized or keto-enol form) may provide a molecular cage (analogous in structure to the vicinal dicarboxylic acid groups of hydrolyzed 4-META) that can chelate calcium ions (*10*). However, it has been shown that reversible adsorption of 4-META and NPG on hydroxyapatite occurs, suggesting hydrogen bonding of these molecules with hydroxyapatite rather than chemical bonding by chelation (*11,12*). Similar reversible, hydrogen bonding reactions have been reported to occur with hydroxyapatite and AA (*13*). In the presence of acidic adhesive monomers, e.g., 4-META, demineralization of the dentin surface occurs with penetration of the monomer into collagen (*13*). This impregnation of modified dentin by polymerized resin forms a hybrid resin reinforced layer, which also can be considered a biocomposite of modified dentin and polymerized resin.   AA, whether it is demineralizing the dentin surface by virtue of its moderate acidity ($pK_a = 4.17$) and/or by its ability to chelate calcium ions, effectively conditions intertubular dentin and opens dentinal tubules so that infiltration, diffusion and polymerization of the adhesive components occurs in the altered dentinal interphase, thereby promoting adhesion by micromechanical retention.

TBS measurements also demonstrate consistent dentin bonding using NPG and MMEP with AA pretreatment. These values compare favorably with the TBSs, 7.4-8.7 (2.2) MPa, obtained previously for NPG-MMEP systems that used an acidified FO pretreatment (*15*). TBSs obtained using BPDM and DSDM as bonding resins further demonstrate the efficacy of using AA as a dentin etchant/conditioner. Significantly, DSDM appeared to give higher TBS with two coats of the monomer solution compared to one coat.

Also of potential significance for resin bonding is the fact that the auto-oxidation of AA yields radicals that can initiate polymerization (*8*). Thus, both AA and NPG, acting as reducing agents, can reduce the effects of oxygen inhibition and generate free radicals that, acting synergistically, may enhance the polymerization of SAMs.   A mechanism for generating initiating radicals from the interaction of aryl amines, e.g., NPG and SAMs, has been proposed (*15*).

## Acknowledgements

We thank Dr. Agnes K. Ly for her technical assistance, Dr. Cliff Carey for assisting with the statistical analysis, and Dr. Allen D. Johnston for providing high purity NPG. Additional thanks are extended to Rohm Tech, Inc. and Bisco, Inc. for their generous contributions of the surface-active monomers.

This work was partially supported by Interagency Agreement 2Y01 DE 30001 with the National Institute of Dental Research, Bethesda, MD 20892.

Certain commercial materials and equipment are identified in this paper to define adequately the experimental procedure.   In no instance does such identification imply recommendation or endorsement by the National Institute of Standards and Technology, the American Dental Association Health Foundation, and the National Institutes of Health, or that the material or equipment is necessarily the best available for the purpose.

**Literature Cited**

1. Erickson, R.L. In: International Congress on Dental Materials. Joint Meeting of the Academy of Dental Materials and the Japanese Society for Dental Materials and Devices. **1989**, pp 55-69.
2. Asmussen, E.; Munksgaard, E.C. In: *Posterior composite resin dental restorative materials*; G. Vanherle and D.C. Smith, Eds.; Utrecht, Peter Szulc Publishing Co.: 1985, pp 217-230.
3. Beech, D.R. In: *Biocompatibility of dental materials*; D.C. Smith and D.F. Williams, Eds.; CRC Press: FL, 1982, Vol.II; pp 87-100.
4. Nakabayashi, N. *Int. Dent. J.* **1985**, *35*, 145.
5. Bowen, R.L. *Int. Dent. J.* **1978**, *28*, 97.
6. Bowen, R.L.; Cobb, E.; Rapson, J.E. *J. Dent. Res.* **1982**, *61*, 1070.
7. Blosser, R.L.; Bowen, R.L. *Dent. Mater.* **1988**, *4*, 225.
8. Antonucci, J.M.; Grams, C.L.; Termini, D.J. *J. Dent. Res.* **1979**, *58*, 1887.
9. Johnston, A.D.; Asmussen, E.; Bowen, R.L. *J. Dent. Res.* **1989**, *68*, 1337.
10. Birch, G.G.; Parker, K.J. In: Vitamin C, New York, Wiley, **1974**, pp 136-149, 221-252.
11. Misra, D.N. *J. Dent. Res.* **1989**, *68*, 42.
12. Misra, D.N.; Johnston, A.D. *J. Biomed. Mater. Res.* **1987**, *21*, 1329.
13. Misra, D.N. *Langmuir* **1988**, *4*, 953.
14. Nakabayashi, N.; Kojima, K.; Masuhara, E. *J. Biomed. Mater. Res.* **1982**, *16*, 265.
15. Schumacher, G.E.; Eichmiller, F.C.; Antonucci, J.M. *J. Dent. Mater.* **1992**, *8*, 278.

RECEIVED May 13, 1993

# Chapter 13

# Salt Partitioning in Polyelectrolyte Gel–Solution Systems

Yu-Ling Yin and Robert K. Prud'homme

Department of Chemical Engineering, Princeton University,
Princeton, NJ 08544

This chapter studies the salt partitioning behavior in polyelectrolyte gel-solution systems. Nonlinear counterion condensation is obtained by solving the full Poisson-Boltzmann equation numerically. Several electrostatic models for salt partitioning were examined. Katchalsky's theory happens to give the right salt accumulation result although the theoretical basis is incorrect. The counterion condensation does account for a large portion of the non-ideality of the salt absorption in polyelectrolyte gels but the interactions among fixed charges have to be considered to fully describe the salt absorption behavior of the polyelectrolyte gels. A new expression has been developed, which uses the results of nonlinear counterion condensation and also considers other electrostatic interactions. The comparison between theoretical calculations and experimental results shows that the new model works well in predicting the salt absorption behavior of polyelectrolyte gels.

Salt partitioning in polyelectrolyte gel-solutions is important in both applications and theoretical work. Gels swelling in aqueous solution can absorb a large amount of the solvent together with the solute (1-5). Cussler and co-workers showed that gels can be used as extraction solvents to separate small molecules with proteins or other polymers (6-9). If one of the small molecules is a salt, it is important to know how much salt can be absorbed by a given amount of polymer gel. Salt absorption is theoretically important because it is related to all the aspects of electrostatic interactions. Furthermore, salt absorption is a quantity relatively easy to measure experimentally so that it can be used as a probe to test models of electrostatic interactions. Despite the importance of the topics, not much work has been done since Katchalsky's work in the 1950s, partially because of the complexity of the problem. Katchalsky and Michaeli's work was widely cited in the later publications (10-13), although no one examined their results seriously. Many new concepts have been proposed since their work, such as Manning's counterion condensation theory (14,15), and numerical solution of the nonlinear Poisson-Boltzmann equation (16-18). However, no one knows how these results affect the salt absorption behavior of polyelectrolyte gels.

0097–6156/94/0540–0157$06.00/0

Salt partitioning is largely determined by the electrostatic interactions in gels. In this chapter, we solve the nonlinear Poisson-Boltzmann equation first, and then incorporate this result with other electrostatic interactions to obtain a model for salt absorption. Then we compare theoretical predictions with experimental results.

## Solution of Poisson-Boltzmann Equation and Counterion Condensation

Manning proposed a linear counterion condensation theory to account for the low activity of counterions in polyelectrolyte solutions (*14* ). The basic idea of the theory is that there is a critical charge density on a polymer chain beyond which some counterions will condense to the polymer chain to lower the charge density, otherwise the energy of the system would approach infinite. The concept of this theory has been widely accepted. The shortcoming of the linear counterion condensation is that it predicts that counterion condensation is independent of ionic strength in the solution, which is not in agreement with experimental observations. Counterion condensation can be obtained directly by solving the nonlinear Poisson-Boltzmann equation.

Consider the interaction between polyions and small ions. For a polyelectrolyte chain, the charges on the polymer chains repel each other so that the polymer chain tends to assume a more extended configuration. Because the diameter of a polymer chain is very small compared with it length, in many applications, a polyelectrolyte chain can be treated as a charged cylinder. The interaction between small ions and a charged cylinder of infinite length can be described by the Poisson-Boltzmann equation (*16-18*)

$$\frac{d}{dr}(r\frac{d\psi}{dr}) = \sinh{(\psi)} \tag{1}$$

where $\psi$ is the electric potential scaled on $kT/e$, k is Boltzmann constant, T is absolute temperature and e is the electronic charge. The radius, r, is scaled on $\kappa^{-1}$, the Debye length. Equation (1) is a second order, nonlinear differential equation. There is no analytical solution for it in a cylindrical coordinate. However, when $\psi > 5$, equation (1) can be approximated by

$$\frac{\partial}{\partial r}(r\frac{\partial \psi}{\partial r}) = e^\psi / 2 \tag{2}$$

Fuoss showed that this equation has the following solution

$$\psi(r) = -\ln\left[\frac{r^2 \sinh^2(\pm \sqrt{D} \ln r + \sqrt{D} \ln c_2)}{4D}\right] \tag{3}$$

where the negative sign corresponds to $\partial\psi/\partial r > -2$ and the positive sign corresponds to $\partial\psi/\partial r < -2$. This solution is valid near the polymer surface. Different forms of the solution result depending on the sign of D. If we rewrite D as

$$\sqrt{D} = \pm c_1 \qquad \text{for D > 0} \tag{4a}$$

and

$$\sqrt{D} = \pm \, i c_1 \qquad\qquad \text{for } D < 0 \qquad (4b)$$

then

$$\psi(r) = - \ln \left[ \frac{r^2 \sinh^2(\pm \, c_1 \ln r + c_1 \ln c_2)}{4 \, c_1^2} \right] \qquad \text{for } D > 0 \qquad (5a)$$

$$\psi(r) = - 2 \ln \left[ \frac{\pm r \ln r + r \ln c_2)}{2} \right] \qquad \text{for } D = 0 \qquad (5b)$$

$$\psi(r) = - \ln \left[ \frac{r^2 \sin^2(\pm \, c_1 \ln r - c_1 \ln c_2)}{4 \, c_1^2} \right] \qquad \text{for } D < 0 \qquad (5c)$$

In order to determine the values of the constants $c_1$ and $c_2$, these equations must be matched to a solution valid in the outer region (i.e., $\psi < 5$). We followed Russel and Stigter's approach in calculating $\psi$ in the outer region numerically and matched $\psi$ to the Fuoss solution for $\psi > 5$ to obtain a complete solution of the full nonlinear Poisson-Boltzmann equation for a arbitrary surface charge density

Stigter showed that equation (1) can be rewritten as (*16*)

$$\frac{d\psi}{d\rho} = X \qquad (6a)$$

$$\frac{dX}{d\rho} = e^{2\rho} \sinh(\psi) \qquad (6b)$$

with the following variable substitutions

$$\rho = \ln r \qquad (7a)$$

and

$$X = \frac{\partial \psi}{\partial r} \qquad (7b)$$

The substitutions translate the second order, nonlinear Poisson-Boltzmann equation into two coupled first order differential equations. The boundary conditions are:

$$\psi \to 0 \qquad\qquad \text{when} \qquad r \to \infty \qquad (8a)$$

$$\frac{\partial \psi}{\partial r} = - \frac{2}{a\kappa} \left( \frac{L_b}{L_\sigma} \right) \qquad \text{when} \qquad r = a\kappa \qquad (8b)$$

where $(L_b/L_\sigma)$ is the charge density on the polymer chain and a is the radius of the polymer cylinder. The boundary condition (8b) was obtained by applying Gauss' Law to the cylinder surface.

Stigter showed that when $\rho > 2$, the Poisson-Boltzmann equation can be linearized to obtain

$$\psi = -2 \left(\frac{L_b}{L_\sigma}\right)_{eff} \frac{K_o(e^\rho)}{(a\kappa)K_1(a\kappa)} \qquad (9)$$

where $(L_b/L_\sigma)_{eff}$ is the effective charge density that a test charge would experience if it approaches the charged cylinder whereas the true experimental charge density is $(L_b/L_\sigma)$. $K_0$ is the zeroth order modified Bessel function of the second kind, $K_1$ is the first-order modified Bessel function of the second kind. Differentiating equation (9) provides an equation for the potential gradient which is also needed to start the numerical integration:

$$\frac{\partial \psi}{\partial r} = -2 \left(\frac{L_b}{L_\sigma}\right)_{eff} \frac{e^\rho K_1(e^\rho)}{(a\kappa)K_1(a\kappa)} \qquad (10)$$

Equations (6) are solved with a forth-order Runge-Kutta routine combined with a shooting procedure to find the proper $(L_b/L_\sigma)_{eff}$. When the potential is low ($\psi < 5$), $(L_b/L_\sigma)_{eff}$ is obtained in the following way: for a solution with given ionic strength, we guess a starting value of $(L_b/L_\sigma)_{eff}$ and obtain initial values of $\psi$ and $\partial\psi/\partial r$ from (9) and (10) evaluated at $\rho = 2$. The computed derivative $\partial\psi/\partial r$ at the surface, $\rho = \ln a\kappa$, is compared with the boundary condition (8b). If the two values do not agree closely, a modified $(L_b/L_\sigma)_{eff}$ value is used and the Runge-Kutta integration is repeated.

When potential is high ( $\psi > 5$), the numerical routine diverges. In this case, the Runge-Kutta integration is matched to the Fuoss's solutions (5a,5b and 5c). The first constant in the solutions is found by

$$D = \left(1 + \frac{X}{2}\right)^2 - \frac{e^{(\psi + 2\rho)}}{4} \qquad (11)$$

If $D > 0$, then $c_1 = D^{1/2}$ and the second constant $c_2$ is given by

$$\ln c_2 = \pm \rho + \frac{1}{2c_1} \ln \left[\frac{\pm (1 + \frac{X}{2}) + c_1)}{\pm (1 + \frac{X}{2}) - c_1)}\right] \qquad (12)$$

If $D = 0$

$$\ln c_2 = \pm \rho + 2e^{-(\frac{X}{2} + \rho)} \qquad (13)$$

If $D < 0$, then, $c_1 = (-D)^{1/2}$ and

$$\ln c_2 = \pm \rho + \tan^{-1}\left[\frac{\pm c_1}{(1 + \frac{X}{2})}\right] \qquad (14)$$

For $X > -2$, the plus sign is used in (5a), (5c), (11-13), and the negative sign is used in (5b). For $X < -2$, the signs are reversed. For $X = -2$, either sign is appropriate in the matching equations (12) and (13), but (11) is singular. This is not important because X is never less than -2 in actual calculations. Once $c_1$ and $c_2$ are determined, then $\psi$ and $\partial\psi/\partial r$ are computed from equation (5). As with the low potential case, if the computed surface gradient X does not match the specified gradient (8b), a new value of $(L_b/L_\sigma)_{eff}$ is guessed and the combined Runge-Kutta integration-analytical matching procedure is repeated.

The surface potential and effective charge density were computed for $0 < L_b/L_\sigma < 2.86$, corresponding to degree of neutralization from 0 to 100% for polyacrylic acid (PAA). Effective degree of neutralization was calculated based on the effective charge density for different ionic strengths as shown in Figure 1. Also shown in the figure is the prediction from Manning's linear counterion condensation theory.

From Figure 1 we can see that linear counterion condensation theory predicts that counterion condensation is only a function of charge density or titrated degree of neutralization, while nonlinear theory predicts that counterion condensation is not only a function of titrated degree of neutralization, but also a function of ionic strength. The lower is the ionic strength, the more obvious is the counterion condensation. In the later discussion, we will use nonlinear counterion condensation as the basis to calculate other electrostatic interactions.

For a polyelectrolyte gel, however, the ionic strength or salt concentration in the gel is not known. In this case, usually we first use linear counterion condensation theory to find the counterion concentration, then we use Donnan equilibrium to calculate the salt concentration. From the counterion concentration and salt concentration we can find the ionic strength. Then the ionic strength is used to solve the nonlinear Poisson-Boltzmann equation to find the real counterion condensation and effective degree of

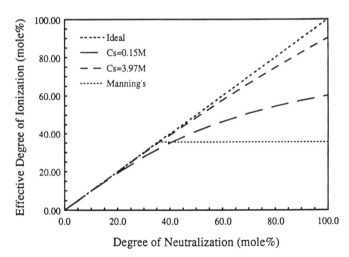

Degree of Neutralization (mole%)

Figure 1. Effective degree of neutralization versus experimental degree of neutralization. The two lines in the middle represent the predictions from nonlinear counterion condensation at two different ionic strengths.

ionization. Since the counterion condensation is not a strong function of ionic strength, we expect that the error introduced by this procedure is small.

## Salt Partitioning in a Gel and Its Solution

**Katchalsky's Theory.** Suppose there is a gel swelling in an aqueous solution. At equilibrium, the chemical potentials of the water in the gel and in the external solution are equal, $\mu_w^{gel} = \mu_w^{ext}$, as are the chemical potentials of the salt in the gel and in the external solution, $\mu_s^{gel} = \mu_s^{ext}$. From thermodynamics it can be seen:

$$\mu^{ext} = \mu_w^0 - RT \sum X_i^{ext} + RT \sum \ln \gamma_w^{ext}$$
(15a)

$$\mu^{gel} = \mu_w^0 - RT \sum X_i^{gel} + RT \sum \ln \gamma_w^{gel}$$
(16b)

where $X_i$ is the molar fraction of the salt and $\gamma_w$ is the activity coefficient of water. At equilibrium,

$$\Delta \ln \gamma_w = \ln \gamma_{gel} - \ln \gamma_{ext} = - \sum X_i^{ext} + \sum X_i^{gel}$$
(17)

If denoting the activity coefficient of the salt by $\gamma_s$, the following Donnan equilibrium will hold for equilibrium swelling:

$$RT \ln X_{cat}^{ext} X_{an}^{ext} \gamma_s^{ext} = RT \ln X_{cat}^{gel} X_{an}^{gel} \gamma_s^{gel}$$
(18)

or:

$$\Delta \ln \gamma_s = \ln \gamma_s^{gel} - \ln \gamma_s^{ext} = \ln X_{cat}^{ext} X_{an}^{ext} - \ln X_{cat}^{gel} X_{an}^{gel}$$
(19)

Katchalsky assumed that the difference in the activity coefficients, $\Delta \ln \gamma_s$, is due only to the free energy of the elasticity of the polymer network and to the electrostatic free energy of the charged network. The small difference in the activity coefficients due to Debye-Hückel interactions among the free ions largely cancels each other, as the salt concentration in the gel and in the external solutions are almost equal. The free energy of elasticity was obtained by using Gaussian distribution and the electrostatic free energy was obtained by summing up all the Debye-Hückel interactions among the fixed charges on the polymer chain (*19*). Their final results are shown in the following (*10*):

$$2X_s^{ext} = \frac{P}{Q} - \frac{0.42}{zQ^{1/3}} - \frac{\alpha^2}{8(pQ)^{1/2}} \left\{ \frac{2.5}{1+\chi} - \frac{\ln(1+\chi)}{\chi} \right\}$$
(20)

and

$$\ln X_s^{ext} = \frac{1}{2}\ln \frac{p^2 - \alpha^2}{4Q^2} - \frac{3\alpha^2}{16(1+\chi)} \frac{Q^{1/2}}{p^{3/2}}$$

(21)

where,

$$p = \frac{\Sigma n_i}{v_e z} = 2\frac{n_{an}}{v_e z} + \alpha$$

(22)

having the meaning of the number of free-ions-per-monomer in the gel; $\chi$ is given by:

$$\chi = \frac{0.3}{z^{1/2}} \frac{Q^{5/6}}{p^{1/2}},$$

(23)

and $Q = n/v_e z$, is the reciprocal of the molar fraction of the monomer unit in the gel. The conventional degree of swelling, q, which is the ratio ($V/V_0$), is related to $Q$ by the equation

$$Q = q(\overline{V}_m/\overline{V}_w)$$

(24)

where $\overline{V}_m$ and $\overline{V}_w$ are the partial molar volume of the monomer and water. $\alpha$ is degree of neutralization, and z is the number of repeat units of a polymer chain between crosslinks.

Theoretically, Equation (20) and (21) should permit the evaluation of both the equilibrium swelling q and the salt accumulation (p-a)/2 from the known data of the external solution. In Katchalsky's work, however, instead of solving the two equations simultaneously, they substituted the experimental swelling ratios into the equation (21) and deduced from it the salt accumulation data. Then they compared the deduced salt accumulation with experiments and found they agreed well. They concluded that their theory worked well. However, we found that equation (20) could not predict the correct swelling if the salt absorption is substituted into it.

The merit of Katchalsky's approach is that it allows us to calculate the salt absorption without knowing the form of the activity coefficient of the free ions in the gel explicitly. But, Katchalsky's theory does not consider counterion condensation. The interactions between various charge species are over-simplified.

**Prediction from Counterion Condensation Alone.** Katchalsky's theory does not consider counterion condensation. The question then is how important is the counterion condensation in determining the salt partitioning? Can the counterion condensation alone account for all the non-ideality of the Donnan equilibrium? To examine this problem, we will calculate the salt absorption by only considering the counterion condensation from both linear and nonlinear counterion condensation theories and compare the predictions with experimental results.

From linear counterion condensation, if the charge density parameter is defined:

$$\xi = e^2/\varepsilon kT b \tag{25}$$

where $\varepsilon$ is the dielectric constant and b the space between two charged groups, then the activity coefficient of the counterion in the a polyelectrolyte gel, $\gamma^{gel}$, can be expressed as

$$\gamma^{gel} = \begin{cases} 1 & \xi < 1 \\ \dfrac{1}{\xi} & \xi > 1 \end{cases} \tag{26}$$

By using a Donnan equilibrium, we found:

$$C_s^{gel} = \sqrt{\left(\frac{\rho\alpha}{2qMw}\gamma^{gel}\right)^2 + \left(C_s^{ext}\right)^2} - \frac{\rho\alpha}{2qMw}\gamma^{gel} \tag{27}$$

where $C_s^{gel}$ is the salt concentration in the gel, $C_s^{ext}$ is the salt concentration in the external solution, $\rho$ is the density of the polymer at dry state, $\alpha$ is the degree of neutralization, Mw is the molecular weight of the monomer and q is the swelling ratio. Once $\gamma^{gel}$ is known, then we can calculate salt absorption from equation (27) and compare the calculations with experimental results. Similarly, $\gamma^{gel}$ can also be calculated by using nonlinear counterion condensation.

**New Expression for Salt Partitioning.** As mentioned before, Katchalsky's theory does not consider the counterion condensation. We will show later that the counterion condensation theories alone, both linear and nonlinear theories, cannot account for all the non-ideality of the Donnan equilibrium. Therefore, we seek an expression that can combine both contributions. We follow a similar approach of Katchalsky by writing the Donnan equilibrium as:

$$2\Delta\ln\gamma_t = 2(\ln\gamma_s^{gel} - \ln\gamma_s^{ext}) = \ln X_{cat}^{ext} X_{an}^{ext} - \ln X_{cat}^{gel} X_{an}^{gel} \tag{28}$$

where $\gamma_t$ is the total activity coefficient of the small ions in the gel. We assume that the interactions of different charge species are independent. Thus the total activity coefficient of the mobile ions is the product of the activity coefficient from the electrostatic interactions of the fixed charges on the polymer chains and those from the interactions of the fixed charges and mobile ions

$$\gamma_t = \gamma_{ff} \cdot \gamma_{fm} \tag{29}$$

Define:

$$P_{eff} = \frac{\sum n_i}{v_e z} = 2\frac{n_{an}}{v_e z} + \alpha_{eff} \tag{30}$$

where $p_{eff}$ is the number of active-free-ions-per-monomer in the gel, $\alpha_{eff}$ is the effective degree of neutralization. Note that our definition is different from Katchalsky's. The molar fraction of the anions and cations can be expressed as:

$$X_{an}^{gel} = \frac{p_{eff} - \alpha_{eff}}{2Q}$$

(31)

$$X_{cat}^{gel} = \frac{p_{eff} + \alpha_{eff}}{2Q}$$

(32)

For the interactions of the fixed charges and mobile ions, we treat the polyions as charged cylinders. We use nonlinear theory to calculate the counterion condensation. After the counterion condensation, the interaction between the free counterions and polyions can be calculated by using an equation suggested by Manning (14). The final result is

$$\ln X_s^{ext} = \frac{1}{2} \ln \frac{p_{eff}^2 - \alpha_{eff}^2}{4Q^2} - \frac{2.85}{2} \frac{\alpha_{eff}^2}{p_{eff}} f - \frac{3\alpha_{eff}^2}{16(1 + \chi_{eff})} \frac{Q^{1/2}}{p_{eff}^{3/2}}$$

(33)

In this equation there are three unknowns: the molar swelling ratio Q, the effective degree of neutralization $\alpha_{eff}$ and the number of active-free-ions-per-monomer $p_{eff}$. Swelling ratio Q can be measured experimentally. The effective degree of neutralization $\alpha_{eff}$ can be found by solving equation (1). The parameter f can be found by fitting the experimental data at one experimental condition and it was found f=1/2 for polyacrylic acid gel. If Q, $\alpha_{eff}$ and f are known, equation (33) can be solved for $p_{eff}$ and salt absorption then can be deduced from $p_{eff}$:

$$C_s / C_p = \frac{1}{2} (p_{eff} - \alpha_{eff})$$

(34)

where $C_p$ is the "molar" concentration of polymer, i.e., the molar concentration of the repeat unit.

**Comparison with Experiments**

**Salt Absorption Experiments.** Gels were synthesized in acid form using the procedure described in the reference (5). Gel samples were washed with deionized water for four weeks to remove extractable components and measure the polymer content. Then, gel samples were neutralized to the desired degrees of neutralization by titrating with standardized sodium hydroxide solution. Salt and deionized water then were added to make the final solution having the desired salt concentration. The volume of the polymer is much less than the volume of the solution so the salt concentration can be treated as a constant. After the equilibrium was reached, the swelling ratios of the samples were determined gravimetrically. Then the samples were dried in a vacuum oven at 100°C. The weight of the samples were measured until constant weight was reached. Salt accumulation was deduced from these measurements.

**Results and Discussion.** Table I gives experimental salt accumulation data for the gels with different crosslink densities. The gels were in swelling equilibrium in 0.1538M (0.9% wt) NaCl solution. In the tables DN denotes the degree of neutralization, Mw the molecular weight of the neutralized monomer, Wg the weight of the gel samples before swelling (polymer content is about 23.5%), Wsg the weight of the swelling gel samples, Ws the weight of the dried samples, q the swelling ratio and Cs/Cp the ratio of the salt concentration to polymer concentration in the gels. These tables show that salt accumulation increases monotonically with the degree of neutralization and swelling ratio.

In Figures 2 and 3, we compare some of the experimental results with the theoretical prediction purely based on counterion condensation theories, Katchalsky's theory and ideal Donnan equilibrium. The following facts are revealed from the plot: 1) At a given crosslink density, all theories agree well with the experimental results and theoretical predictions when the degree of neutralization is low. When degree of neutralization is high, the predictions from counterion condensation theories alone under-estimate the salt absorption; 2) Out of all these theories, Katchalsky's theory seems to agree best with the experiments. Ideal Donnan equilibrium always predicts a lower salt absorption than the experiments.

It is obvious that the predictions purely based on counterion condensation theories under-estimate the salt accumulation. This is due to the fact that the calculations neglect the interactions between the fixed charges on polymer chains. However, it should be noted that counterion condensation alone does account for a large portion of the non-ideality of the Donnan equilibrium. We compare the predictions from our new expression, Katchalsky's theory, ideal Donnan equilibrium with experimental results in Figure 4 and 5. The comparison shows that our new model works well in predicting the salt absorption in polyelectrolyte gels. Although our theory and Katchalsky's theory predict about the same salt absorption, the predicted free-ions-per-polymer-unit is different, which means that the two theories will predict different osmotic pressure. By combining the salt partition model with other models for polyelectrolyte gels, we have developed a comprehensive model which can predict salt partitioning and swelling behavior simultaneously (*21*), while Katchalsky's model does not work in predicting the swelling, especially when ionic strength is low. Since the free ion concentration is different from that in the corresponding solution, many researchers predicted that the pH

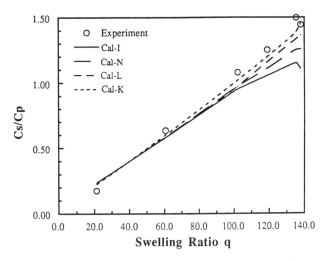

Figure 2. Comparison of experimental salt absorption with theoretical predictions based on ideal Donnan equilibrium (Cal-I), nonlinear counterion condensation theory (Cal-N), linear counterion condensation theory (Manning's theory) (Cal-L) and Katchalsky's theory (Cal-K). The gel has crosslink density of $v = 3.5 \times 10^{-6}$ moles (polymer chains)/cm$^3$, swelling in 0.1538 M NaCl solution.

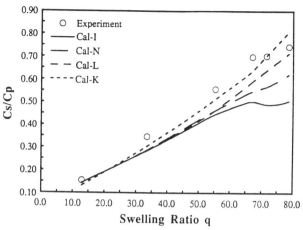

Figure 3. Comparison of experimental salt absorption with theoretical predictions based on ideal Donnan equilibrium (Cal-I), nonlinear counterion condensation theory (Cal-N), linear counterion condensation theory (Manning's theory) (Cal-L) and Katchalsky's theory (Cal-K). The gel has crosslink density of $\nu = 6.1 \times 10^{-6}$ moles (polymer chains)/cm$^3$, swelling in 0.1538 M NaCl solution.

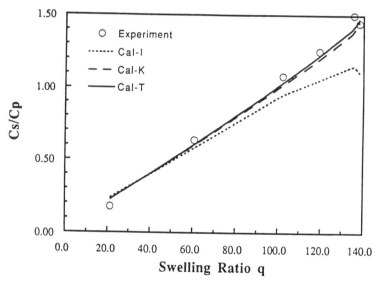

Figure 4. Comparison of experimental salt absorption with theoretical predictions based on ideal Donnan equilibrium (Cal-I), Katchalsky's theory (Cal-K), and our theory (Cal-T). The gel has crosslink density of $\nu = 3.5 \times 10^{-6}$ moles (polymer chains)/ cm$^3$, swelling in 0.1538 M NaCl solution.

Table I.  Salt Absorption of Polyacrylic Acid Gels

(a) Gels with crosslink density of $3.5 \times 10^{-6}$ (mole/cm$^3$)

| samples | DN | MW | Wg | Wsg | q | Ws | Cs | Cs/Cp |
|---|---|---|---|---|---|---|---|---|
| 1-1 | 0 | 72.00 | 0.1990 | 0.9279 | 21.31 | 0.0062 | 0.1168 | 0.1750 |
| 2-1 | 20 | 76.40 | 0.1739 | 2.3088 | 60.71 | 0.0196 | 0.1490 | 0.6340 |
| 3-1 | 40 | 80.80 | 0.1600 | 3.5794 | 102.29 | 0.0307 | 0.1495 | 1.0790 |
| 4-1 | 60 | 85.20 | 0.1861 | 4.8561 | 119.31 | 0.0414 | 0.1485 | 1.2510 |
| 5-1 | 80 | 89.60 | 0.1938 | 5.7293 | 135.45 | 0.0515 | 0.1565 | 1.4980 |
| 6-1 | 100 | 94.00 | 0.2053 | 6.1974 | 138.03 | 0.0527 | 0.1480 | 1.4440 |

(b) Gels with crosslink density of $6.1 \times 10^{-6}$ (mole/cm$^3$)

| samples | DN | MW | Wg | Wsg | q | Ws | Cs | Cs/Cp |
|---|---|---|---|---|---|---|---|---|
| 1-3 | 0 | 72.00 | 0.2030 | 0.6167 | 12.94 | 0.0060 | 0.1663 | 0.1540 |
| 2-3 | 20 | 76.40 | 0.2013 | 1.5817 | 33.48 | 0.0133 | 0.1497 | 0.3460 |
| 3-3 | 40 | 80.80 | 0.1940 | 2.5249 | 55.42 | 0.0206 | 0.1436 | 0.5570 |
| 4-3 | 60 | 85.20 | 0.2004 | 3.1612 | 67.21 | 0.0267 | 0.1482 | 0.6990 |
| 5-3 | 80 | 89.60 | 0.2004 | 3.4421 | 71.75 | 0.0278 | 0.1417 | 0.7030 |
| 6-3 | 100 | 94.00 | 0.2007 | 3.7050 | 78.71 | 0.0285 | 0.1348 | 0.7440 |

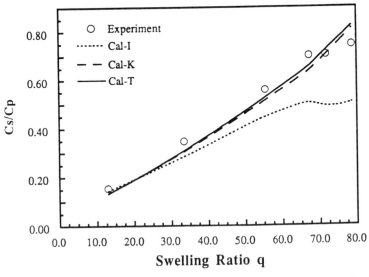

Figure 5.  Comparison of experimental salt absorption with theoretical predictions based on ideal Donnan equilibrium (Cal-I), Katchalsky's theory (Cal-K), and our theory (Cal-T). The gel has crosslink density of $v = 6.1 \times 10^{-6}$ moles (polymer chains) / cm$^3$, swelling in 0.1538 M NaCl solution.

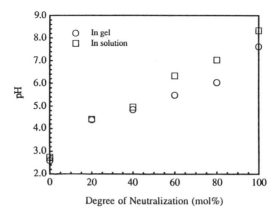

Figure 6. pH in polyacrylic acid gels and their solution as a function of degree of neutralization.

in polyelectrolyte gels and their solution should be different. This prediction has been verified experimentally by using a new procedure (*22*). It was found that the pH in polyelectrolyte gels are substantially lower than that in the solutions as shown in Figure 6.

## Summary

Nonlinear counterion condensation is obtained by solving the full Poisson-Boltzmann equation numerically. Several electrostatic models were examined. Katchalsky's theory happens to give the right salt accumulation result although the theoretical basis is incorrect. The predicted total number of active mobile ions by the theory is incorrect. The counterion condensation does account for a large portion of the non-ideality of the salt absorption in polyelectrolyte gels but the interactions among fixed charges has to be considered to fully describe the salt absorption behavior of the polyelectrolyte gels. A new expression has been developed, which considers both the counterion condensation and the electrostatic interaction of the fixed charges, and the new expression agrees very well with the experimental results.

## Acknowledgments

We would like to acknowledge financial support from the Dow Chemical Company.

## Literature Cited

1. Tanaka, T., Scientific American, **1981**, *244* (1), 124.
2. Oppermann, W.; Rose, S.; and Rehage, G., British Polymer J., **1985**, *17*, 175.
3. Nossal, R.; Jolly, M., J. Appl. Phys., **1982**, *53*, 5518.
4. Prange, M. M.; Hooper , H. H.; Prausnitz, J. M., AIChE Journal, **1989**, *35*, 803.
5. Yin, Y. L.; Prud'homme, R. K. and Stanley, F., *In Polyelectrolyte Gels*, Harland, R.;S.; Prud'homme, R. K., Ed., ACS Symposium Series 480, ACS, Washington D.C.,1992, pp91-113.
6. Cussler, E. L.; Stoker, M. R.; Varberg, J. E., AIChE Journal, **1984**, *30*, 578.
7. Freitas, R. F. S.; Cussler, E. L., Chemical Engineering Science, **1987**, *42*, 97.
8. Gehrke, S. H. and Cussler, E. L., Chemical Engineering Science,**1989**,*44*, 559.

9.  Gehrke, S. H.; Anderews, G. P.; Cussler, E. L., Chemical Engineering Science, **1986**,*41*, 2153.
10. Katchalsky, A.; Micheali, I.,. J. Polym. Sci., **1955**, 9, 69.
11. Hasa, J.; Ilavsiky, M.; Dusek, K., J. Polym. Sci., **1975**, *13*, 253.
12. Hasa, J.; Ilavsky, M., J. Polym. Sci., **1975**, *13*, 263 .
13. Konak, C.; Bansil, R., Polymer, **1989**, *30*, 677.
14. Manning, G., J. Chem. Phys., **1969**, *51*, 924.
15. Manning, G., J. Chem. Phys., **1969**, *51*, 934.
16. Stigter, D., J. Colloid Interface Sci., **1975**, *53*, 296.
17. Russel, W. B., J. Polym. Sci., Phys. Ed., **1982**, *20*, 1233.
18. Davis, R. M., PhD Thesis, Princeton University, Princeton ,1984.
19. Lifson, S.; Katchalsky, A., J. Polym. Sci., **1954**, *8*, 43.
20. Katchalsky, A.; Lifson, S., J. Polym. Sci., **1953**, *5*, 409.
21. Yin, Y.L.; Prud'homme, R. K., Macromolecules, submitted.
22. Yin, Y.L.; Prud'homme, R. K.; Warr, G. G., to be published.

RECEIVED April 22, 1993

Chapter 14

# Ring-Opening Polymerization of a 2-Methylene Spiro Orthocarbonate Bearing a Pendant Methacrylate Group

Jeffrey W. Stansbury

Dental and Medical Materials Group, Polymers Division, National Institute of Standards and Technology, Gaithersburg, MD 20899

A methacrylate-substituted spiro orthocarbonate monomer was synthesized and evaluated in polymerizations using radical and/or cationic initiators. The monomer contains an exocyclic double bond on the spiro group for radical addition and ring opening independent of the remote methacrylate functionality. Crosslinked polymers were obtained by all modes of initiation with mixed radical and cationic giving optimum conversions and ring opening. The incorporation of the pendant methacrylate group minimizes concerns of leachable products generated by polymerization mechanisms involving single ring opening with elimination of a cyclic carbonate. The spiro vinyl ether-type double bond appears to activate the monomer toward cationic polymerization. The ring-opening polymerization of spiro orthocarbonate monomers can yield expansion in volume and may improve a variety of dental and medical materials such as composites, adhesives and coatings.

Unlike conventional 1,2-vinyl polymerizations, the double ring opening of spiro orthoester and orthocarbonate monomers, which have strained tri- and tetra-oxaspiro linkages, respectively, provides access to volume expansion as polymers are formed (1,2). This phenomenon results from the cleavage of two covalent bonds for each addition of a monomer unit to the polymer chain. The reduction or elimination of polymerization shrinkage is highly desirable in a number of polymer applications. In dentistry, a gap-free interface between polymer-based restorative materials and the tooth is difficult to obtain due to the contraction on polymerization. The use of a free radically polymerizable spiro orthocarbonate to counter the unwanted shrinkage in dental resins was first investigated by Thompson et al. (3). Subsequent development of asymmetric oxaspiro monomers which were liquid at room temperature simplified their incorporation into dental resin formulations (4,5).

While these approaches did provide for reductions in polymerization shrinkage and improvements in adhesion, the oxaspiro monomers were more sluggish in free radical polymerizations than the methacrylate comonomers in the resins. Efforts to increase the radical reactivity of these monomers have involved the attachment of the exocyclic double bond to one of the spiro-fused oxygens (6,7) and the use of conjugated 1,3-diene-based oxaspiro compounds (7).

Another alternative to enhance the free radical reactivity of potential ring-opening monomers has been to append readily polymerizable functional groups to the spirocyclic frame. In this manner, allyl- and styryl-modified symmetric spiro orthocarbonates have been evaluated (8). Similarly, spiro orthoesters (1; Figure 1) bearing acrylate or methacrylate substituents have also been examined (9,10). This approach involves the sequential polymerization of the vinyl group by free radical initiation followed by cationic ring-opening polymerization of the pendant oxaspiro functionality to produce expansion in the final crosslinking stage. Somewhat different acrylate-containing spiro orthoesters (2; Figure 2) have also been devised (11). Bulk polymerizations of monomer 2 under free radical conditions resulted in crosslinked polymer with a small amount of ring opening as a consequence of chain transfer from the spirocyclic portion of the molecule.

The current study describes the synthesis and polymerization of a methacrylate-substituted 2-methylene spiro orthocarbonate monomer [3 (SOCM); Figure 2] which has an exocyclic double bond on the oxaspiro segment for free radical addition and ring opening independent of the pendant methacrylate functionality. The difunctional nature of the monomer in terms of radical addition allows high modulus, crosslinked polymers to be formed in concert with the ring opening. The presence of a polymerizable vinyl group on each of the spiro rings also serves to minimize concerns of single ring opening with elimination of a cyclic carbonate fragment. This elimination pathway, which also serves to cleave two bonds for each monomer addition, can be a significant pathway in both free radical and cationic polymerizations of oxaspiro monomers (7,12).

**Materials and Methods**

The SOCM monomer was synthesized according to the following multi-step procedure (Figure 3):

Dry glycerol (5.53 g, 60 mmol) and dibutyltin oxide (14.94 g, 60 mmol) in 120 mL of toluene were refluxed for 4 h under argon. The theoretical amount of water was collected in a Dean-Stark sidearm. The reaction mixture was cooled to 70°C and 4-chloromethyl-1,3-dioxolane-2-thione (5; 9.16 g, 60 mmol) (7) was added to the cyclic tin compound 4 in one portion. After 5 h, the solvent was removed under reduced pressure to leave a two-phase liquid residue. The dibutyltin sulfide by-product was removed by several extractions with hexane to provide the crude chloromethyl-hydroxy-substituted oxaspiro intermediate 6 as a pale yellow oil in 99% yield.

An argon-blanketed solution of compound 6 (12.54 g, 60 mmol) in 80 mL of dry tetrahydrofuran was cooled to 5°C in an ice bath while potassium t-butoxide (16.84 g, 150 mmol) was added. The dark reaction mixture was vigorously stirred at room temperature for 18 h. The solvent was removed under reduced pressure and the semi-solid residue was partitioned between dichloromethane and water. The organic layer along with two additional dichloromethane extracts of the aqueous phase were

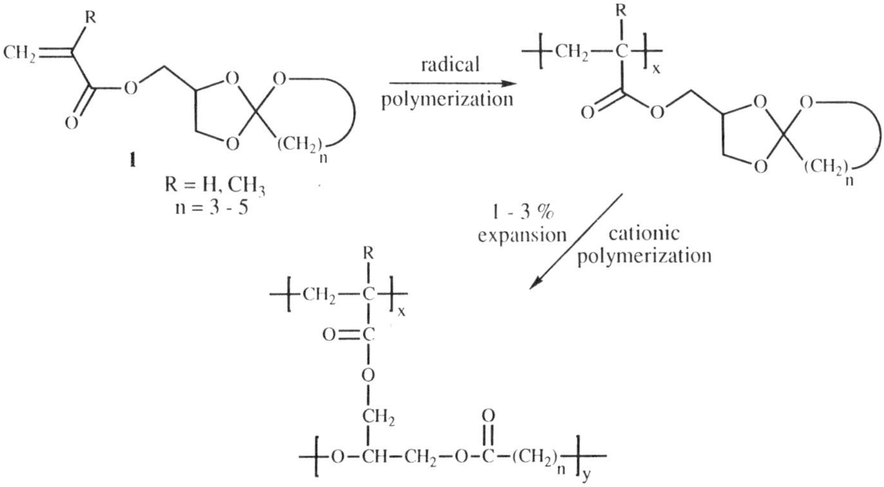

Figure 1.   Polymerization of a methacrylate-substituted spiro orthoester (1) by free radical addition to the methacrylate group followed by cationic ring opening.

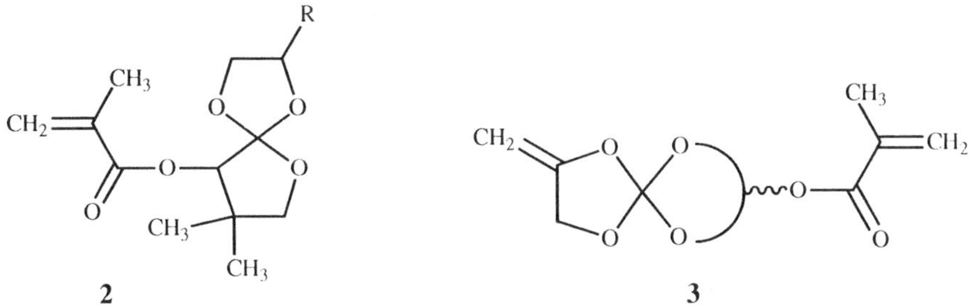

Figure 2.   Structures of an alternate spiro orthoester methacrylate (2) and the spiro orthocarbonate methacrylate (SOCM, 3) monomer.

combined and dried over anhydrous sodium sulfate. The solvent was evaporated to give the exocyclic methylene oxaspiro intermediate 7 as a yellow liquid in 78%. The hydroxy oxaspiro intermediate 7 (3.28 g, 20 mmol) was combined with triethylamine (2.23 g, 22 mmol) in 30 mL of toluene. The solution was cooled to 5°C under argon and freshly distilled methacryloyl chloride (2.09 g, 20 mmol) diluted with 7 mL of toluene was added dropwise. The mixture was stirred at room temperature for 18 h and then filtered to remove the amine hydrochloride precipitate. The majority of the solvent was removed under reduced pressure and the residue was eluted through a small silica gel pad with hexane-ethyl acetate (1:1). Evaporation of the solvent under reduced pressure provided the SOCM monomer 3 as a pale yellow liquid in 93% yield.

The overall yield of the four-step synthesis of SOCM from glycerol was 72%. The monomer was characterized by IR, $^1$H and $^{13}$C NMR and by HPLC (Whatman Partisil 10 silica column with 9:1 hexane-ethyl acetate as eluant) analyses. The IR spectra of intermediates 6 and 7 as well as the SOCM monomer 3 are provided in Figure 4. The $^{13}$C NMR spectrum of monomer 3 is shown in Figure 5.

Polymerization behavior of the SOCM monomer was investigated under a variety of conditions with a number of free radical and cationic initiators. The polymerization initiators, their sources and their modes of action are given in Table I.

Table I.   Initiators and and Their Effects on SOCM Polymerization

| Initiator | Abbrev | Source[*] | Conversion[†] MA | SV | Ring opening |
|---|---|---|---|---|---|
| azobis(isobutyro-nitrile) | AIBN | PB | good | fair | fair |
| di-*tert*-butyl peroxide | DTBP | PB | good | fair | poor |
| camphorquinone | CQ | A | fair | fair | fair |
| benzoin methyl ether | BME | A | excel | good | good |
| boron trifluoride etherate | BTFE | A | poor | excel | excel |
| 4-diazo-*N,N*-diethyl-aniline fluoroborate | DDFB | A | poor | excel | excel |
| Cyracure UVI-6974 (UV initiator for cycloaliphatics) | UVI | UC | good | excel | excel |

[*] PB = Pfaltz and Bauer, Waterbury, CT; A = Aldrich, Milwaukee, WI; UC = Union Carbide, Danbury, CT.
[†] MA = methacrylate double bond; SV = spiro vinyl double bond.

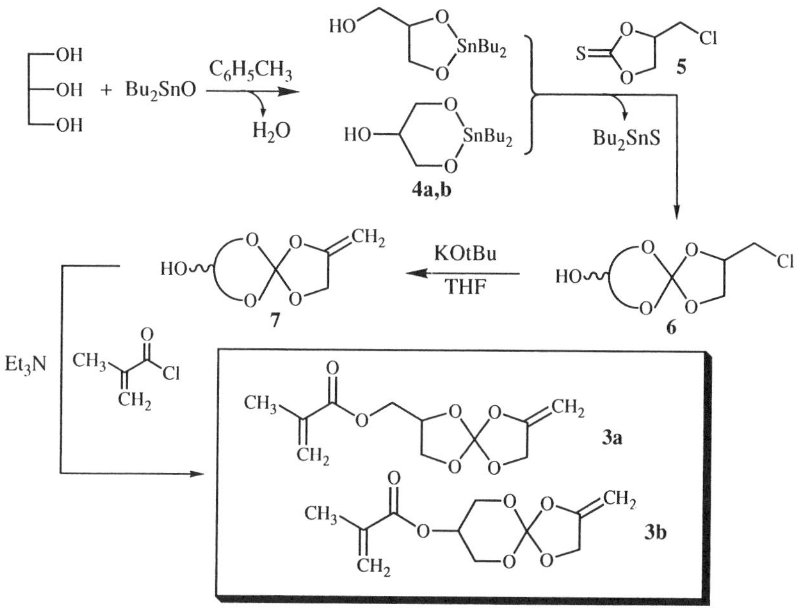

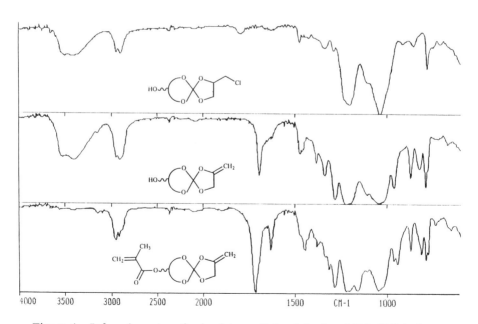

Figure 3.  Synthesis of SOCM starting from glycerol.

Figure 4.  Infrared spectra of spiro intermediates 6 (top) and 7 (middle) along with that of SOCM 3 (bottom).

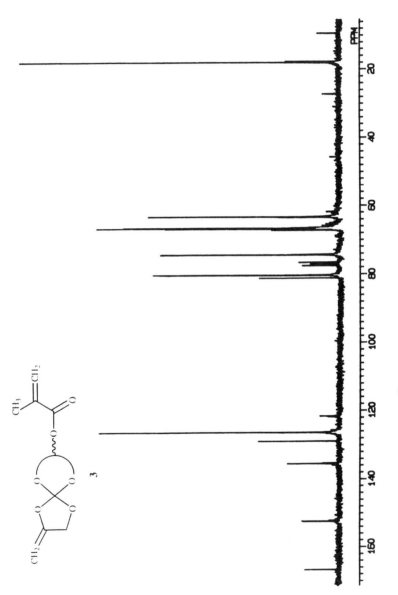

Figure 5.   $^{13}$C NMR spectrum of the SOCM monomer 3.

The cationic UV photo-initiator (UVI) is a triarylsulfonium hexafluoroantimonate salt in propylene carbonate. This initiator can undergo photolysis to yield radical species as well as a strong acid for cationic polymerization initiation (*13*). The relative degrees of conversion and ring opening of the resultant polymers were obtained from the IR spectra with emphasis on the carbonyl and vinyl absorption regions.

Dental resin formulations were prepared by the combination of SOCM with an ethoxylated bisphenol A dimethacrylate (EBPAD, Diacryl 101, Akzo Chemie America) in a 26.6:73.4 wt ratio which corresponds to a 41.5:58.5 mol ratio. Resin samples were activated by the addition of various photo-initiators and were polymerized as unfilled films between Mylar sheets. The cured samples were stored for 24 h at 37°C and then removed from the Mylar. The degree of conversion of the combined methacrylate groups of SOCM and EBPAD as well as the conversion of the spiro vinyl group of SOCM were determined by comparison of the IR spectra of the uncured resin with those of the polymerized films. The methacrylate double bond absorption at 1637 $cm^{-1}$ and the spiro vinyl absorption at 880 $cm^{-1}$ were monitored with the aromatic band at 1585 $cm^{-1}$ utilized as an internal standard reference (*14*).

**Results and Discussion**

The IR spectra in Figure 4 trace the synthetic pathway to the SOCM monomer 3. The entire series is dominated by the strong IR bands characteristic of the spiro orthocarbonate group. IR absorptions associated with the shortened $CO_4$ central ether bonds occur near 1200 $cm^{-1}$ while the external ether bands are found near 1050 $cm^{-1}$. Intermediates 6 and 7 produce a strong OH band around 3400 $cm^{-1}$ and 7 also yields peaks due to the oxaspiro-based double bond at 1698 and 880 $cm^{-1}$. Addition of the methacrylate group in monomer 3 eliminates the OH band and gives rise to the carbonyl and methacrylate vinyl absorptions at 1720 and 1637 $cm^{-1}$, respectively.

The initial synthesis of the SOCM monomer was attempted via the reaction of 1-methacryloxy-2,3-propanediol with dibutyltin oxide followed by treatment with the cyclic carbonate 5. However, this process failed to provide any significant quantity of the chloromethyl-methacrylate-substituted intermediate. If this route had been successful, the final SOCM monomer would have been comprised solely of the spiro-fused five-membered rings (3a). Instead, the synthesis of SOCM as shown in Figure 3 yielded a mixture of the 5-5 and 5-6-membered ring systems (3a and 3b, respectively). Each of these different structural isomers is also further subdivided into a pair of syn and anti diastereomers as shown in Figure 6. The complex eutectic mixture which results is a liquid of low viscosity that is well suited for use as a diluent and wetting agent.

HPLC fractionation of a SOCM sample into its individual components provided information about the relative proportions of the various isomers. The ratio of 3a to 3b was approximately 3:1 as determined by peak areas (UV detection at 254 nm) of the HPLC chromatograms. This should be a reasonably good estimation of product ratio since the pendant methacrylate functionality is the only UV active group in these compounds. The spiro-fused five-membered rings were characterized by a $CO_4$ resonance at 135 ppm in the $^{13}C$ NMR spectra. By contrast, the compounds of type 3b, with mixed five- and six-membered ring sizes, produced $CO_4$ signals at 122 ppm. A small amount of an oxaspiro dimethacrylate (8, Figure 7) was also noted. This

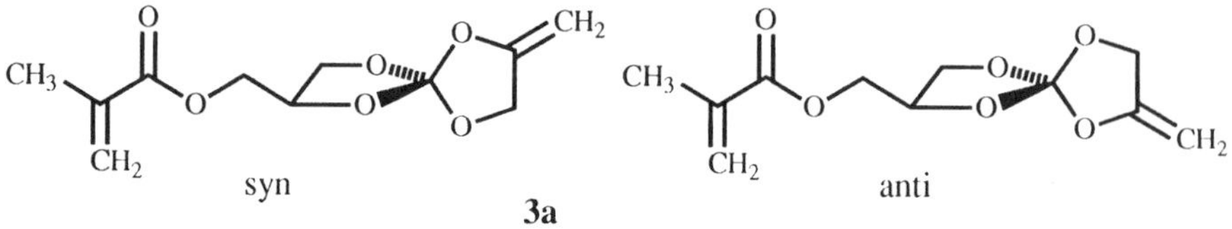


**3a**

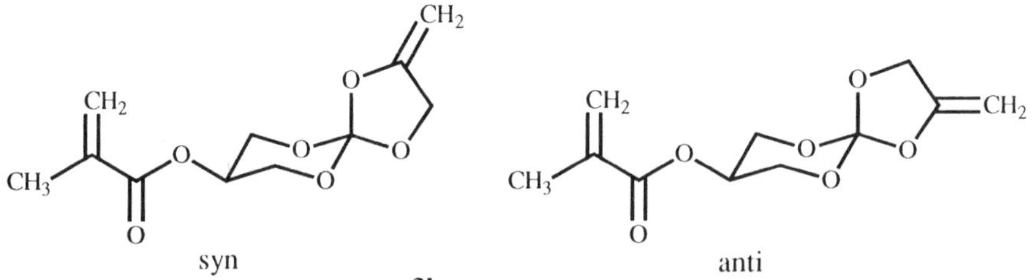


**3b**

Figure 6. Various isomers of SOCM 3 separated by HPLC.

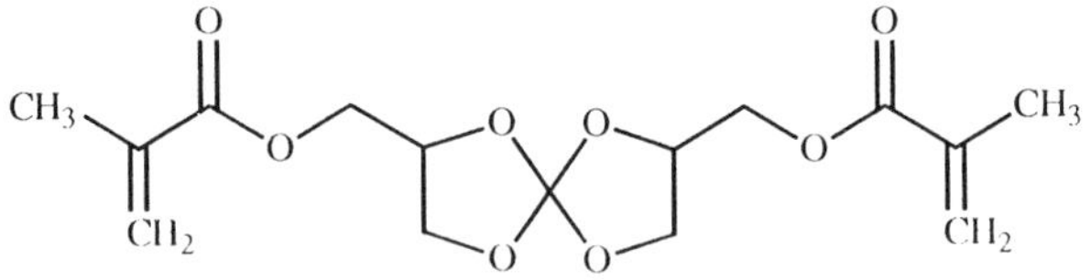


**8**

Figure 7. Structure of the symmetric dimethacrylate trace impurity isolated from the SOCM product by HPLC.

structure was proposed since signals assigned to the spiro and methacrylate groups were present, but the exocyclic double bond resonances at 80 and 152 ppm were absent from the $^{13}$C NMR spectrum.

Initial polymerization studies of the SOCM monomer (Table I) utilized thin films cured between Mylar sheets. Free radical polymerization with AIBN (1 wt%) at 65°C resulted in lower conversion but more extensive ring opening than a comparable polymerization at 120°C with DTBP (1 wt%). Similar polymerizations conducted open to the air gave appreciably more carbonate formation due to more extensive ring opening. Polymerization under a sunlamp (General Electric 275 W, 350-600 nm) with BME (1 wt%) as the radical initiator produced a polymer with excellent methacrylate conversion and moderate conversion of the spiro vinyl group. The SOCM monomer was also polymerized in solution (3% in benzene) with 1 wt% AIBN at 65°C. The majority of the polymer was crosslinked and precipitated from solution. However, a small amount of non-crosslinked polymer was isolated by addition of the benzene to hexane. Analysis of the soluble polymer fraction by $^1$H NMR indicated that there was approximately equal incorporation of the SOCM monomer via the methacrylate and spiro vinyl groups (Figure 8). The soluble polymer was apparently generated with significant ring opening since strong carbonate carbonyl absorptions were present in the IR spectrum while the spiro-based $CO_4$ band was greatly reduced.

The possibility of cationic ring-opening polymerization of SOCM either alone or in concert with a free radical polymerization was explored with a number of cationic initiators. BTFE was added to an argon flushed sample of the SOCM monomer at room temperature. An immediate dark brown viscous mass was produced at the point of contact. After several hours, the discoloration extended throughout the sample which had solidified. The majority of the material was crosslinked polymer with virtually no intact spiro rings remaining. The IR spectrum of the polymer had strong carbonate-related bands at 1800 and 1750 cm$^{-1}$. The methacrylate group did not appear to be involved in this polymerization. A significant amount of the methacrylate-substituted, five-membered cyclic carbonate (9, Figure 9) was isolated from the crude polymer. This was apparently produced by a single ring-opening/elimination polymerization pathway that has been observed previously in cationic polymerizations of oxaspiro monomers with five-membered rings (*12*). Treatment of the non-vinyl-containing bis(chloromethyl) spiro orthocarbonate (10, Figure 9) with BTFE yielded a soluble polymer with a significantly decreased rate of formation compared with the SOCM polymerization. This polymer also appeared to result mainly from the single ring-opening with cyclic carbonate elimination.

Polymerization of SOCM with the UVI cationic photo-initiator (1.5 wt%) under the sunlamp resulted in high conversion of the spiro vinyl group and modest conversion of the methacrylate group. The IR spectrum of the polymer contained more carbonate-based absorptions than was found for the BME-cured material and much more than for the AIBN-cured polymer. Linear carbonate from the double ring-opening pathway appeared to be the dominant structure in the polymer. These results indicate that the UVI initiator produces efficient cationic polymerization of the SOCM monomer and generates sufficient radical species upon photolysis to affect free radical polymerization to at least a limited degree. However, the attempted UV polymerization of the non-vinyl-containing oxaspiro compound 10 with UVI produced no polymer and no degradation of the monomer. The slow rate of the BTFE-induced polymerization and

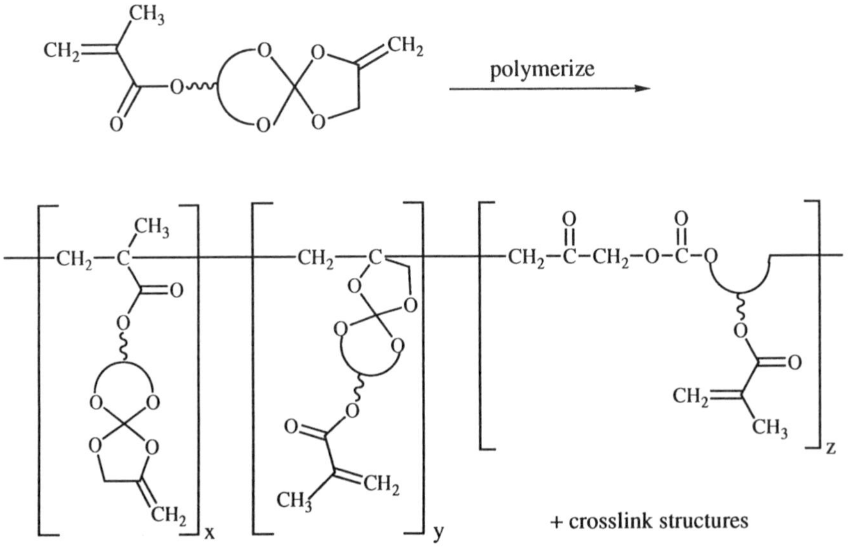

Figure 8. Idealized structure of the polymer obtained by the free radical polymerization of SOCM by non-ring-opening and double ring-opening pathways.

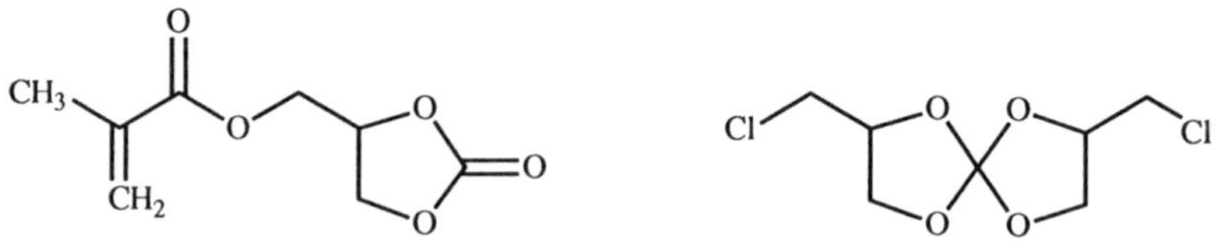

**9**                    **10**

Figure 9. Structure of the cyclic carbonate elimination product (9) isolated from the BTFE cationic polymerization of SOCM. Symmetrical bis(chloromethyl) spiro model monomer (10) used in comparative cationic polymerizations.

the lack of any polymerization with UVI for this monomer indicates that the presence of a vinyl ether-type group within the spiro linkage, such as in the SOCM monomer, may be beneficial to the cationic ring-opening polymerization efficiency.

Another compound, DDFB, was also investigated to determine its potential as a dual radical/cationic polymerization initiator with the SOCM monomer. The dark DDFB solid had only limited solubility in the monomer (approximately 0.5 wt%). The activated SOCM sample was heated to 100°C for 10 min which produced a darkened polymer that was predominantly crosslinked. An unactivated SOCM control sample also received the same heat treatment and was recovered unchanged. The IR spectrum of the polymer showed extensive carbonate formation in the 1800 and 1750 cm$^{-1}$ regions and a nearly complete disappearance of the spiro absorption bands. This initiator appeared to have little affinity for the methacrylate double bond since a strong 1637 cm$^{-1}$ band was still present. The attempted polymerization of oxaspiro monomer 10 with DDFB at 100°C yielded no polymer and no reaction. The DDFB-containing SOCM monomer sample was then irradiated under the sunlamp with no polymer formation observed after 30 min.

Preliminary evaluations of the SOCM polymers have shown that the mixed radical-cationic polymerizations gave significantly less shrinkage than free radical alone. The polymers obtained from cationic initiation alone exhibited only a small amount of shrinkage which can probably be attributed to the contribution of the single ring-opening/elimination pathway.

Unfilled SOCM/EBPAD resin films activated with a conventional CQ (0.2 wt%) and ethyl 4-*N,N*-dimethylaminobenzoate (EDMAB, 0.8 wt%) visible light photo-initiator system were cured with either a blue light (Primetics, LD Caulk, Div of Dentsply) for 40 s or under the sunlamp for 5 min. The samples polymerized under the sunlamp reached a temperature of approximately 45°C. The results shown in Table II demonstrate a relatively high conversion of the methacrylate groups with the standard blue light cure technique compared with the sunlamp application while utilization of the spiro vinyl group was approximately the same for both techniques. The addition of UVI to the CQ/EDMAB-activated resin produced improved conversions for both types of double bonds with the sunlamp curing procedure. In contrast, the addition of DDFB to the CQ/EDMAB resin caused a large reduction in the polymerization efficiencies of both the methacrylate and spiro vinyl groups.

In a move away from the traditional CQ radical polymerization initiator, the SOCM/EBPAD resin was activated with BME and irradiated under the sunlamp. As shown in Table II, the methacrylate conversion obtained was similar to that characteristic of the CQ-initiated blue light polymerizations. However, the degree of cure of the spiro vinyl group in these films was poor. The combination of UVI and BME in a mixed initiator system provided the optimum conversions observed for both the methacrylate and spiro vinyl groups. Since the spiro vinyl conversion did not increase with extended exposure to the sunlamp, it is likely that the irradiation time could be reduced to less than 5 min. The comparable polymerization efficiencies of the two types of double bonds indicate that this cure process has potential to significantly reduce the levels of resin shrinkage. It may also allow higher proportions of the ring-opening oxaspiro monomer to be utilized in the resin composition which should permit greater control of the polymerization shrinkage.

Table II.    Degree of Conversion of SOCM/EBPAD Resins

| Initiator | Wt% of resin | Cure conditions | Conversion*% MA | SV |
|---|---|---|---|---|
| CQ/EDMAB | 0.2/0.8 | blue light/40 s | 77 | 53 |
| CQ/EDMAB | 0.2/0.8 | sunlamp/5 min | 41 | 50 |
| CQ/EDMAB/UVI | 0.2/0.8/1.5 | sunlamp/5 min | 77 | 71 |
| CQ/EDMAB/D-DFB | 0.2/0.8/0.5 | sunlamp/5 min | 23 | 29 |
| BME | 0.7 | sunlamp/5 min | 73 | 33 |
|  |  | 1 h | 85 | 48 |
| BME/UVI | 0.7/1.5 | sunlamp/5 min | 81 | 83 |
|  |  | 1 h | 87 | 84 |

* MA = methacrylate double bond; SV = spiro vinyl double bond.

## Acknowledgements

This work was supported by Interagency Agreement 01-DE 30001 with the National Institute of Dental Research, Bethesda, MD 20892.

Certain commercial materials and equipment are identified in this paper for adequate definition of the experimental procedure. In no instance does such identification imply recommendation or endorsement by the National Institute of Standards and Technology, or that the material or equipment is necessarily the best available for the purpose.

## Liturature Cited

1. Endo, T.; Okawara, M.; Yamazaki, N.; Bailey, W.J. J. Polym. Sci., Polym. Chem. Ed. 1981, 19, 1283.
2. Bailey, W.J.; Endo, T. J. Polym. Sci., Polym. Chem. Ed. 1976, 14, 1735.
3. Thompson, V.P.; Williams, E.F.; Bailey, W.J. J. Dent. Res. 1979, 58, 1522.
4. Endo, T.; Bailey, W.J. J. Polym. Sci., Polym. Chem. Ed. 1975, 13, 2525.
5. Stansbury, J.W.; Bailey, W.J. J.Dent. Res. 1986, 65, 219, Abst. No. 452.
6. Tagoshi, H.; Endo, T. J. Polym. Sci., Polym. Chem. Ed. 1989, 27, 1415.
7. Stansbury, J.W. J. Dent. Res. 1992, 71, 1408.
8. Endo, T. Nippon Setchaku Kyokaishi 1986, 22, 437.
9. Endo, T.; Kitamura, N.; Takata, T.; Nishikubo, T. J. Polym. Sci., Polym. Lett. 1988, 26, 517.
10. Kitamura, N.; Takata, T.; Endo, T.; Nishikubo, T. J. Polym. Sci., Polym. Chem. Ed. 1991, 29, 1151.

11. Haase, L.; Klemm, E. *Makromol. Chem.* **1990**, *191*, 549.
12. Sakai, S.; Fujinami, T.; Sakurai, S. *J. Polym. Sci., Polym. Lett.* **1973**, *11*, 631.
13. Crivello, J.V.; Lee, J.L.; Conlon, D.A. *J. Rad. Curing* **1983**, *10*, 6.
14. Rueggeberg, F.A.; Hashinger, D.T.; Fairhurst, C.W. *Dent. Mater.* **1990**, *6*, 241.

RECEIVED March 30, 1993

Chapter 15

# Ring-Opening Dental Resin Systems Based on Cyclic Acetals

B. B. Reed[1], Jeffrey W. Stansbury[2], and Joseph M. Antonucci[2]

[1]Paffenbarger Research Center, American Dental Association Health Foundation, Gaithersburg, MD 20899
[2]Dental and Medical Materials Group, Polymers Division, National Institute of Standards and Technology, Gaithersburg, MD 20899

For monomers of comparable size, ring-opening polymerization results in less shrinkage than that which accompanies 1,2-vinyl addition polymerization. Two monomer types were synthesized, nonvinyl (NVCA) and vinyl (VCA) cyclic acetals. The goals of this study were to assess the potential for reduced shrinkage through free radical ring-opening polymerization of NVCA and VCA type monomers, and to test the mechanical strength of dental resin composites formulated with these novel monomers. Homo- and copolymerizations were conducted with several NVCAs and VCAs to evaluate their potential as comonomers in dental polymeric composites. Composite specimens were formulated with PBMD, a VCA derived from terephthaldehyde, and EBPADM, an ethoxylated bisphenol A dimethacrylate, and tested for their mechanical strength. Three control formulations were tested, one containing 100% EBPADM, the second containing PBD, a NVCA derived from terephthaldehyde, and the last containing p-dimethoxybenzene (DMOB). The composites containing PBMD and PBD gave mechanical strength values similar to the EBPADM control, while the control containing DMOB had much lower strength.

Polymerization shrinkage in dental resin composite systems causes numerous undesirable results: internal stresses, micro-cracks, debonding at the filler particle-resin interface that leads to accelerated degradation through reduced mechanical strength and abrasion resistance, and external loss of adhesion that produces marginal gaps at the composite-tooth interface and ultimately results in secondary carries and staining. To counteract the shrinkage caused when chain growth polymerization occurs, monomers capable of free radical ring-opening polymerization have been designed (1). Ring-opening monomers have the potential for polymerization with less volume change than noncyclic vinyl monomers such as methyl methacrylate or styrene. During 1,2-vinyl addition polymerization, monomer units go from Van der Waals to covalent bond distances. In contrast, during ring-opening polymerization, volume contraction is offset as some covalent bonds are cleaved to give near Van der Waal bond distances (Figure 1a and 1b).

0097–6156/94/0540–0184$06.00/0
© 1994 American Chemical Society

In this investigation two classes of monomers were synthesized, nonvinyl cyclic acetals and vinyl cyclic acetals (NVCA and VCA, respectively (Figure 2)). All monomers were evaluated for their ability to homo- and copolymerize under free radical conditions. The monomers that displayed sufficient reactivity to warrant further study were then incorporated into dental composite resin systems. The diametral tensile strength (DTS) of several experimental composites were determined and compared with various controls.

## Materials and Methods

Unless otherwise stated, the reagents used in the syntheses of NVCA and VCA monomers and the monomers used in the copolymerization studies were obtained commercially (Aldrich Chemical Co.) and used as received. All monomers and non-crosslinked polymers were characterized by interpretation of their respective $^1$H and proton-decoupled $^{13}$C NMR spectra which were obtained on a JEOL GSX-270(FT) spectrometer operated at 270 and 68.1 MHz, respectively. All NMR samples were analyzed in CDCl$_3$. IR spectra were obtained on a Ratio Recording Perkin Elmer 1420 instrument controlled through a data station.

Nonvinyl cyclic acetal synthesis was carried out according to the following general scheme (Figure 3). Stoichiometric amounts of the appropriate aryl aldehyde and diol were combined in a single-neck, round bottom flask with toluene as the solvent and 1 mole % of $p$-toluene sulfonic acid (PTSA), recrystalized from methanol, as the catalyst. The flask was fitted with a condenser and a Dean-Stark trap to collect, by azeotropic distillation, the water generated in the reaction. An inert atmosphere (argon) was maintained throughout all steps of the procedure. After refluxing for several hours, the mixture was allowed to cool to room temperature. The mixture was then extracted three times with a 5 wt/vol% of an aqueous sodium bicarbonate solution. The organic layer was dried with anhydrous sodium sulfate, filtered and concentrated under reduced pressure via rotary evaporation.

PBD $^1$H NMR: δ 4.00 (m, CH$_2$CH$_2$), 5.80 (s, OCHO), 7.45 (s, arom). $^{13}$C NMR: δ 65.0 (CH$_2$CH$_2$) 103.0 (OCHO) 126.5 (arom C$_{2,3,5,6}$) 139.0 (arom C$_{1,4}$).

Synthesis of a vinyl cyclic acetal monomer is similar to the synthesis of a NVCA but includes a dehydrohalogenation step as shown in Figure 4. Stoichiometric amounts of the appropriate aryl aldehyde and 3-chloropropane-1,2-diol were reacted and worked up as previously described. The chloromethyl cyclic acetal intermediate was dissolved in toluene and added slowly to three equivalents of potassium *tert*-butoxide in toluene in a round bottom flask cooled in an ice bath. When the addition was complete, the mixture was allowed to warm to room temperature and stirred vigorously overnight. The dark reaction mixture was vacuum filtered through a paper filter covered with a small amount of silica gel. The filtrate was dried and concentrated as before.

MPD $^1$H NMR: δ 4.00 (s, C=CH$_E$), 4.45 (s, =CH$_Z$), 4.60 (d d, CH$_2$), 6.15 (s, OCHO), 7.40-7.55 (m, arom). $^{13}$C NMR: δ 67.5 (OCH$_2$C=), 78.5 (=CH$_2$), 106.0 (OCHO), 126.5 (arom C$_4$), 128.5 (arom C$_{2,6}$), 130.0 (arom C$_{3,5}$), 136.5 (arom C$_1$), 156.0 (C=CH$_2$).

PBMD $^1$H NMR: δ 4.00 (s, =CH$_E$), 4.45 (s, =CH$_Z$), 4.55 (d d, OCH$_2$C=), 6.15 (s, OCHO), 7.55 (s, arom). $^{13}$C NMR: δ 67.0 (OCH$_2$C=), 78.5 (=CH$_2$), 105.0 (OCHO), 126.5 (arom C$_{2,3,5,6}$), 138 (arom C$_{1,4}$), 155.5 (C=CH$_2$).

Homopolymerizations of each monomer were attempted in bulk and in solution. Bulk polymerizations were conducted at 60°C for 24 h with vacuum degassed monomers containing 2,2'-asobis(isobutronitrile) (AIBN). Solution polymerizations were conducted by placing an aliquot of the monomer in a screw cap vial along with 0.5-1.0 wt% AIBN and ≈90 w/w% benzene. Argon was then bubbled through the mixture for 10 s before the cap was secured. The sealed vial was placed in the 60°C

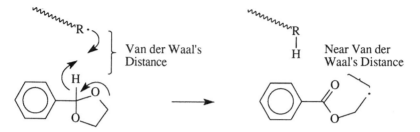

Figure 1a. Free radical ring opening of a NVCA type monomer to an ester intermediate.

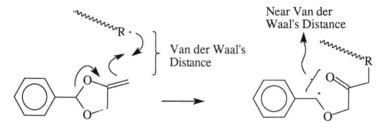

Figure 1b. Free radical ring opening of a VCA type monomer to a keto ether intermediate.

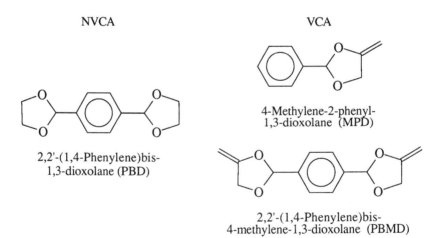

NVCA                                    VCA

4-Methylene-2-phenyl-
1,3-dioxolane (MPD)

2,2'-(1,4-Phenylene)bis-
1,3-dioxolane (PBD)

2,2'-(1,4-Phenylene)bis-
4-methylene-1,3-dioxolane (PBMD)

Figure 2. Monomers used in this study.

oven for 24 h.  Copolymerizations were conducted by transfering equimolar amounts of the experimental monomer and either methyl methacrylate (MMA) or styrene along with the AIBN initiator into a glass vial.  The vial was degassed, vacuum sealed, and placed in the 60°C oven for 24 h.

After 24 h the vial was removed from the oven, allowed to cool to room temperature, and the contents of the vial dissolved into a minimal amount of dichloromethane.  The solution was added dropwise to ≈20 fold v/v hexane to precipitate polymers of significant molecular weight present in the sample.  The precipitate was filtered, dried and dissolved into CDCl$_3$ for analysis by $^1$H and $^{13}$C NMR.  If the combined NMR spectra were not sufficient to determine the structure of the polymers, then IR spectral analysis was used to aid in the elucidation of the structure.

The monomers that yielded homo- or copolymers were formulated into composites.  Three comonomers, PBD, PBMD and *p*-dimethoxybenzene (DMOB), were combined individually with EBPADM, an ethoxylated bisphenol A dimethacrylate (Diacryl 101, Noury Chemicals), in various ratios to form the resin components of the composites.  The initiator system used was 0.4 wt% camphorquinone (CQ) and 0.8 wt% ethyl 4-*N,N*-dimethylaminobenzoate (4-EDMAB).

The respective resins were mixed in a wt/wt powder to liquid ratio of 5 with a barium oxide containing glass powder (Corning Glass 7724), which was silanized according to a previously reported procedure (2).  The resulting paste was deaerated under a slight vacuum (≈ 175 Pa) for 2 h.  Samples were made by hand packing the composite paste into cylindrical stainless steel molds, 3 mm H by 6 mm D, which were then covered with 1 mm thick glass plates and afixed with spring clips.  Each sample was irradiated with a visible light (Prismetics Lite, Caulk/Dentsply) for 60 s *per* side and placed in a 37°C oven for 15 min.  They were then removed from oven, ejected from their molds and stored in deionized water at 37°C for 24 h before being tested.  A universal testing machine (United Calibration Corp.) was used to measure the diametral tensile strength (DTS) of each sample at a crosshead speed of 1 cm/min.  Five or more samples were tested from each composite formulation.  The stress-strain curves and the broken samples were inspected to determine if the sample failure was primarily due to tensile stress without significant plastic deformation.

## Results and Discussion

VCAs homopolymerized under the conditions previously described; therefore they were considered promising candidates for inclusion in the resin phase of composites.  MPD yielded  homopolymer which was dissolved in CDCl$_3$ for NMR analysis. The most distinguishing features of its $^{13}$C NMR spectrum were peaks around 103 and 206 ppm.  A peak at 103 ppm is corresponds to a cyclic acetal which indicates that 1,2-addition polymerization occurred, leaving the ring intact.  The peak at 206 ppm is characteristic of a ketone; which indicates ring-opening to the keto-ether after radical attack on the exomethylene group (Figure 5).  PBMD, the difunctional analog of MPD, yielded crosslinked homopolymer, which could not be characterized with solution NMR.  Presumably PBMD polymerized through similar mechanistic pathways as MPD.  PBMD was formulated into composites because its difunctional nature offers the following possible advantages: (1) the potential for two rings to be opened during copolymerization, thus further counteracting the shrinkage caused by the dimethacrylate comonomer conversion (EBPADM), and (2) the potential for enhancing the degree of crosslinking, thereby strengthening the composite.

It was hypothesized that a monomer of the first type (NVCA) could undergo free radical polymerization through abstraction of its tertiary hydrogen (Figure 1a). Results of this study indicate this did not occur to a measurable extent under the reaction conditions given.  NVCAs proved to be not only stable under the circumstances described in the previous section but also unreactive even under harsher conditions (2 wt% *t*-butylperoxide at 126°C for 24 h).  Polymerization studies of the NVCA

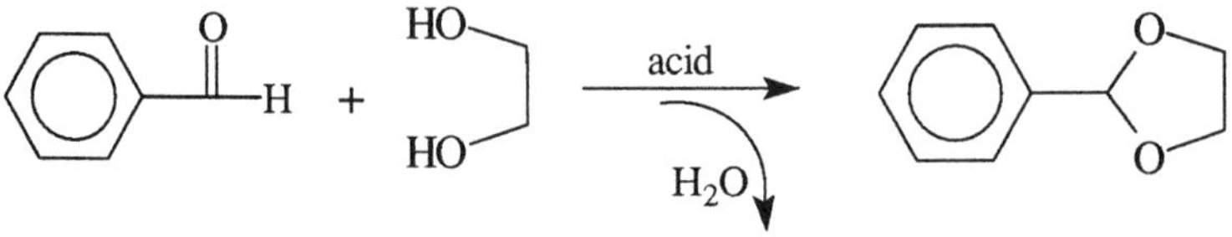

Figure 3. Synthesis of a NVCA type monomer.

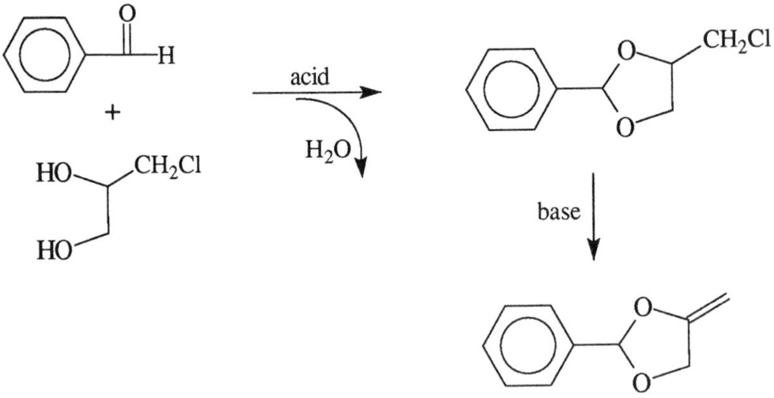

Figure 4. Synthesis of a VCA type monomer.

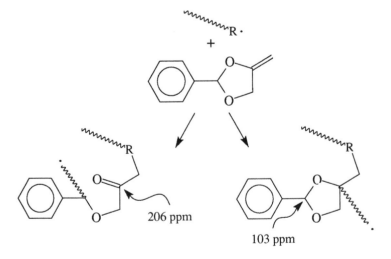

Figure 5. Polymerization pathway via ring-opening and vinyl addition.

compounds with monofunctional comonomers such as MMA indicated no incorporation of the cyclic acetal or any ring-opened products into the polymer (PMMA). Therefore, PBD was considered an excellent candidate for the formulation of a negative control for the composite study; it is approximately the same size, shape, and molecular weight as PBMD, yet seemingly unreactive under free radical conditions.

One composite formulation was made with 100 wt% EBPADM to serve as a positive control. A second negative control using DMOB as a nonreactive component for EBPADM was also formulated into a composite. DMOB was chosen because it has even less chemical potential for incorporation into the polymeric matrix than PBD. It contains no vinyl group or labile hydrogen, yet it has comparable size and molecular weight to both PBD and PBMD.

The results of the DTS tests are shown in Table I. The values given are in MPa with the standard deviation indicated by the number in parentheses. All statistical information was generated with the general linear model program of the Statistical Analysis System software (*3*). Comparisons of the data were made with Duncan's Multiple Range Test (modified for unequal sample sizes) at p<0.5 (*4-5*). Values preceded by an asterisk indicate no significant difference from each other.

## Table I. Results of DTS Tests

| Monomer(s) | Molar Ratio | DTS Average |
|---|---|---|
| EBPADM | -------- | *50.97 (2.65) |
| EBPADM / PBMD | 2.12 / 1.00 | 56.03 (1.09) |
| EBPADM / PBMD | 1.39 / 1.00 | *52.03 (3.45) |
| EBPADM / PBD | 1.45 / 1.00 | 46.95 (1.80) |
| EBPADM / DMOB | 1.40 / 1.00 | 34.60 (2.52) |

The DTS value for the PBMD formulation with the higher molar ratio (2.12/1.00) is greater than the DTS value associated with the more stoichiometric equivalent formulation (1.39/1.00). The explanation for the difference in DTS values lies in the relative reactivities of methacrylate groups to vinyl acetal groups. The methacrylate groups on EBPADM are considerably more reactive than the vinyl groups on PBMD. During polymerization, propagating polymer radicals are consumed by addition of monomer until they become highly immobilized, and further addition favors the more reactive EBPADM over PBMD. That is probably why the the formulation with the higher PBMD content had a somewhat lower DTS value; the unpolymerized PBMD may be acting as a plasticizer.

The table shows the DTS value for PBMD (with the lower ratio) to be only marginally higher than that with PBD. This prompted the formulation of a composite with DMOB as a negative control that could act only as a plasticizer. DMOB appears to have fulfilled this role since the DTS value obtained with this EBPADM formulation was drastically less than the values from the PBD and PBMD based resins. Therefore, apparently PBD does not function simply as a plasticizer in this system.

The fact that PBD containing composites had strengths significantly higher than DMOB containing composites, and yet relatively close to PBMD containing composites, leads to speculation that the labile hydrogen, present in PBD but not in DMOB, may have a role in these crosslinking polymerizations. This speculation seems to contradict the conclusions derived from the attempted copolymerization of PBD with the monofunctional MMA or styrene, which indicated that there was no detectable abstraction of the labile hydrogen of PBD. A possible explanation for these contradictory results may be the fact that crosslinking monomers such as EBPADM rapidly undergo gelation and form a relatively immobile network structure. The more restrictive matrix of EBPADM also contains immobilized radical sites that may not be in a proximate position to react with a methacrylate group but rather can now only react with an available labile hydrogen by a slower radical abstraction reaction route. The spectroscopic data currently available neither support nor contradict this hypothesis.

## Acknowledgments

Supported by NIDR Grant DE09322, ADAHF, and NIST. Certain commercial materials and equipment are identified in this paper to define adequately the experimental procedure. In no instance does such identification imply recommendation of endorsement by the National Institute of Standards and Technology or by the National Institutes of Health or that the material or equipment is necessarily the best available for the purpose.

## References

1.  Bailey, W.J. and Endo, T. *J Poly Sci.* **1978,** *Poly Symposium 64,* 17.
2.  Chen, T.M. and Brauer, G.M. , *J Dent Res.* **1982,** *61,* 1439.
3.  SAS Institute Inc. *SAS/STAT User's Guide, Release 6.03 Edition.* Cary, NC: SAS Institute Inc., 1988, pp 549-637.
4.  Duncan, D.B. *Biometrics.* **1955,** *11,* 1.
5.  Wall, F.J. *Statistical Data Analysis Handbook*; McGraw-Hill Inc: New York, NY, 1986; pp. 4.10-4.11 and 15.3.

RECEIVED March 30, 1993

# Chapter 16

# Synthesis of Novel Hydrophilic and Hydrophobic Multifunctional Acrylic Monomers

Joseph M. Antonucci[1], Jeffrey W. Stansbury[1], and G. W. Cheng[2]

[1]Dental and Medical Materials Group, Polymers Division, National Institute of Standards and Technology, Gaithersburg, MD 20899
[2]West China University of Medical Sciences, People's Republic of China

A unique formaldehyde/acrylate insertion-condensation reaction can be used to form novel difunctional and multifunctional hydrophilic and hydrophobic monomers and oligomers. Hydrophilic polyethylene glycol diacrylates, formed more viscous, water soluble oligomeric products without the need for a solvent. Because of the predominant 1,6-arrangement of double bonds in these oligomers, they have a propensity to undergo cyclopolymerization as well as the usual crosslinking associated with the free radical polymerization of multifunctional monomers and oligomers. However, the synthesis of analogous hydrophobic difunctional monomers and multifunctional oligomers derived from highly fluorinated and siloxane-containing acrylic monomers required the use of a dipolar aprotic solvent such as dimethyl sulfoxide. The siloxane oligomers had the expected predominance of 1,6-diene structure. In contrast, the fluorinated difunctional monomers and multifunctional oligomers were characterized by a predominantly 1,4-diene structure. These novel oligomeric monomers have potential utility in a wide spectrum of dental and medical applications.

Novel difunctional acrylic monomers and multifunctional acrylic oligomers with a propensity for intra-intermolecular addition polymerization (cyclopolymerization) have been conveniently synthesized from conventional acrylates or diacrylates and paraformaldehyde under neat conditions using 1,4-diazabicyclo[2,2,2]octane (DABCO) as the catalyst (1-5).

The mechanism for the formation of oxybismethacrylates such as 2 (Figure 1) from monoacrylates (and presumably the oligomeric monomers derived from diacrylates as well) appears to involve an unusual base catalyzed self etherification of the intermediate $\alpha$-hydroxymethyl acrylate 1 (1,6). In addition to products 1 and

0097–6156/94/0540–0191$06.00/0
© 1994 American Chemical Society

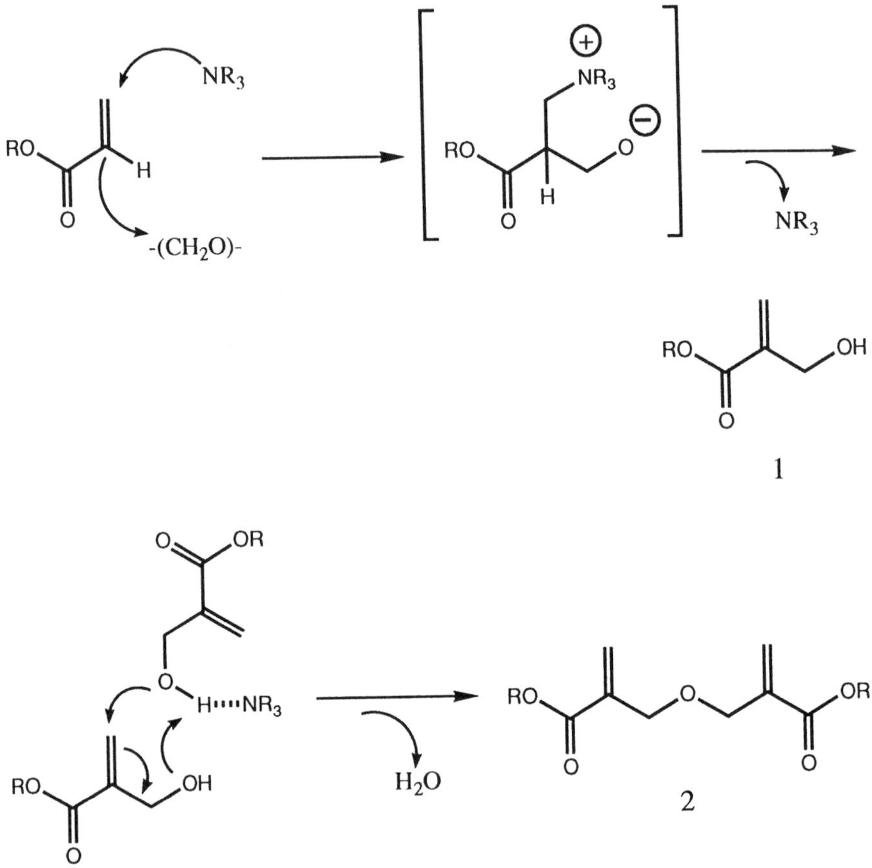

Figure 1. Mechanism for the formation of 1,6-oxybismethacrylates from the base-catalyzed insertion/condensation reaction of monofunctional acrylates and formaldehyde.

2, minor amounts of a methylenebisacrylate 3 were also observed from a competing alternate pathway (Figure 2).

The scope of this interesting reaction recently has been extended to include the synthesis of similar multifunctional hydrophilic and hydrophobic monomers and oligomers derived from ethylene oxide-based acrylates and fluorinated and siloxane-based acrylates, respectively *(7,8)*. This paper describes the synthesis, characterization and possible biomedical applications of these new acrylic monomers and oligomers.

## Materials and Methods

**Synthesis of Ethylene Oxide-based Multifunctional Vinyl Oligomers (Figure 3).** Polyethylene glycol (PEG) 400 diacrylate (average molecular weight = 508, Scientific Polymer Products, Inc., Ontario, N.Y.), and paraformaldehyde (95%, Aldrich Chemical Co., Milwaukee, WI), 2 mmol (1.016g) and 4 mmol (0.126g), respectively, were combined with 0.2 mmol (0.0224g) DABCO (Aldrich Chemical Co., Milwaukee, WI) in a sealed vial which was heated (oil bath) at 90-95°C for 6 h. The clear, viscous liquid was isolated in ca. 85% yield by column chromatography (silica gel) using methanol as the eluant. As determined by $^1$H NMR spectroscopy (JEOL GSX-270, Peabody, MA), the oligomeric product consisted mainly of segments having in-chain 1,6-diene and 1,4-diene linkages as well as $\alpha$-hydroxymethylacrylate end groups in an overall ratio of 6:3:1.

In a slightly modified preparation the same starting mixture of reagents was reacted at 90-95°C for 20 h. The viscous, crude product contained no measurable amount of the PEG 400 diacrylate starting material as determined by $^1$H NMR analysis. The following alternate product isolation procedure was used. The oligomer was washed with several portions of carbon tetrachloride to remove any low molecular weight product as well as the DABCO reaction catalyst. The oligomer was then dissolved in chloroform and dried over anhydrous sodium sulfate. The solvent was removed under reduced pressure (to ca. 5Pa) to provide the PEG 400 oligomer as a viscous pale yellow oil in ca. 63% yield. Characterization of the product by $^1$H NMR indicates no unreacted acrylate end groups and an average of 3.7 repeat units per oligomer. This corresponds to an average molecular weight of ca. 2700 for the oligomer. The repeating diene structure of the oligomer was predominantly 1,6; about 8% of the diene units had the less common 1,4 orientation.

IR (neat) 3465, 2868, 1720, 1636, 1109 cm$^{-1}$. $^1$H NMR (CDCl$_3$) $\delta$ 3.65 (internal OCH$_2$CH$_2$O), 3.77 (OC$\underline{H}_2$CH$_2$O$_2$C), 4.25 (C$\underline{H}_2$OH), 4.28 (CH$_2$OCH$_2$), 4.33 (CH$_2$O$_2$C), 5.84 (HOCH$_2$C=C$\underline{H}_E$), 5.92 (CH$_2$OCH$_2$C=C$\underline{H}_E$), 6.29 (HOCH$_2$C=C$\underline{H}_Z$), 6.34 (CH$_2$OCH$_2$C=C$\underline{H}_Z$)

The ethoxylated oligomeric multifunctional vinyl monomer derived from polyethylene glycol (PEG) 200 diacrylate (molecular weight = 302, Scientific Polymer Products, Inc., Ontario, NY) was synthesized and characterized in a similar manner. The PEG 200 oligomer, prepared under the same conditions described above (90-95°C for 20 h), contained an average of 4.2 repeat units per oligomer having mainly 1,6-diene units with only 6% 1,4-diene units. The average molecular weight of the PEG 200 oligomer was only ca. 1900 due to the shorter ethylene oxide

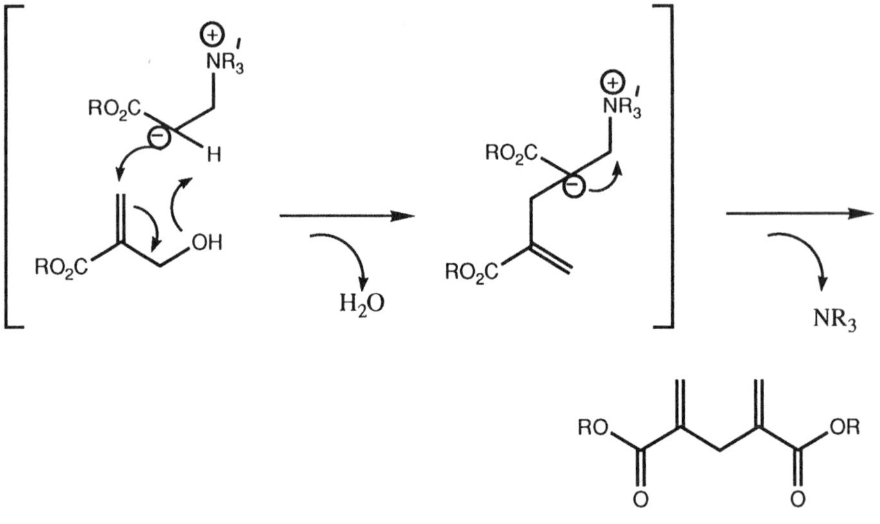

**3**

Figure 2. Mechanism for the formation of 1,4-methylenebisacrylates from the base-catalyzed insertion/condensation reaction of monofunctional acrylates and formaldehyde.

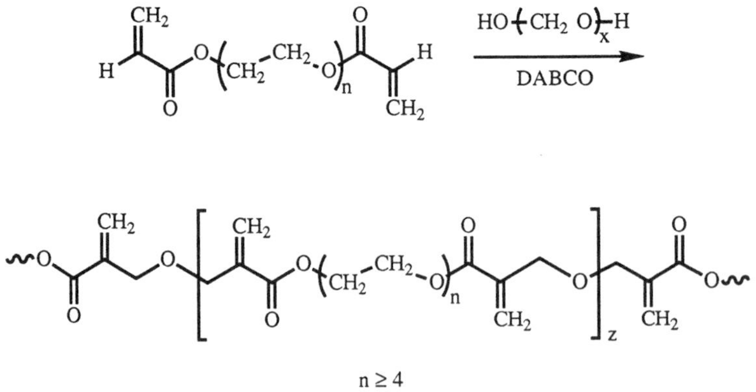

$n \geq 4$

Figure 3. Synthesis of ethylene oxide-based multifunctional acrylic vinyl oligomers.

chain length compared with that in the PEG 400 material; however, it was slightly more viscous than the PEG 400 oligomer.

**Synthesis of Highly Fluorinated Difunctional Monomers and Multifunctional Oligomers.** A commercially available monofunctional fluoroacrylate, 2,2,3,3,4,4,4-heptafluorobutyl acrylate (4, Figure 4), was used as received in this study. The initial experiment essentially followed published procedures and involved the neat bulk reaction of 5 mmol of 4 with 5 mmol of paraformaldehyde in the presence of 10 mole % (0.5 mmol) of DABCO at $23 \pm 2°C$ for long periods of time (1-5 months) in a sealed vial or ampule equipped with magnetic stirring. A second synthesis procedure involved both heating and stirring the above mixture at 85-90°C in an oil bath for 30 h. Neither of these approaches lead to significant product development. A variation of the latter procedure involved the addition of 0.6 mL of dimethyl sulfoxide (DMSO) to the above charge and heating the sealed ampule with stirring for 30 h.    This technique resulted in near complete conversion of the fluoroacrylate 4 to a complex mixture of products dominated by methylenebisacrylate 5, which was isolated by column chromatography (silica gel) with an elutant consisting of 10% ethyl acetate in hexane: IR (neat) 1739, 1632, 1228, 1126. $^1$H NMR (CDCl$_3$) $\delta$ 3.38 (s, 2H, C-CH$_2$-C), 4.64 (t, J= 13.7 Hz, 4H, CH$_2$O), 5.77 (s, 2H, = CH$_E$), 6.40 (s, 2H, = CH$_z$). $^{13}$C NMR (CDCl$_3$) $\delta$ 33.5 (C-$\underline{C}$H$_2$-C), 59.8 (t, J- 27.4 Hz, CH$_2$O), 129.6 (= CH$_2$), 135.8 (C=), 164.6 (C=O).

The fluorinated diacrylate 6 (Figure 5) used in this study was prepared in 91% yield by the esterification of the commercially available fluorodiol, 2,2,3,3,4,4-hexafluoropentane-1,5-diol, in dichloromethane using acryloyl chloride with triethylamine as the catalyst/acid acceptor. After isolation, purification and characterization (IR, $^1$H NMR), the reaction of 6 with two equivalents of paraformaldehyde was conducted in DMSO at 85-90°C in the presence of catalytic amounts of DABCO as described in the last experiment with 4. The resulting oligomeric product 7 was dissolved in dichloromethane and the resulting solution was extracted with several portions of dilute aqueous HCl to remove the DABCO and DMSO. The solution was dried over anhydrous sodium sulfate and the solvent was removed under reduced pressure. Oligomer 7, as a viscous pale yellow oil residue, was charcterized by $^1$H NMR. The 1,4- and 1,6-diene linkages were present in a ratio of 3:1.

**Synthesis of Tetramethyldisiloxane-based Multifunctional Vinyl Oligomers.** The starting siloxane diacrylate 8 (Figure 6), 1,3-bis(4-acryloxybutyl)tetramethyl-disiloxane, was synthesized from the corresponding diol, 1,3-bis(4-hydroxybutyl)-tetramethyldisiloxane (Columbia Organic Chemicals Co., Columbia, S.C.) as described previously. The siloxane diacrylate 8 was isolated in 85% yield: IR (neat) 2945, 1718, 1631, 1401, 1264, 1254, 1060, 950, 840 cm$^{-1}$. $^1$H NMR (CDCl$_3$) $\delta$ 0.00 (s, Si-CH$_3$), 0.51 (t, Si-CH$_2$), 1.40 (m, Si-CH$_2$-C$\underline{H}_2$), 1.66 (m, O-CH$_2$-C$\underline{H}_2$), 4.12 (t, O-CH$_2$), 5.78 (d, =CH$_E$), 6.09 (q, CH$_2$=C$\underline{H}$), 6.38 (d, =CH$_Z$).

Diacrylate 8 was oligomerized according to the following procedure. A neat mixture of the siloxane diacrylate and DABCO in a molar ratio of 1:0.4 was heated for 30 min at 80-85°C and then transferred to a vial containing 2 equivalents of

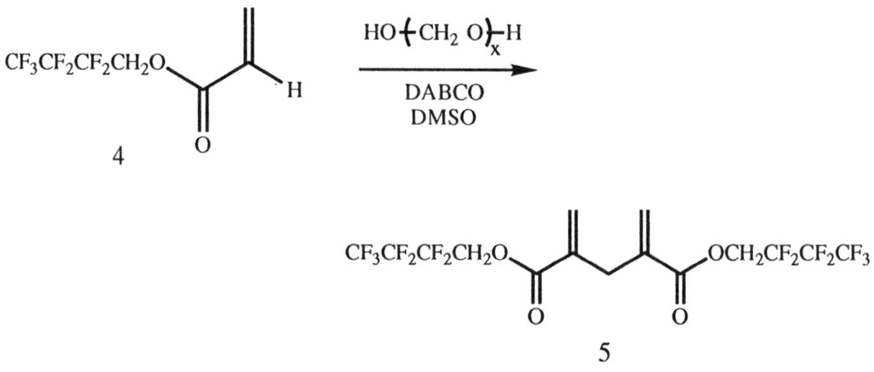

Figure 4. Synthesis of highly fluorinated difunctional acrylic vinyl monomers.

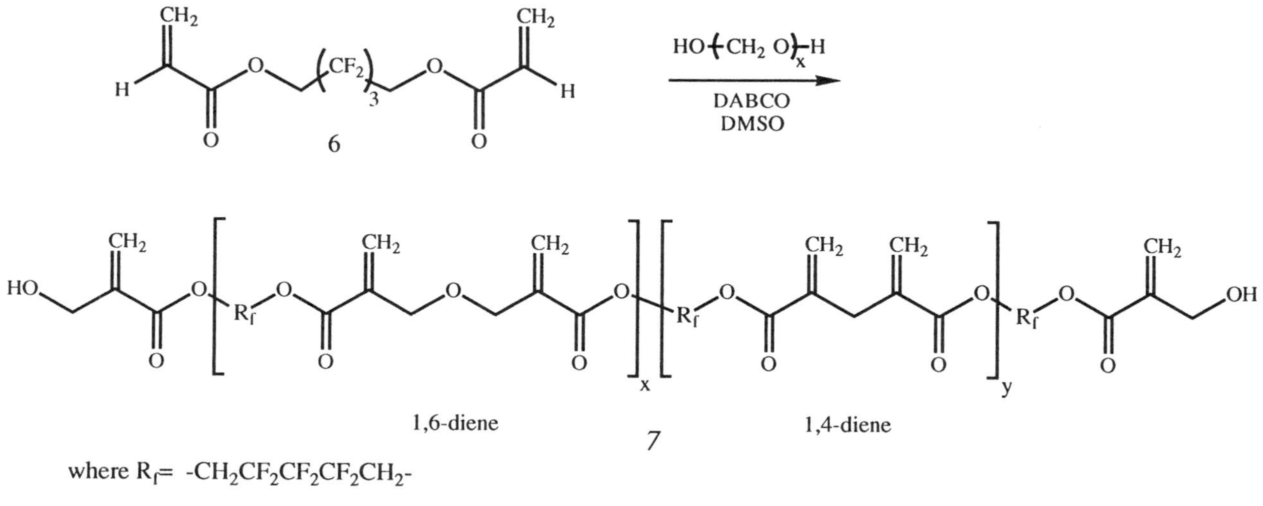

Figure 5. Synthesis of highly fluorinated multifunctional acrylic vinyl oligomers.

where $R_f = -CH_2CF_2CF_2CF_2CH_2-$

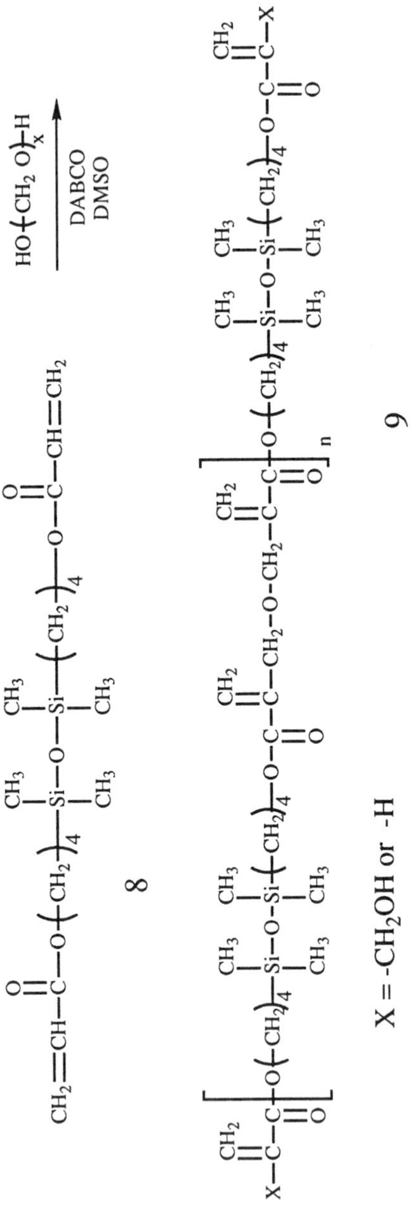

Figure 6.  Synthesis of tetramethyldisiloxane-based multifunctional acrylic vinyl oligomers.

paraformaldehyde and DMSO (50% by weight based on the diacrylate). The vial was sealed and heated with stirring (magnetic) at 95°C (oil bath) for 20 h. The workup of the siloxane-based oligomeric product followed that described for the fluorinated oligomer 7. A small amount of residual, unreacted diacrylate 8 was chromatographically separated on silica gel using 20% ethyl acetate in n-hexane. Ethyl acetate was used to elute the oligomeric product 9 which was isolated as a near-colorless, slightly viscous oil: IR (neat) 3465, 1718, 1634, 1254, 1164, 1060 cm$^{-1}$. $^1$H NMR (CDCl$_3$) $\delta$ 0.01 (s, Si-CH$_3$), 0.48 (t, Si-CH$_2$), 1.36 (m, Si-CH$_2$C$\underline{H}_2$), 1.64 (m, Si-CH$_2$CH$_2$C$\underline{H}_2$), 4.12 (t, CH$_2$O$_2$C), 4.22 (s, CH$_2$OCH$_2$), 4.30 (s, C$\underline{H}_2$OH), 5.80 (d, HOCH$_2$C=C$\underline{H}_E$), 5.87 (d, CH$_2$OCH$_2$C=C$\underline{H}_E$), 6.21 (d, HOCH$_2$C=C$\underline{H}_Z$), 6.38 (d, CH$_2$OCH$_2$C=C$\underline{H}_Z$). $^{13}$C NMR (CDCl$_3$) $\delta$ 0.2 (Si-CH$_3$), 17.8 (Si-CH$_2$), 19.6 (Si-CH$_2$C$\underline{H}_2$), 31.9 (Si-CH$_2$CH$_2$C$\underline{H}_2$), 62.0 (HOCH$_2$), 64.4 (CO$_2$C$\underline{H}_2$), 68.7 (CH$_2$OCH$_2$), 125.0 and 125.3 (C=C$\underline{H}_2$, internal and terminal), 137.0 (CH$_2$OCH$_2$C=CH$_2$), 139.6 (HOCH$_2$C=CH$_2$), 165.7 (CH$_2$OCH$_2$C(=CH$_2$)C$\underline{}$=O), 166.2 (HOCH$_2$C(=CH$_2$)C$\underline{}$=O).

The $^1$H NMR spectrum of oligomer 9 also showed a small peak at 5.54 ppm which corresponds to the 1,4-diene structure, but this amounted to only about 3% of the linkages in the oligomer. In addition, the $^{13}$C NMR spectrum shows a very small peak at 94.5 ppm which indicates that a trace of a 1,8-diene linkage is also present in the oligomer. The ratio of 1,6-diene to terminal hydroxymethyl acrylates is ca. 2:1 which gives a value of n=2 on average for the oligomer. The composition of end groups in the oligomer is X = -CH$_2$OH (95%) and X = -H (5%).

## Results and Discussion

This investigation demonstrates that the unique insertion/ condensation reaction of formaldehyde with mono- or difunctional acrylates can be extended to the synthesis of hydrophilic polyethylene oxide and hydrophobic siloxane multifunctional vinyl monomers. The preparation of the hydrophilic oligomers was similar to the procedures used to convert conventional hydrocarbon mono- and diacrylates to the corresponding difunctional and multifunctional vinyl monomers. As expected the hydrophilic oligomers had a predominantly 1,6-diene structure similar to that obtained from oligomers derived from hydrocarbon diacrylates. The siloxane oligomer also had mainly the 1,6 arrangement of double bonds but required the use of the compatibilizing additive or solvent, DMSO, in its synthesis. Previously it was found that DMSO also was necessary to effect the synthesis of analogous highly fluorinated difunctional monomers and oligomers from the corresponding fluoroacrylates (7).

Because of favorable proximity of many of their double bonds (i.e. the predominance of 1,6-diene units), these oligomers should have a propensity for cyclopolymerization via six membered cyclic rings (9,10). Under bulk free radical conditions, which is the common mode of polymerization in dental resin applications, crosslinking also results. The degree of cyclopolymerization versus crosslinking will depend on the 1,4-diene content and other structural features of the oligomer, the type and concentration of comonomers and the conditions of polymerization. Generally, cyclopolymerization will aid in both reducing residual

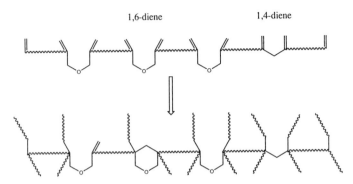

Figure 7. Idealized representation of a vinyl oligomer showing intramolecular cyclization and crosslinking.

vinyl unsaturation and polymerization shrinkage but will tend to lower the crosslink density of the polymer (2). The formation of in-chain cyclic structures as shown in Figure 7 is expected to aid in raising the glass transition temperature of these polymers which may somewhat offset the effects of lower degrees of crosslinking.

The hydrophilic and hydrophobic characters of these oligomers should allow the formation of unique homopolymers and copolymers. The poly(ethylene oxide) oligomers are novel hydrogels which should find application in composites, cements and other resin-based materials. Their large bulk, cyclopolymerizability and potential for hygroscopic expansion after water uptake may aid in moderating the effects of polymerization contraction. The siloxane oligomers, which combine large bulk, comparatively low viscosity, high flexibility and cyclopolymerizability, offer a facile means of achieving low shrinking, hydrophobic, tough polymeric structures suitable for use in a variety of dental and medical applications.

## Acknowledgment

The authors gratefully acknowledge access to the American Dental Association/Paffenbarger Research Center multinuclear NMR facility (obtained under NIH shared instrument grant #RR03266) located at NIST.

This work was partially supported by Interagency Agreement Y01-DE 30001 with the National Institute of Dental Research, Bethesda, MD  20892.

Certain commercial materials and equipment are identified in this paper to define adequately the experimental procedure. In no instance does such identification imply recommendation or endorsement by the National Institute of Standards and Technology or that the material or equipment is necessarily the best available for the purpose.

## Literature Cited

1. Mathias, L.J.; Kusefoglu, S.H. *Macromolecules* **1987**, *20*, 2039.
2. Stansbury, J.W. *J. Dent. Res.* **1990**, *69*, 844.
3. Mathias, L.J.; Dickerson, C.W. *J. Polym. Sci., Polym. Lett. Ed.* **1990**, *28*, 175.

4. Stansbury, J.W. *Macromolecules* **1991**, *24*, 2029.
5. Stansbury, J. W. *J. Dent. Res.* **1992**, *71*, 434.
6. Mathias, L.J.; Kusefoglu, S.H.; Ingram, J.E. *Macromolecules* **1988** *21*, 545.
7. Antonucci, J.M.; Stansbury, J.W.; Cheng, G.W. *Polym. Prepr.* **1990**, *31*, 320.
8. Antonucci, J.M.; Stansbury, J. W.; Cheng, G.W. *J. Dent. Res.* **1991**, *70*, 527.
9. Marvel, C.S. *J. Polym. Sci.* **1960**, *48*, 101.
10. Butler, G.B. In: *Proceedings of the International Symposium on Macromolecules*; E.B. Mano, Ed.; Elsevier Scientific Publishing Company: New York, NY, 1975; pp 57-76.

RECEIVED May 13, 1993

Chapter 17

# Effect of Structure on Properties of Absorbable Oxalate Polymers

Russell A. Johnson and Shalaby W. Shalaby

Department of Bioengineering and Materials Science and Engineering Program, Clemson University, Clemson, SC 29634–0905

Absorbable copolyesters entailing hexamethylene oxalate repeat units were synthesized and studied to access the effect of incorporating isomorphic sequences on the properties of these absorbable systems. The chemical identity and physical properties were determined using infrared spectroscopy (IR), nuclear magnetic resonance (NMR), X-ray diffraction, and differential scanning calorimetry (DSC). The effect of compositional changes on thermal transitions, percent crystallinity, and absorption profiles was emphasized. Copolymers with non-isomorphic sequences with aromatic moieties were studied in the same manner. It was noted that the isomorphic systems maintained a relatively high crystallinity throughout the entire range of composition. Furthermore, hydrolytic stability increased with changes in composition, particularly with the increase in cyclic content and presence of aromatic moieties. Histological evaluations of 3 mm diameter transcortical pins at 4 and 8 weeks post-operatively did not reveal any significant differences (p=0.10) between the oxalate-based polymers and poly(p-dioxanone) (PDS) controls. None of the implants elicited a toxic or severe hard or soft tissue reaction as implanted.

The adverse tissue reactions and uncontrollable absorption rates associated with natural absorbable polymers, such as collagen, stimulated the search for synthetic systems. Limitations identified with the lactone systems that are commercially available, including the presence of impurities (i.e. monomers), narrow windows of absorption profiles for structurally useful materials, and sterilization constraints, have lead to investigations of different compositions (1,2). Among those are oxalate-based polymers which are a relatively new system of absorbable polymers. They are synthesized by step-growth polymerization, and therefore, avoid monomer complications associated with ring-opening polymerization. Oxalates were developed by Shalaby and coworkers in the 1970s, who initially focused on the simplest form of alkylene oxalate polymers for suture coating purposes (3). The hydrolytic stability of these systems was reported to decrease primarily with an increase in methylene fraction in the repeat unit, and the melting temperature was found to approach that of polyethylene. To increase the Tm of alkylene oxalate polymers, copolymers of cyclic and aromatic diols were synthesized. In particular, polyesters based on 1,4-*trans*-

0097–6156/94/0540–0202$06.00/0
© 1994 American Chemical Society

cyclohexane-dimethanol (1,4-HDM) were reported to have desirable melting temperatures and when synthesized in conjunction with 1,6-hexanediol, isomorphic copolymers were formed. Surprisingly, the crystallinity of these copolymers displayed substantial degree of crystallinity throughout the entire composition range and the melting temperature increased progressively with increases in cyclic fraction(4).

$$\text{--}\!\!\left[\text{O--C--C--O--(CH}_2)_n\right]_p$$

poly(alkylene oxalate)

poly(*trans*-1,4-cyclohexyldicaronyl-co-hexamethylene oxalate)

The purpose of this investigation was to determine the effects of changing the chemical composition of hexamethylene-based oxalate copolymers on their thermal properties, hydrolytic stability, and tissue reaction in bone. Eight compositions, including the homopolymer poly(hexamethylene oxalate), three isomorphic copolyoxalates, and three copolymers containing aromatic moieties, were synthesized and characterized. The aromatic sequences based on dimethyl terephthalate and dimethyl isophthalate were included for subsequent studies of radiostability against gamma radiation.

dimethyl terephthalate          dimethyl isophthalate

## Materials and Methods

**Polymer Synthesis.** Oxalate-based polyesters are synthesized by the step-growth polymerization involving the interchange of an ester (diethyl oxalate) with a diol (1,6-hexanediol or 1,4-*trans*-cyclohexanediol). For the aromatic systems, dimethyl terephthalate or dimethyl isophthalate were used as comers at variable mole percentages with diethyl oxalate. All reactants, which were supplied by Aldrich, were distilled and/or dried prior to synthesis. All glassware was flame dried before charging the reactants. Diethyl oxalate and the selected diol(s), and in three compositions, aromatic moieties were heated in a mechanically stirred reactor using stannous octoate and/or dibutyltin oxide as the catalyst(s). The prepolymerization (formation of the hydroxyl-capped oligomers) was conducted in a dry nitrogen environment at 120° C and 160° C for two and three hours, respectively, or until the precalculated amount of ethanol had collected in the receiver. After cooling the mixture under nitrogen, post-polymerization was performed at reduced pressures using suitable heating schemes. When the polymer melt had reached a highly viscous state, polymerization was concluded. Each polymer was cooled to room temperature, isolated, dried for 2 to 3 days, and stored under a vacuum of less than 0.1 mmHg. The polymerization charges for every composition are shown in Table I and corresponding polymerization schedules are given in Table II.

Table I. Chemical compositions of polymerization charges (in mol %)

| Polymer number | Desired composition | DEO | 1,6-Hex | 1,4-HDM | DI | DT | Catalyst(s) |
|---|---|---|---|---|---|---|---|
| I | 100 Ox-6 | 0.25 | 0.27 | 0 | 0 | 0 | SO |
| II | 20/80 Ox-c/Ox-6 | 0.25 | 0.216 | 0.054 | 0 | 0 | SO, DBTO |
| III | 50/50 Ox-c/Ox-6 | 0.25 | 0.13 | 0.14 | 0 | 0 | SO, DBTO |
| IV | 80/20 Ox-c/Ox-6 | 0.25 | 0.054 | 0.216 | 0 | 0 | SO, DBTO |
| V | 90/10 Ox-6/T-6 | 0.225 | 0.26 | 0 | 0 | 0.025 | SO, DBTO |
| VI | 90/10 Ox-6/I-6 | 0.225 | 0.27 | 0 | 0.025 | 0 | SO, DBTO |
| VII | 85/15 Ox-6/I-6 | 0.213 | 0.27 | 0 | 0.037 | 0 | SO, DBTO |
| VIII | 50/50/15 Ox-c/ Ox-6/I-6 | 0.213 | 0.14 | 0.13 | 0.037 | 0 | SO, DBTO |

| | | | |
|---|---|---|---|
| DEO | = Diethyl oxalate | DT | = Dimethyl terephthalate |
| 1,6-Hex | = 1,6-hexanediol | SO | = Stannous octoate |
| DI | = Dimethyl isophthalate | DBTO | = Dibutyltin oxide |
| 1,4-HDM | = 1,4-*trans*-cyclohexanedimethanol | | |

Table II. Postpolymerization schedules for all polymer compositions

| Polymer number | Desired composition | Temperatures (° C) | Time at each temp. (hrs) |
|---|---|---|---|
| I | 100 Ox-6 | 100, 115, 150, 175, 195 | 2, 2, 3, 4, 9 |
| II | 20/80 Ox-c/Ox-6 | 155, 180, 190, 200 | 5, 4.5, 2, 2 |
| III | 50/50 Ox-c/Ox-6 | 170, 190, 200, 205 | 4, 2, 1.5, 3.5 |
| IV | 80/20 Ox-c/Ox-6 | 205, 215, 230, 235 | 2, 2.5, 1, 0.5 |
| V | 90/10 Ox-6/T-6 | 120, 150, 160, 200, 210 | 2.5, 2, 11, 2.5 |
| VI | 90/10 Ox-6/I-6 | 125, 150, 180, 200, 210 | 1.5, 2, 1.5, 3, 6.5 |
| VII | 85/15 Ox-6/I-6 | 150, 170, 180, 200, 205 | 2, 2, 4, 2.5, 0.5 |
| VIII | 50/50/15 Ox-c/ Ox-6/I-6 | 135, 160, 185, 205, 225 | 1.5, 2, 5.5, 2, 0.5 |

**Polymer Characterization.** The inherent viscosity of each composition was determined at a 0.1 percent concentration in chloroform at 31° C. IR spectrometry was used to verify the presence of aromatics. To determine the actual composition of each polymer composition, NMR spectra of polymer solutions in deuterated chloroform were recorded on a Bruker AC300 NMR spectrometer. Both $^1H$ and $^{13}C$ spectra were prepared and computational analysis of the copolymers was done using the Inverse Gating procedure provided by the system. A DuPont 990 DSC was used to discern the thermal characteristics of each polymer in a nitrogen environment at a heating rate of 20° C/min. The melting temperature (Tm) and the area of the melting peak (ΔH) were recorded for the initial heating and reheating schemes after quenching the molten sample with liquid nitrogen. Finally, polymers were scanned with a Scintag XCD 2000 X-ray diffractometer to obtain the diffraction patterns characteristic of semicrystalline materials. The recorded diffraction patterns were resolved to separate the amorphous halo from the crysalline-originated reflections. The percent crystallinity was calculated by dividing total areas under the crystalline reflections by the total area of the curve.

*In Vitro* **Studies** A physiological environment was simulated using a phosphate buffer solution of pH = 7.28 at 37° C. Films with dimensions 4 x 2 x 0.174 cm weighing an average of 150 mg were constructed via compression molding using a Carver Laboratory Press. At least two samples were applied for each time period and were individually placed in a 50 mL, preweighed, screw-capped polyethylene centrifuge tube containing the buffer. Mass losses were taken at the first, second, and fourth weeks, and every month thereafter. The data were compared statistically using a 2-tailed t-test to discern percent decrease in original mass between the isomorphic compositions, and the aromatic systems at concurrent time periods. Comparisons were considered highly significant if their significance values were less than 0.05 and moderately significant if less than 0.10. Similarly, an accelerated absorption study was carried out at 55° C using 2 x 2 x 0.175 cm films weighing and average of 70 mg.

Two samples were made for each time period, consisting of weekly intervals for the first three weeks, and one final time period.

*In Vivo* **Study**  To investigate the hard-tissue response to all compositions of oxalate-based polymers, a total of four polymer systems were chosen as implant materials including: poly(hexamethylene oxalate), 50/50 poly(*trans*-1,4-cyclohexyldicarbonyl-co-hexamethylene oxalate), 90/10 poly(hexamethylene oxalate-co-terephthalate), and PDS as a control. The polymers were converted to 3 mm diameter pins by compression molding using a specially constructed mold. The pins implanted transcortically in the femur of goats. Prior to implantation, the implants were scoured with a 70% ethanol solution, sterilized with ethylene oxide, and stored under vacuum for at least 4 days. Two time periods, one and two months, were selected for implantation since bone requires approximately four weeks to regain mechanical integrity from a fracture, and furthermore, the implants were expected to show at least partial absorption by the end of eight weeks. Two animals were studied per time period so that four examples of each composition could be observed (one of each implant was placed in both femurs). Histological sections were harvested, prepared, and analyzed qualitatively for comparison of composition and tissue response evaluation. The evaluation focused on four main parameters: 1.) bone apposition, 2.) bone remodelling, 3.) periosteal and endosteal bone formation, and 4.) the soft tissue reaction in the marrow.

**Results and Discussion**

The IR spectra were obtained to assure that the groups of primary importance were present. Frequencies characteristic of the functional groups relevant to the investigation were as follows: carbonyl (C=O), 1750; methylene (CH$_2$), 2870 and 2940; and aromatic (C=C) and (C-H), 1600 and 3100, respectively (4). All compositions were consistent with the desired compositions. Both proton ($^1$H) and carbon-13 ($^{13}$C) spectra prepared for of each polymer, and an Inverse Gating program was applied to attenuate first order coupling and allow integration. The compositions of copolymers was determined by relating representative carbons for each particular repeat unit to those of the co-repeat unit. Results of the characterization including the final compositions, viscosities, melting temperatures and corresponding heat of fusion, and crystallinities of all polymers are shown in Table III. The melting peaks were recorded for each composition: some compositions exhibited two adjacent melting peaks in a single scan, indicating differences in crystallite sizes. The isomorphic copolyoxalates displayed a slight initial decrease in Tm with increase in cyclic content, and a progressive increase thereafter. A progressive increase in Tm was expected throughout since the cyclic moieties increase chain stiffness, a phenomena which Shalaby and Jamiolkowski observed (4,6). However, these investigators did not examine a copolymer composition consisting of 85% cyclic and 15% acyclic chain, which may be an "eutectic-like" composition for the system. The crystallinity of each composition was determined predominantly by X-ray diffraction, with some consideration given  to heat of fusion data from DSC. Having distinct crystalline reflections in the diffraction patterns and presence of melting temperature in all polymers indicated that all examined systems are crystalline. The observed crystallinity throughout the range of isomorphic compositions is an unusual phenomena for copolymers (Fig. 1), and was previously demonstrated by Shalaby and Jamiolkowski (4). Similarities in length of co-repeat units must have been conducive for crystallite growth.

Table III.  Properties of oxalate-based polymers

| Poly. No. | Identity | Composition Initial | Final[a] | $\eta_{inh}$[b] | Thermal data[c] #1 (°C) | #2 (°C) | % crystal.[d] |
|-----------|----------|---------|----------|--------|---------|---------|-------------|
| I | Ox-6 | 100 | 100 | 0.65 | 76 | 77 | 48  (11.5) |
| II | Ox-c/Ox-6 | 20/80 | 20/80 | 0.31 | 69 | 71 | 43  (7.2) |
| III | Ox-c/Ox-6 | 50/50 | 51/49 | 0.48 | 120,130 | 101,121 | 34  (8.3) |
| IV | Ox-c/Ox-6 | 80/20 | 85/15 | 0.45 | 188 | 180,192 | 67  (7.4) |
| V | Ox-6/T-6 | 90/10 | 89/11 | 0.31 | 60,66 | 65 | 38  (9.8) |
| VI | Ox-6/I-6 | 90/10 | 90/10 | 0.29 | 60 | 62 | 30  (6.5) |
| VII | Ox-6/I-6 | 85/15 | 84/16 | 0.31 | 61 | 55 | 37  (10.0) |
| VIII | Ox-c/Ox-6/I-6 | 50/50/15 | 52/48/14 | 0.29 | 105 | 101 | 44  (9.2) |

[a] The final composition (mol %) was determined by NMR.

[b] The viscosity is based on the average of 10 repetitions of one solution.

[c] DSC heating of the polymer was initiated at room temperature and heated to above melting (20° C/min).  It was reheated at the same rate and to the same maximum.  Two melting temperatures represents two melting peaks.

[d] The percent crystallinities and errors were determined by curve resolving X-ray diffraction patterns.

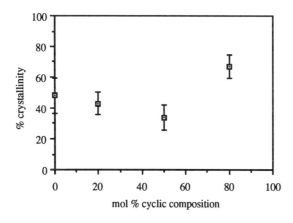

Figure 1.  Percent crystallinities of isomorphic copolyoxalates determined by X-ray diffraction.

***In Vitro* Absorption** The initial absorption profiles for each system are shown in Fig. 2. Three distinct behaviors were apparent from the absorption profiles and there significances were verified statistically. First, all compositions had a 20 to 45% mass loss within the first few weeks. This may be related to earlier reports describing oxalates as having a substantial decrease in the *in vitro* and *in vivo* strength retention as fibers in a few to several days (4). Second, in general, aromatic systems increased the hydrolytic stability of oxalate compositions, which was evident form the larger mass retained in comparison to the homopolymer at both 4 and 16 weeks. These groups may have increased the hydrophobicity of the polymer and provided a barrier to water. Third, increasing the cyclic content of the isomorphic copolyoxalates increased the hydrolytic stability at 4 and 8 weeks. However, the composition based on 20% cyclic 80% cyclic diols had a higher mass loss than the homopolymer at 4, 16, and 20 weeks. This may be related to disorder caused by the small amounts of cyclic moieties or low molecular weight (indicated by the low solution viscosity). Under accelerated conditions, most of the compositions had completely absorbed by the end of three weeks.

**Biological Performance of Transcortical Pins** Implanting the absorbable oxalate polymers as transcortical pins was a unique study and was considered an evaluation of the efficacy of the model as well as an evaluation of the tissue response. During implantation, many of the pins were loosely fit in the drilled hole so that micro motion was possible. This may have affected the reliability of the evaluation, however, the partial absorption of the polymers would compromise the fit and mechanical properties (if simultaneous bone growth did not prevail) as time passed. Thus, the differences in hard tissue response between securely and loosely fit absorbable implants may not have been significant. Furthermore, the aromatic composition was very brittle, breaking into several pieces during placement in at least half of the designated drilled holes. The fragments of those implants which had broken and were located in the marrow of the femur were evaluated for soft tissue response only. Scores assigned to specimens for each of the four factors were evaluated using the Kruskal-Wallis analysis of variance test for nonparametric data with p = 0.05. According to the comparisons, the relevant scores from the oxalate-

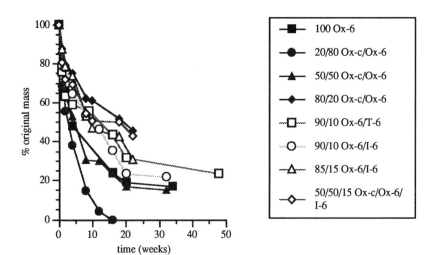

Figure 2. *In vitro* absorption profiles of all compositions at 37° C.

based polymers were not significantly different from those of PDS. Previous investigations have indicated that PDS elicits a minimal foreign body tissue reaction (7,8) and was supported by this research for both hard and soft tissues. The portions of material at the site of implantation of all oxalate-based polymers had good hard tissue compatibilities; none of the sections demonstrated large areas of bone resorption. When viewed microscopically with a polarized light, all oxalate compositions had birefringent particles in the tissue outside the initial area of implantation, which were surrounded by inflammatory cells (foreign body giant cells, macrophages, etc.). The homopolymer (Ox-6) had the highest amount of absorption, and consequently, the largest occurrence of these birefringent particles; their composition remains unknown.

## Literature Cited

1. Lerwick, E. *Surg. Gynecol, Obstet.* **1983**, *156*, p. 51.
2. Shalaby, S. W.; Johnson, R. A. In *Biomedical Polymers: Designed to Degrade Systems*; Shalaby, S. W., Ed.; Hanser Publishers: Brooklyn, NY(in press).
3. Shalaby, S. W.; Jamiolkowski, D. D., U. S. Pat. (to ETHICON, Inc.) 4,105,034 **1978**.
4. Shalaby, S. W.; Jamiolkowski, D. D., U. S. Pat. (to ETHICON, Inc.) 4,140,678; 4,141,087 **1979**.
5. Sorrell, T. N. *Interpreting Spectra of Organic Molecules,* University Science Books: Mill Valley, **1988**; pp. 11-115.
6. Shalaby, S. W.; Jamiolkowski, D. D., U. S. Pat. (to ETHICON, Inc.) 4,205,399 **1980**.
7. Makela, E. A. Dissertation, Department of Orthopaedics and Traumatology, Helsinki University, **1989**.
8. Ray, J. A.; Doddi, N.; Regula, D.; Williams, J. A.; Melveger, A. *Surg. Gynecol, Obstet.* **1981**, *153*, p.497.

RECEIVED June 21, 1993

Chapter 18

# Fluorescent Cure Monitoring of Dental Resins

**Spurgeon M. Keeny III, Joseph M. Antonucci, Francis W. Wang, and John A. Tesk**

**Polymers Division, National Institute of Standards and Technology, Gaithersburg, MD 20899**

Fluorescent probes are increasingly being used for *in situ* cure monitoring of resins. Benefits of this type of cure monitoring are that it allows for real time measurements and holds potential for remote sensing of the quality of the cure. In this study ethyl 4-*N*,*N*-dimethyl-aminobenzoate (4EDMAB) and its isomers (2EDMAB and 3EDMAB) were used in experimental, visible-light-activated dental monomer systems as both the photoreductant and fluorescent probe. These probes are sensitive to changes in the microenvironment of the curing resin. Compared to 4EDMAB, the ortho and meta isomers showed much less sensitivity to viscosity changes. The fluorescence changes of 4EDMAB were correlated with degree of cure (DC) measurements from IR spectroscopy. The relationship between fluorescence intensity changes and DC indicates a monotonically increasing but nonlinear function.

Recently fluorescent probes have been used for *in situ* monitoring of the cure kinetics of resin systems (*1-6*). The fluorescent probes are sensitive to changes in the microenvironment that depend on the degree of cure of the resin (i.e. the micropolarity and microviscosity) as shown by changes in the intensity or position of their excitation frequencies. The benefits of fluorescence spectroscopy for cure monitoring of dental resins are that it allows for real time measurements and holds the potential for remote sensing of the resin cure as a function of depth. Visible-light-activated resin based dental materials employ a photooxidant, e.g., camphorquinone (CQ), and a photoreductant, typically a tertiary amine. One such tertiary aryl amine, ethyl 4-*N*,*N*-dimethylaminobenzoate (4EDMAB), is not only an excellent photoreductant for CQ but its fluorescence behavior is sensitive to both the local polarity and the local viscosity of its surroundings. This and similar probes have been used to study the polymerization of methyl methacrylate (*4*), polymer-solvent interactions (*7*), and free volume effects in both dilute and concentrated polymer solutions (*8*). In this paper, 4EDMAB, its isomers and several other aryl amines were evaluated as fluorescent probes for cure monitoring of dental resins. The fluorescence intensity changes of 4EDMAB in an experimental vinyl monomer system were correlated with degree of cure measurements obtained from IR spectroscopy.

## Experimental

### Resin Materials and Curing Conditions
The resin used in this study consisted of a 1/1 by weight mixture of 2,2,-bis[p-(2'-hydroxy-3'-methacryloxypropoxy)phenyl]propane (BIS-GMA) and triethylene glycol dimethacrylate (TEGDMA). The resin was activated for visible light cure with CQ (0.05 and 0.2%, w/w%) and a tertiary aryl amine prior to irradiation using a visible (blue) light source (MAX LIGHT, CAULK/DENTSPLY). Several aryl amines were evaluated for their effectiveness as probes including 6-propionyl-2-$N,N$-dimethylaminonaphthalene (PRODAN), which was used in typical fluorescent probe concentrations ($\sim 10^{-5}$ mol/L). The other aryl amines were used at much greater concentrations than typical fluorescent probes, i.e., $4.14 \times 10^{-2}$ mol/L (typical photoreductant concentrations), and included the three isomers (para, meta, and ortho) of ethyl-$N,N$-dimethylaminobenzoate (4EDMAB, 3EDMAB, and 2EDMAB, all at 0.8% w/w%) and $N,N$-dimethylamino-p-toluidine, DMPT at 0.56% w/w%. Some of the tertiary aryl (Ar) amine, $(CH_3)_2$-N-Ar, is transformed by the photoredox reaction with CQ to the initiating radical, $\cdot CH_2$-N-$(CH_3)$-Ar, which becomes incorporated into the polymer. This and other 4EDMAB derived byproducts from the breakdown of the CQ-amine exiplex forms products which probably have fluorescent properties similar to that of the starting tertiary aryl amine.

To simulate clinical conditions resins specimens were cured in air under ambient conditions prior to the fluorescence or IR measurements. Since these photoactivated dental thermosetting resins undergo almost instantaneous polymerization on irradiation at 450-500 nm, samples were cured under four different sets of conditions designed to yield significantly different conversions of the vinyl groups (see Table I). Conditions 2 and 3 were designed to yield less than optimal conversions. Cure condition 1a mimicked typical clinical conditions (CQ = 0.2%, 60 second light exposure). Condition 2 used reduced light exposure (CQ = 0.2%, 18 second exposure) and condition 3 employed both reduced CQ concentration and reduced light exposure (CQ = 0.05%, 6 second exposure). Condition 1b was similar to 1a but was designed to enhance cure by an immediate post cure at 60°C for 1 hour rather than the usual post cure at 20°C. The broad range of degree of cure (DC) values obtained by IR were then correlated with the corresponding fluorescence intensity changes.

### Fluorescence and IR Measurements
Fluorescence spectra were obtained as a function of time from mostly single resin

samples contained in 1.5 mm thick rectangular quartz cells with internal dimensions of 8 x 30 mm for the front face. For conditions Ia and Ib several resins were tested at least in triplicate (n ≥ 3). The front surface fluorescence spectra were obtained at 20°C (± 2°C) in a SPEX FLUOROLOG series 2 fluorometer. The fluorescence changes with time were measured over a range of 340-540 nm with excitation at 345 nm for 4EDMAB and 3EDMAB, 370 nm for 2EDMAB, and 360 for PRODAN. The IR spectra were obtained from thin polymer films with a ratio recording transmission IR spectrophotometer (Perkin-Elmer 1420). Spectra for the uncured resin samples and for the poorly cured samples (conditions 2 and 3, which did not produce strong coherent polymer films immediately after photocuring), were obtained between KBr windows. Polymer films (n ≥ 3) for conditions 1a and 1b were obtained by photocuring between removable mylar films. The DC was determined directly by measuring the ratio of the peak height of the $C=C$ double bond absorption band at 1637 cm$^{-1}$ to the internal standard aromatic absorption band at 1582 cm$^{-1}$ and comparing this value with that of the uncured sample. Baseline subtraction was used since both of these peaks are shoulders of the aromatic peak at 1608 cm$^{-1}$ (baselines were drawn from 1660 cm$^{-1}$ and 1550 cm$^{-1}$ to the shoulder at 1590 cm$^{-1}$).

## Results and Discussion

**Fluorescence of PRODAN.** Resins with PRODAN in low concentrations ($10^{-5}$ mol/L) in combination with 4EDMAB (0.8 w/w%) and CQ (0.2 w/w%) on excitation at 360 nm, exhibited a broad fluorescence emission peak between 430 and 450 nm in the uncured resin. A blue shift in the fluorescence peak to 414-418 nm occurred within 2 minutes after photocuring the resin for 60 seconds. The peak position did not change with subsequent post cure fluorescence measurements, indicating that PRODAN was only sensitive to changes in the initial curing stages while the resin was still relatively fluid. Thus, PRODAN was not suitable for the determination of cure in rapidly setting photocured resins with significant post cures.

**Fluorescence of EDMAB isomers and DMPT.** The fluorescence changes of the photoreductant 4EDMAB after a 60 second photocure of the CQ activated resin were quite dramatic and demonstrated more than a 20 fold increase after one hour when measured at 360 nm (Figure 1). The initial fluorescence of 4EDMAB in a 1/1 mixture of BIS-GMA and TEGDMA was very weak and showed two peaks (360 and 460 nm). Theoretical explanations of the fluorescence spectra of 4EDMAB and similar aromatic tertiary amines have been proposed (*4, 7-9*).

Compared to 4EDMAB the ortho and meta isomers (2EDMAB and 3EDMAB) showed much less sensitivity to viscosity changes. Figure 2 shows the fluorescence changes with post cure of 2EDMAB which amounted to about a 15% increase after one hour. The very strong initial fluorescence of 2EDMAB and the subsequent minimal changes with post cure are consistent with the concept of enhanced restricted internal rotation due to the steric hinderance imposed on this molecule by ortho substitution. The fluorescence changes for 3EDMAB are shown

Table I. Correlation of Fluorescence Intensity Ratio,
$I_t/I_o$, with Infrared Degree of Cure, DC (IR)

| Post Cure Time at 20°C (min) | Cure Conditions | | | | | |
|---|---|---|---|---|---|---|
| | 1a &1b* | | 2 | | 3 | |
| | t = 60 sec CQ = 0.2% | | t = 18 sec CQ = 0.2% | | t = 6 sec CQ = 0.05% | |
| | $I_t/I_o$ (s.d.) | DC (IR) | $I_t/I_o$ | DC(IR) | $I_t/I_o$ | DC(IR) |
| 1 | 18.5 (0.4) | 54.2 (1.6) | | | | |
| 5 | 20.2 | | 10.0 | 43.0 | 3.2 | 13.5 |
| 10 | 20.4 (0.3) | 60.5 (3.5) | 11.9 | 45.0 | 4.4 | 23.6 |
| 20 | 20.6 (0.3) | | 13.5 | 48.5 | 5.8 | 30.3 |
| 30 | 21.0 | | 15.0 | 50.0 | 6.7 | 36.3 |
| 60 | 21.5 (0.2) | 63.8 (0.7) | | | | |
| *60 min at 60°C | 25.3 (0.2) | 75.0 (1.8) | | | | |

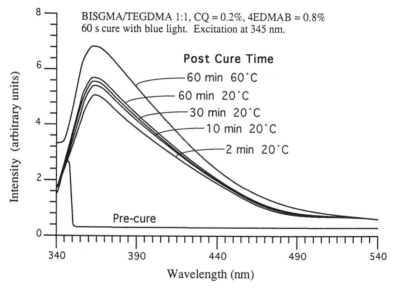

Figure 1. Fluorescence changes of 4EDMAB with post cure time of samples irradiated for 60 seconds at 20°C and at 60°C.

in Figure 3 which shows a 2.5 fold increase after 2 minutes of post cure and almost no change thereafter. Neither of the 4EDMAB isomers showed solvent quenching and thus are highly fluorescent in the uncured state. DMPT, which was only slightly fluorescent in the uncured resin, increased in intensity but only by a factor of 2 after photocuring. These results suggest the importance of both the nature and position of the aryl substituents in this type of probe. Because of the involvement of both CQ and 4EDMAB in photoinitiation, it was important to ascertain whether the change in fluorescence intensity might have been due to redox chemical reactions involving 4EDMAB and CQ. Significantly, in a thermally initiated polymerization (azobisisobutyronitrile, 24 h at 60°C) of the same resin containing only 0.8 w/w% of this amine, 4EDMAB exhibited similar fluorescence intensity changes. Therefore, photocuring is not the cause of the fluorescence behavior of 4EDMAB. In addition, an analogous unpolymerizable BIS-GMA/TEGDMA formulation containing CQ and 4EDMAB in which the vinyl groups were saturated via hydrogenation was used as a negative control. In contrast to the non-hydrogenated (polymerizable) resins, the fluorescence intensity of the hydrogenated resin did not increase after light exposure. This implies that the fluorescence changes of 4EDMAB were due to microviscosity changes related to the polymerization of the vinyl resin.

**Fluorescence changes of 4EDMAB vs IR degree of cure.** The fluorescence intensity ratio of the emission at 360 nm of the uncured, $I_{o,360}$, and cured samples at various time intervals, $I_{t,360}$, for the various curing conditions are given in Table I and are shown graphically in Figure 4. Similar measurements of DC determined by IR spectroscopy are shown in Figure 5. Assuming similar cure conditions at each time interval an empirical correlation between fluorescence and DC (IR) appears to exist (Figure 6). Thus, since the fluorescence intensity increases monotonically with the degree of cure, this emperical correlation can be used to at least estimate the degree of cure from fluorescence measurements.

## Conclusion

In addition to its role as a photoreductant in light activated dental resin systems, 4EDMAB is also a microviscosity sensitive fluorescent probe. Its fluorescence emission spectra is strongly quenched by the uncured, polar liquid resin. After fully curing the resin the fluorescence intensity of 4EDMAB increases dramatically (over 20 fold after 1 hour). The relationship between fluorescence intensity changes and DC indicates a monotonically increasing but nonlinear function.

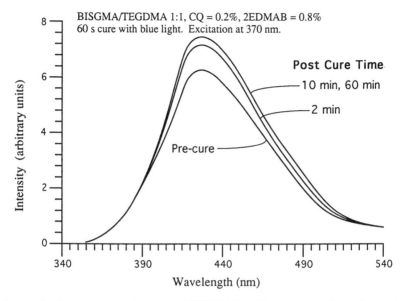

Figure 2.   Fluorescence changes of 2EDMAB with post cure time of samples irradiated for 60 seconds at 20°C.

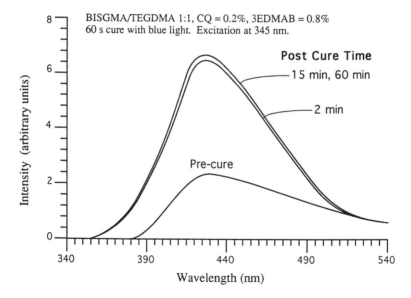

Figure 3.   Fluorescence changes of 3EDMAB with post cure time of samples irradiated for 60 seconds at 20°C.

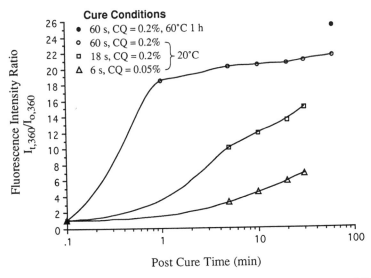

Figure 4. Fluorescence intensity ratio changes with post cure time for different curing conditions, using 4EDMAB as the photoreductant/fluorescent probe.

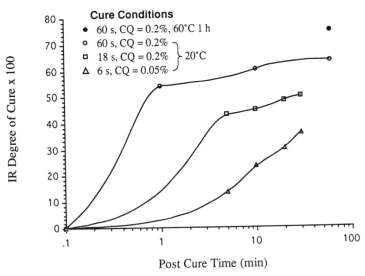

Figure 5. Infrared degree of cure changes with post cure time for different curing conditions, using 4EDMAB as the photoreductant.

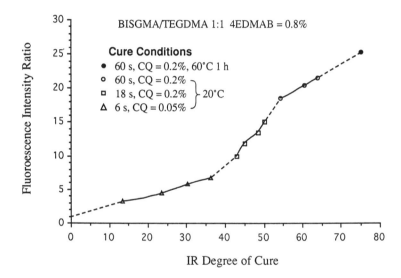

Figure 6. Correlation of fluorescence intensity ratio and infrared degree of cure, using 4EDMAB as the photoreductant.

## Acknowledgments

This work was partially supported by the National Institute of Dental Research (#Y01 DE30001). Thanks are extended to ESSCHEM CO. (Essington, PA) for providing us with samples of BIS-GMA and TEGDMA, and to Mr. Robert E. Lowry and Dr. Luanne Tilstra for their helpful discussions and to Dr. Jeffrey Stansbury for providing the hydrogenated BIS-GMA/TEGDMA.
Certain commercial materials and equipment are identified in this paper for adequate definition of the experimental procedure. In no instance does such identification imply recommendation or endorsement by NIST or that the material or equipment is necessarily the best available for the purpose.

## Literature Cited

1. Wang, F.W.; Lowry, R.E.; Fanconi, B.M. *Polymer* **1986**, *27*, 1529.
2. Sung, C.S.P.; Pyun, E.; Sun, H.L. *Macromolecules* **1986**, *19*, 2922.
3. Itagaki, H.; Horie, K.; Mita, I. *Prog. Polym. Sci.* **1990**, *15*, 361.
4. Paczkowski, J. and Neckers, D.C. *Chemtracts-Macromol. Chem.* **1992**, *3*, 75.
5. Paczkowski, J. and Neckers, D.C. *Macromolecules* **1992**, *25*, 548.
6. Paczkowski, J. and Neckers, D.C. *J. Polym. Sci.: Part A: Polym. Chem.* **1993**, *31*, 841.
7. Ledwidth, A. *Makromol. Chem. Suppl.* **1981**, *5*, 42.
8. Hayashi, R.; Tazuke, S.; Frank, C.W. *Macromolecules* **1987**, *20*, 983.
9. Loufty, R.O.; Law, Y. *J. Phys. Chem.* **1980**, *84*, 2803.

RECEIVED May 13, 1993

# BIOLOGICAL EFFECTS RELATED TO SPECIFIC PHYSICOCHEMICAL FACTORS

# Chapter 19

# Activation of Leukocytes by Arg–Gly–Asp–Ser-Carrying Microspheres

Keiji Fujimoto[1], Y. Kasuya[1], M. Miyamoto[2], and H. Kawaguchi[1]

[1]Department of Applied Chemistry, Faculty of Science and Technology, Keio University, 3–14–1 Hiyoshi, Kohoku-ku, Yokohama 223, Japan
[2]Department of Research, Central Blood Center, Japanese Red Cross, 4 Hiroo, Shibuya, Tokyo 150, Japan

It is known that a tetrapeptide, Arg–Gly–Asp–Ser (RGDS) has cell adhesion activity and plays an important role in cell–cell and cell–extracellular matrix interactions. RGDS-carrying microspheres, RGES-carrying microspheres, and parent microspheres were prepared to elucidate the RGDS-mediated activation of leukocytes. The ingestion of microspheres by leukocytes, the rate of oxygen consumption, the production of superoxide anion and hypochlorite ion and the inhibition of these products were examined. There was no difference in the number of microspheres ingested by leukocytes among the three kinds of microspheres. Leukocytes appeared specific to have an increase in oxygen consumption when they were mixed with RGDS-carrying microspheres. Moreover, this specificity was lowered by adding soluble RGDS or RGES-carrying microspheres. These findings indicate that the specific increase in oxygen consumption is caused by a signal transmission induced by immobilized RGD-integrin binding. Furthermore, it was found that the profiles of liberation of superoxide anion were similar to those of oxygen consumption. It seems probable that leukocytes are activated to produce active oxygens by interaction with RGDS peptides. In addition, specific oxygen consumption was inhibited by inhibitors of the growth of the actin filament. Immobilized RGDS may adhere to integrins and lead to the clustering of integrins and a change in the actin filament, and consequently cells may recognize the signal of activation through the actin filament. These results suggest that the microsphere with immobilized RGDS can be utilized as an activator of neutrophils.

Microspheres containing immobilized bioactive molecules are used not only in cell targetting and separation, but in cell activation as well, by taking advantage of their biospecific affinity. In this work, we focused on a tetrapeptide, Arg-Gly-Asp-Ser (RGDS), which is well known to be located within fibronectin and is able to partially substitute its function(*1*). The adhesive interaction of cells with extracellular matrix components such as fibronectin plays an important role in not only cell adhesion but also many other cellular functions, i.e.: cell morphology, growth, differentiation, and proliferation(*2*). As previously reported, immobilized RGDS played a role in mediating the adherence and activation of platelets (Kasuya,Y. et al. *J. Biomater. Sci. Polymer Edn,* in press.). Leukocytes also express a glycoprotein, "integrin", as a cell surface adhesion receptor. This glycoprotein recognizes the RGDS sequence(*3*) and affects the leukocyte's functions(*4,5*). Yet little is known about the molecular events. In this work, therefore, we tried to prepare RGDS tetrapeptides and immobilize them on the polymeric microspheres and intended to investigate the RGDS-mediated activation of leukocytes using RGDS-carrying microspheres.

Experimental

Preparation of RGDS-carrying microspheres
    Z-Arg(Mbs)-Gly-Asp(OBz)-Ser(Bzl)-OH (protected RGDS-OH) was synthesized by solution method(*6*). Substrate microspheres were prepared by soap free emulsion copolymerization of styrene, acrylamide and divinylbenzene. RGDS-carrying microspheres were prepared in a manner depicted in Figure 1. Amino groups were produced on the substrate microspheres by carrying out the Hofmann reaction (parent microspheres) and protected RGDS was coupled onto the amino groups with carbodiimide, followed by deprotection. As a control, RGES-carrying microspheres, which have no cell adhesion activity, were prepared. The amount of peptide immobilized on the microspheres was determined from the amino acid analysis of the supernatant after hydrolysis of the peptide bonds. Glycyl glycine (GG) was used as a spacer.

Evaluation of phagocytosis
    Polymorphonuclear leukocytes were isolated from fresh human blood by the Dextran-Ficoll method at 25 °C. They were suspended in phosphate buffered saline (pH 7.4) at $10^7$ cells ml$^{-1}$. Two milliliters of this suspension was mixed with 100 $\mu$ l of microsphere dispersion. The interaction between leukocytes and various microspheres was first evaluated by measuring oxygen consumption. The concentration of oxygen in the media was determined with an oxygen electrode. Soluble RGDS and RGES (Bachem Fine Chemicals Inc., USA) were used to examine the competitive reaction between soluble peptide and immobilized one. Colchicine and cytochalasin D (Cyt.D) (Sigma Chemical Co. Ltd., USA) were used to inhibit the signal transmission induced by Integrin.
    Thin cross-sections of microspheres internalized by a leukocyte were observed with transmission electron microscopy (TEM).

Evaluation of activation of leukocytes
    The luciferin- and the luminol-enhanced chemiluminescence were measured by a multichannel luminescence analyzer (BIOLUMAT LB 9505, Berthold, Wilbad, Germany) in order to determine active oxygens liberated from leukocytes. Luciferin derivative and luminol are reagents unique to superoxide anion ($O_2^-$) and hypochlorite ion ($OCl^-$), respectively.

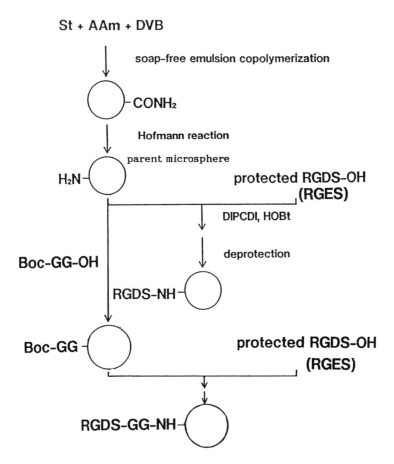

Figure 1. Preparation of RGDS-carrying microspheres. St: styrene, AAm: acrylamide, DVB: divinylbenzene, DIPCDI: diisopropylcarbodiimide, HOBt: 1-hydroxybenzotriazole.

Results & Discussion

The microspheres obtained by soap-free emulsion polymerization were monodisperse and 294 nm in diameter. The amount of amino groups on the amino microspheres was $4.3 \pm 0.4$ units $nm^{-2}$. The amount of immobilized peptide and the zeta potentials of the microspheres are shown in Table 1. Figure 2 shows the oxygen consumption by leukocytes phagocytosing parent microspheres and RGDS- and RGES-carrying microspheres as a function of incubation time. At the beginning of incubation, both RGDS- and RGES-carrying microspheres were phagocytosed to a lesser extent than parent ones. This is probably because immobilization of peptides renders microspheres hydrophilic, independent of the affinity of peptides. In the RGES-carrying microsphere system, oxygen consumption proceeded slowly and remained stable at a low level. On the other hand, leukocytes consumed oxygen very uniquely when they were mixed with RGDS-carrying microspheres. Phagocytosis was accelerated at around 5 min, gradually inclined and leveled off after 20 min (second stage). When only soluble RGDS was added to leukocytes, oxygen consumption could not be detected. These results indicate that RGDS-mediated phagocytosis takes place with a short lag and immobilization of RGDS leads to unique oxygen consumption of leukocytes. Moreover, it can be observed from Figure 2 that oxygen consumption at the second stage was suppressed by doses of free RGDS. Evidently, this unique oxygen consumption is associated with the specificity of the immobilized RGDS for integrin (VLA-5) of leukocytes.

Furthermore, it was observed from Figure 2 that the introduction of a spacer GG accelerated phagocytosis of RGDS-carrying microspheres by a factor of about 1.3. This indicates that greater flexibility in orientation of the RGDS peptide allows better recognition. But there is no denying the possibility that GG altered the conformation of RGDS so as to increase the affinity for integrin.

Transmission electron microscopic observation was performed to investigate the relation between the ingestion and the unique oxygen consumption. There was no difference in the number of microspheres ingested by leukocytes among the three kinds of microspheres. This finding strongly suggests that the unique oxygen consumption is not related to ingestion but caused by the signal transmission induced by immobilized RGDS-integrin binding.

Production of active oxygens was measured by the luciferin- and the luminol-enhanced chemiluminescence to make clear the cause of oxygen consumption at the second stage. As can be seen in Figure 3, the profiles of liberation of superoxide anion were similar to those of oxygen consumption. This suggests that leukocytes might be activated by immobilized RGDS to produce active oxygens. On the other hand, the luminol-dependent chemiluminescence revealed little difference in the time-course of active oxygen production among three kinds of microspheres as shown in Figure 4. Probably, this is because the formation of $OCl^-$ might be much more strongly affected by various enzymes released from leukocytes than by the extent of activation via RGDS-integrin binding.

The effect of inhibitors of cytoskeletons on the total amount of

Table 1. The amount of peptide immobilized on the microspheres and zeta potentials of the microspheres

| immobilized chain | immobilized peptide (unit $nm^{-2}$) | zeta potential (mV) |
|---|---|---|
| parent microspheres | – | 13.1 |
| RGDS | 4.2±0.4 | 1.7 |
| RGES | 2.6±0.2 | 14.4 |
| GG | 5.1±0.5 | 13.8 |
| RGDS-GG | 4.6±0.2 | 1.4 |
| RGES-GG | 2.9±0.1 | 13.2 |

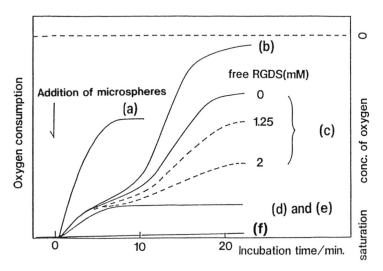

Figure 2. Phagocytosis of peptide-carrying microspheres by leukocytes. (a) parent, (b) RGDS-GG-carrying, (c) RGDS-carrying, (d) RGES-carrying, (e) RGES-GG-carrying microspheres, and (f) soluble RGDS.

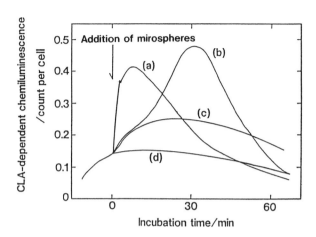

Figure 3. Time course of luciferin-dependent chemiluminescence of leukocytes stimulated by microspheres. (a) parent, (b) RGDS-GG-carrying, (c) RGES-GG-carrying microspheres and (d) absence of microspheres.

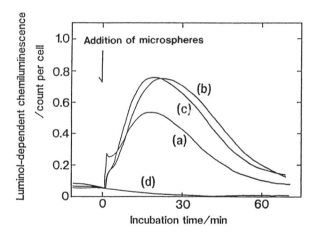

Figure 4. Time course of luminol-dependent chemiluminescence of leukocytes stimulated by microspheres. (a) parent, (b) RGDS-GG-carrying, (c) RGES-GG-carrying microspheres, and (d) absence of microspheres.

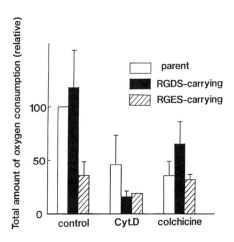

Figure 5. Effect of inhibitors on oxygen consumption of leukocytes induced by microspheres (Cyt.D: 100 $\mu$g ml$^{-1}$, colchicine: 500 $\mu$M). Error bars show standard deviations.

oxygen consumption is shown in Figure 5. Cyt.D binds to the actin filament, resulting in the inhibition of the growth of the actin filament. Figure 5 shows that oxygen consumption in the RGDS-carrying microsphere system seemed to be more effectively inhibited by Cyt.D than that of the others. However, there was no significant difference in the reduction of oxygen consumption among three kinds of microspheres when leukocytes were pretreated by colchicine, which inhibits the polymerization of microtubules. Moreover, it could be observed that the second stage of RGDS-carrying microspheres disappeared by the addition of Cyt.D. It is well known that the intracellular domains of integrins interact with proteins such as talin, vinculin, and $\alpha$-actinin. These proteins link an integrin to the actin filaments. Therefore, it seems that immobilization of RGDS leads to a change in the actin filament and that this change is related to clustering of integrins.

We imagine that specific activation of cells requires the affinity and the microenvironment according to each signal transduction, because the signal is transduced through the conformational changes and clustering of receptors[7]. As polymeric microspheres are the same size as cells, they are expected to be suitable supports for the signal transduction. In this study, an increase in activity was not seen in the presence of soluble RGDS or RGES-carrying microspheres. Moreover, the presence of microspheres alone or the presence of microspheres and soluble RGDS did not produce an increase in activity. Only the microspheres with attached RGDS produced an increase in activity. These indicate that the mechanism of the biospecific activation is synergistic. Probably, the microspheres with immobilized RGDS can adhere to neutrophils via integrins and lead to a change in the actin filament and to clustering of integrins, and consequently cells can recognize these changes as the signal of activation.

Literature Cited

1. Pierschbacher,M.D.; Ruoslahti,E. *Nature*, 1984, *309*, 30.
2. Ruoslahti,E. *Ann. Rev. Biochem.*, 1988, *57*, 375.
3. Brown,E.J.; Goodwin,J. *J. Exp. Med.*, 1988, *167*, 777.
4. Singer,I.I.; Scott,S.; Kawka,D.W.; Kazazis,D.M.; *J. Cell Biol.*, 1989, *109*, 3169.
5. Gresham,H.D.; Goodwin,J.; Allen,P.M.; Anderson,D.C.; Brown,E.J.; *J. Cell Biol.*, 1989, *108*, 1935.
6. *Principles of Peptide Synthesis,;* Bodanszky,M.Ed.; Springer-Verlag, Berlin, Heidelberg, 1984.
7. Kornberg,L.J.; Earp,H.S.; Turner,C.E.; Prockop,C.; Juliano,R.L.; *Proc. Natl. Acad. Sci. USA*, 1991, *88*, 8392.

RECEIVED July 6, 1993

Chapter 20

# Poly(vinyl alcohol) Hydrogels Prepared under Different Annealing Conditions and Their Interactions with Blood Components

Keiji Fujimoto[1], Masao Minato, and Yoshito Ikada

Research Center for Biomedical Engineering, Kyoto University, 53 Kawahara-cho, Shogo-in, Sakyo-ku, Kyoto 606, Japan

Poly(vinyl alcohol) (PVA) hydrogels obtained by annealing in the presence of glycerol exhibited a remarkable decrease in the water solubility and water content because of an increase in the crystallinity. The physical properties of the hydrogels were greatly dependent on the glycerol concentration in the PVA films during annealing. The hydrogels showed reduced protein adsorption in vitro and reduced platelet adhesion both in vitro and ex vivo, compared with that obtained without the addition of glycerol. Characterization of the surface of PVA hydrogels by contact angle and zeta potential suggested that the hydrated PVA chains existed at the surface region. Probably, as the surface coverage with the tethered PVA chains was increased by the addition of glycerol, the blood components were much more protected from direct contact with the material surface because of its steric hindrance effect.

In previous papers(1–3), we have reported that a specially prepared poly(vinyl alcohol) (PVA) film is very low in protein adsorption and platelet adhesion compared with conventional PVA and other polymers. When PVA is used for practical applications as non aqueous solutions, physical or chemical crosslinking or acetalization is introduced to PVA to make it insoluble in water. Acetalization with formaldehyde and physical crosslinking by heat treatment (annealing) are the most common among the insolubilization methods(4). However, for the minimum protein adsorption and platelet adhesion, these physical and chemical modifications have to be avoided because significant interactions with biological components are observed for such PVA, similar to the conventional polymers.

[1]Current address: Faculty of Applied Chemistry, Keio University, 3–14–1 Hiyoshi, Kohoku-ku, Yokohama 223, Japan

The purpose of the present work is to twofold: (1) to clarify why only mild heat treatment of PVA can produce a surface which greatly prevents protein adsorption and platelet adhesion and, (2) to find the optimal annealing condition for production of PVA hydrogels with low water content. In previous studies, we could not obtain high strength PVA hydrogels with low water content because, at that time, a very mild annealing condition (heat treatment at low temperature for a short time ) was chosen without paying attention to the mechanical properties. In the present study, we added glycerol to PVA films to be annealed, since Lee and Hitomi reported about 50 years ago that heat treatment of PVA in the presence of glycerol enhanced crystallization of PVA ( Lee,S.;Hitomi,K. presented at the 2nd meeting of Japanese Association for Synthetic Fiber Research, Oct. 4, 1943, Kyoto, Japan.). To our knowledge, no detailed studies on the effects of glycerol on the PVA crystallization have been reported since then. We describe here the effects of glycerol concentration and annealing temperature on the water content, crystallinity, protein adsorption, and platelet adhesion for PVA hydrogels obtained under different annealing conditions.


Experimental

Preparation of PVA Films.   Commercial PVA with a viscosity- average degree of polymerization of 1,700 and a saponification degree of 99.5 mol% was employed in this study. Glycerol was purchased from Wako Pure Chemical Inc, Osaka, Japan. Both were used as received. Distilled water was added to the PVA powder in a Pyrex flask to prepare a PVA concentration of 10 wt%. Glycerol was added to the mixture by 10-40 wt% of PVA and placed in an autoclave at 110 °C for 30 min to completely dissolve PVA. The resulting PVA aqueous solution containing glycerol was cast on Pyrex Petri dishes and dried at room temperature for a week in a clean room. The PVA films obtained with 80-120 $\mu$m thickness were annealed at temperatures from 70 to 150°C for 10 min, followed by slow cooling, unless otherwise noted. After heat treatment, the PVA films were immersed in methanol to remove glycerol at 37 °C for 48 hr and then air-dried.

Determination of the Sol Fraction and Water Content.   The dried PVA films were immersed in water at 37 °C for 72 hr to determine the sol fraction and the water content of the insoluble PVA hydrogel. They were calculated from equations (1) and (2):

$$\text{Sol fraction} = [(W_F - W_D)/W_F] \times 100 \tag{1}$$

$$\text{Water content} = [(W_S - W_D)/W_S] \times 100 \tag{2}$$

where $W_F$ is the weight of the dry PVA film after glycerol removal, $W_D$ is the weight of the insoluble hydrogel after drying in air, and $W_S$ is the weight of the hydrogel swollen in water at 37°C to equilibrium.

Crystallinity Determination.   The crystallinity of dried PVA hydrogels was evaluated by the density gradient tube method using $CCl_4$-toluene mixtures at 37°C ( *5*). The crystallinity $\chi$, defined as the weight ratio of crystalline PVA to the total PVA, was calculated from equation (3);

$$1/d = \chi/d_C + (1-\chi)/d_A \qquad (3)$$

where d is the density of the dried PVA hydrogel, $d_A$ is the density of 100 % amorphous PVA, and $d_C$ is the density of 100 % crystalline PVA. $d_A$ and $d_C$ are assumed to be 1.27 and 1.34 g cm$^{-3}$, respectively ($6$).

Contact Angle Measurement.    Dynamic contact angle measurement was performed on the dried hydrogels at 25℃ by the Wilhelmy plate technique using equipment manufactured by Shimadzu Inc. (Automated System for Dynamic Contact Angle Measurement, ST-1S type) ($7$). Water used for the measurement was purified by de-ionization after double distillation. Five measurements on different parts of the film were averaged. The movement rate of the water vessel into which the gel specimen was immersed was kept at 0.3 mm sec$^{-1}$.

Zeta Potential Measurement.    Streaming potentials of the hydrogel surface were measured at 25℃ and pH 7.4 using the cell method ($8$). The electric potentials created by streaming an electrolyte solution were recorded on an X-Y recorder. The electrolyte solution was prepared from NaOH and HCl aqueous solutions to have an ionic strength of 6x10$^{-3}$. Five measurements were averaged, and zeta potentials were calculated from the average streaming potentials using the Helmholtz- Smoluchowski equation.

Protein Adsorption.    Bovine $\gamma$-globulin (IgG, crystallized), purchased from Sigma Chemical Co., was used as received. Na$^{125}$I for protein labeling was purchased from Daiichi Pure Chemical Co. Ltd., Tokyo, Japan and the Chloramine-T method was employed for labeling ($9$). Unbound $^{125}$I was removed from the protein solution by applying to an anionic exchange resin (Dowex 1-X8, Dow Chemicals, USA). The aqueous solution of labeled IgG was stored at 4℃ and used for adsorption study within 12 hours after labeling. Protein adsorption was carried out under static conditions at a protein concentration of 1 mg ml$^{-1}$, adjusted with phosphate buffered solution (PBS, pH7.4), at 37 ℃ by immersing five pieces of hydrogel film of 1 cm x 1 cm in 2 ml of protein solution for each experiment. After a 3 hr adsorption, without taking them out of the protein the films were first rinsed with PBS solution and then blotted with tissue paper. Radioactivity of the protein-adsorbed films was measured with a gamma-ray scintillation counter. Five readings were performed on the different pieces cut from the same large film.

Platelet Adhesion.    Venous blood from healthy human volunteers, who did not receive any agents which affect platelet function, was collected with a vacuum syringe containing 5 % citric acid. The blood was centrifuged at 800 rpm for 10 min at 25℃, and the platelet-rich plasma (PRP) was withdrawn with a polyethylene pipette and placed in clean vials at room temperature. A portion of PRP was diluted by adding PBS and centrifuged at 1800 rpm for 10 min to prepare platelet pellets. The residue of the blood was further centrifuged at 3,000 rpm for 10 min to obtain platelet-poor plasma (PPP). The platelet count was determined with Coulter Counter (Type 4) and adjusted to have 150,000 platelets in 1 mm$^3$. Plasma-free platelet solution (washed platelet, WP) and PRP were prepared by adding the platelet pellets to PBS and PPP, respectively. 0.6 ml of WP or PRP was placed on each of the hydrogel films of 1.8 cm$^2$ in a vial and allowed to stand for 1 hr at 37 °C. Then, the films were vigorously washed with PBS and put into 2 ml of 0.1 M phosphate buffer (PB) containing 0.5 % Triton-X100 to lyse the adhered platelets. Lactic acid dehydrogenase (LDH) activity of the lysate was determined with an enzymatic method to count the adhered platelets with the use of a calibration curve of platelet counts(Tamada,Y., Kyoto University, Doctor Thesis, 1989.). To confirm the reproducibility, the platelet adhesion experiment was repeated five times for the same film. A Silastic film, donated by Dow Corning Cor., Ltd., Midland, USA, was used as a reference material.

Ex vivo Arterio-venous Shunt Experiment.    A plasticized poly(vinyl chloride) (PVC) tube, provided by Japan Medical Supply, Co., Ltd., Hiroshima, Japan, was coated with a thin layer of PVA containing 40 wt% glycerol, and heat treatment of this layer was done at 70 °C for 30 min followed by glycerol removal with methanol. Rabbits weighing 2-3 kg were anesthetized and their carotid artery and jugular vein were exposed surgically to make the arterio-venous (A-V) shunt. In the case of venous(V)-venous(V) shunt model, the shunting between the right and the left vein was performed (*10*). The PVA-coated PVC tube was connected to the shunt circuit with the help of a Vessel tip made of fluorine polymer (Quinton Instrument Co., Seattle, WA). After removal of the tube from the shunt, the blood-contacting surface of the tube was rinsed with 100 ml PBS at 100 cmHg and fixed with glutaraldehyde for scanning electron microscopy (SEM).


Results


Bulk Properties of PVA Hydrogels.    Figure 1 shows the effect of annealing of PVA films containing different concentrations of glycerol on the sol fraction. In the absence of glycerol, a considerable sol fraction was observed even by heat treatment at 110 °C for 30 min, whereas annealing at low temperatures such as 30°C could yield PVA gel fractions when glycerol was added. The annealing temperature required for formation of the same amount of gel became lower as the addition of glycerol increased. The water content of PVA gel fractions also decreased with the annealing temperature as shown in Figure 2. Addition of glycerol led to an abrupt decrease in water content when the weight percentage of glycerol added to PVA was above 20 %. A similar tendency was found for the crystallinity of PVA gel. The result is given in Figure 3. Clearly, the crystallinity of the PVA gel film gradually increased as the annealing temperature was higher, regardless of the amount of glycerol added. Glycerol addition always promoted the crystallization of PVA. Figure 3 shows that crystallinity changed discontinuously around Tg of PVA when glycerol was absent, although this phenomenon was not seen in the presence of glycerol probably because of the lowering of Tg. Figure 4 shows the relationship between the water content and the crystallinity of PVA hydrogel. As is seen, the effect of glycerol addition on the water content-crystallinity relationship is not remarkable, but it is apparent that the water content drastically changed in the region of low crystallinity between 0.2 and 0.3.

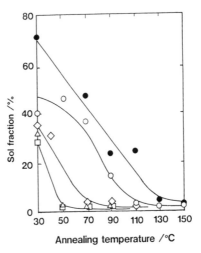

Fig.1  Effect of annealing temperature on the sol fraction of PVA films.  ●; 0 %, ○; 10 %, ◇; 20 %, △; 30 %, □; 40 %  (weight percent of glycerol to PVA).

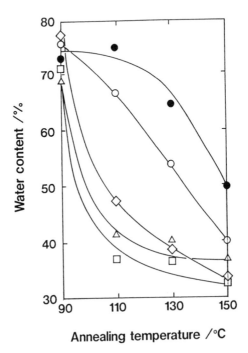

Fig.2  Effect of annealing temperature on the water content of PVA hydrogels.  ●; 0 %, ○; 10 %, ◇; 20 %, △; 30 %, □; 40 %  (weight percent of glycerol to PVA).

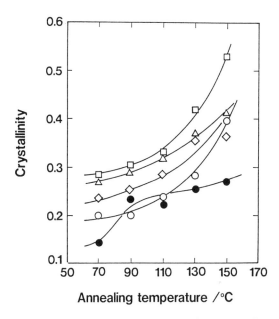

Fig.3  Effect of annealing temperature on the crystallinity of dried hydrogels. ●; 0 %, ○; 10 %, ◇; 20 %, △; 30 %, □; 40 % (weight percent of glycerol to PVA).

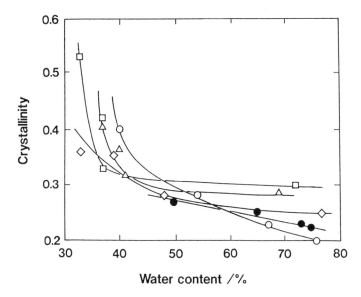

Fig.4  Relation between the water content and the crystallinity of PVA hydrogels. ●; 0 %, ○; 10 %, ◇; 20 %, △; 30 %, □; 40 % (weight percent of glycerol to PVA).

Table I   Water Contact Angles and Zeta Potentials of PVA
Annealed in the Presence of Glycerol by 40 wt% of PVA

| Annealing Temp. (°C) | Contact Angle (deg.) $\theta_a$ | $\theta_r$ | Zeta Potential (mV) |
|---|---|---|---|
| 90 | $29\pm2.5$ | $14\pm1.5$ | $-0.8\pm0.1$ |
| 110 | $35\pm3.1$ | $16\pm1.2$ | $-0.8\pm0.1$ |
| 130 | $37\pm1.8$ | $18\pm0.8$ | $-1.5\pm0.5$ |
| 150 | $42\pm3.1$ | $18\pm0.6$ | $-2.1\pm0.3$ |
| Original[a] | $36\pm1.7$ | $18\pm1.1$ | $-5.7\pm0.6$ |

a) annealed at 150 °C for 30 min without glycerol

Surface Properties of PVA Hydrogels.    Table I shows contact angles and zeta potentials of PVA hydrogels obtained by annealing in the presence of glycerol, together with those obtained in the absence of glycerol. The surface measurements were carried out for the PVA hydrogels in the water-swollen state. It is seen that the receding contact angles ( $\theta_r$ ) were very low in contrast to the advancing contact angles ( $\theta_a$ ), irrespective of the annealing temperature. Both the contact angles $\theta_r$ and $\theta_a$ slightly increased with the annealing temperature. It is also apparent from Table I that the zeta potentials are virtually zero or negative for all samples. Seemingly, annealing of PVA shifted the zeta potentials to a little higher negative values.

In vitro Protein Adsorption.    The results of IgG adsorption onto PVA hydrogels annealed at various temperatures are shown in Figures 5 and 6 as a function of crystallinity and water content of the PVA hydrogels, respectively. The amount of protein adsorbed per unit surface area, as well as platelets adhered, was calculated under the assumption that the surface of the PVA hydrogels was entirely smooth. As seen from Figure 5, IgG adsorption drastically increased as crystallinity increased in the absence of glycerol during annealing; however, it exhibited a slow increase over a wide range of crystallinity when annealed in the presence of glycerol. Protein adsorption leveled off at crystallinity above 0.4 when glycerol was added by 40 % of PVA. Furthermore, the result in Figure 6 reveals that adsorption was remarkably reduced with decreased water content for the hydrogels annealed in the absence of glycerol; whereas, reduced protein adsorption was maintained over a relatively wide range of water content when PVA films were annealed in the presence of glycerol.

In vitro Platelet Adhesion.    Figures 7 and 8 show the time course of platelet adhesion onto Silastic and PVA hydrogels when WP and PRP were used, respectively. PVA hydrogels were prepared under two different conditions: by heat treatment at 150°C in the absence of glycerol and by heat treatment at 70 °C in the presence of glycerol by 40 wt% of PVA. The number of adhered platelets was expressed as a percentage of adhered platelets onto the reference material (Silastic film).   As can be seen in

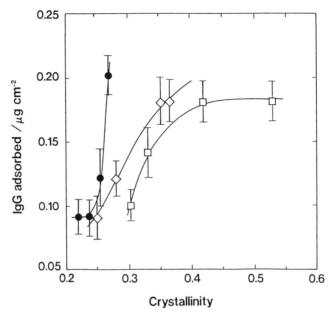

Fig.5  Dependence of IgG adsorption on the crystallinity of PVA hydrogels.  ●; 0 %, ◇; 20 %, □; 40 %  (weight percent of glycerol to PVA).

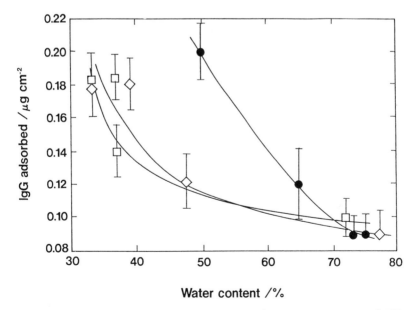

Fig.6  Dependence of IgG adsorption on the water content of PVA hydrogels.  ●; 0 %, ◇; 20 %, □; 40 %  (weight percent of glycerol to PVA).

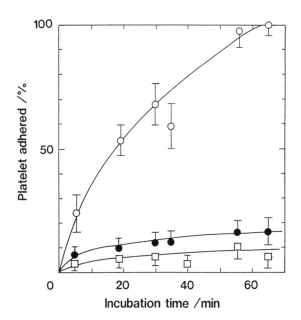

Fig.7  Effect of incubation time on the platelet adhesion to PVA hydrogels and Silastic in WP. ○; Silastic, ●; 0 % (150℃), □; 40 % (70 ℃) (weight percent of glycerol to PVA).

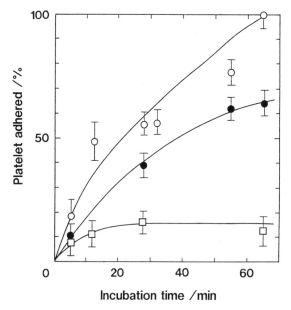

Fig.8  Effect of incubation time on the platelet adhesion to PVA hydrogels and Silastic in PRP. ○; Silastic, ●; 0 % (150℃), □; 40 % (70℃) (weight percent of glycerol to PVA).

Figure 7, platelet adhesion onto Silastic film gradually increased with the incubation time, but platelet adhesion onto PVA hydrogels always remained at low levels. Compared with platelet adhesion in WP, adhesion of platelets from PRP onto the PVA annealed in the absence of glycerol significantly increased with the incubation time as seen in Figure 8. Influences of the crystallinity and water content of PVA hydrogels on platelet adhesion in WP for 60 min are shown in Figures 9 and 10, respectively. The PVA hydrogel prepared by heat treatment at 150 °C in the absence of glycerol was selected as a reference material, to which $6.5 \times 10^3$ platelets were adhered per $mm^2$. As can be seen in Figure 9, the profile of the relation between the crystallinity and the number of platelets adhered is similar to that observed for IgG adsorption. A similar profile is also recognized in Figure 10, where platelet adhesion was found to be greatly suppressed over the wide water content, when PVA was annealed in the presence of glycerol.

Ex vivo Platelet Adhesion.    An ex vivo experiment for platelet adhesion was carried out after coating the surface of a PVC tube with PVA hydrogels. Figure 11 shows SEM photographs of the surface of the PVA hydrogel exposed to blood in the A–V shunt system for different periods of time, together with those of PVC. Platelets adhered to the virgin PVC surface gradually increased in number with the shunting time, whereas platelet deposition was not observed on the surface of PVA–coated PVC even 30 min after exposure to blood. Figure 12 shows SEM photographs of virgin PVC and PVA–coated PVC surfaces 15 min after blood exposure in the V–V shunt system where thrombus was formed more quickly than in the A–V shunt. The surface of PVC was covered with a tightly deposited platelet layer containing red blood cells in fibrin–like network, but very few adherent red blood cells could be observed on the surface of PVA–coated PVC.

Discussion

PVA molecules have one hydroxy group for each of the $-CH_2-CH-$ repeating units. This simple chemical structure allows PVA to readily crystallize to yield high strength fiber and film which are currently used in industry. Once the crystallites are destroyed, PVA becomes soluble in water even at room temperature. Because of these unique properties, several researchers have attempted to utilize PVA for biomedical materials. For instance, Peppas et al. ( *11* ) and Noguchi et al. ( *12* ) tried to use PVA as articular cartilage . Merril et al. ( *13* ) and Sefton ( *14-17* ) studied the interaction of PVA with blood components. Tomita et al. ( *18* ) investigated the possibility of using the PVA fiber as suture material. All of these PVA samples seem to have received severe heat treatment to enhance the mechanical properties, probably resulting in significant interactions with blood components.

    As shown in Figures 1 to 3, annealing in the presence of glycerol gave rise to the enhanced gel fraction and the lowered water content. This must have resulted from promotion of PVA crystallization by glycerol. It is likely that glycerol molecules present in the PVA matrix served as plasticizer to help the Brownian motion of the PVA segments for reorientation. Glycerol is known to become a solvent of PVA when the

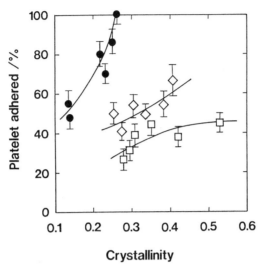

Fig.9 Dependence of platelet adhesion on the crystallinity of PVA hydrogels in WP. ●; 0 %, ◇; 20 %, □; 40 % (weight percent of glycerol to PVA).

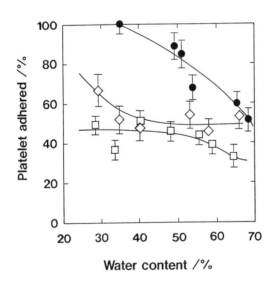

Fig.10 Dependence of platelet adhesion on the water content of PVA hydrogels in WP. ●; 0 %, ◇; 20 %, □; 40 % (weight percent of glycerol to PVA).

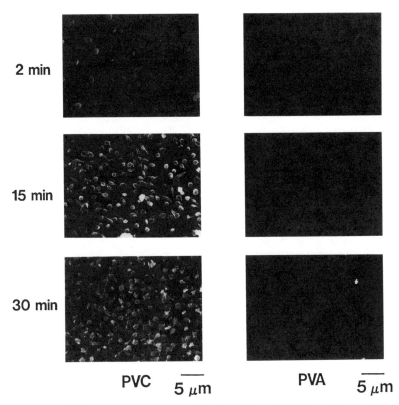

Fig.11 SEM photographs of PVA hydrogels and PVC surfaces after exposure to blood at an A-V shunt system for different times.

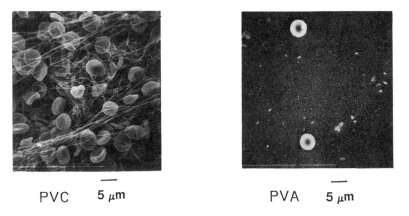

Fig.12 SEM photographs of PVA hydrogels and PVC surfaces 15 min after exposure to blood at a V-V shunt system.

(a) Extremely high water content

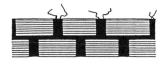

(b) Extremely low water content

(c) Medium water content
with small crystallites

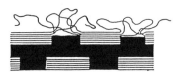

(d) Medium water content
with large crystallites

Fig.13 Assumed models of interfacial structure of a PVA hydrogels
of different water contents in contact with water.

temperature is higher than 100 ℃. It is very reasonable that the promoted crystallization of PVA brought about the increase in gel content and the decrease in water content of PVA gel even at low annealing temperatures.

Since the protein adsorption and platelet adhesion are the surface–associated phenomena, it is obvious that glycerol had also altered the surface structure of PVA hydrogel. Unfortunately, very few analytical methods are currently available for characterizing the surface structure of hydrogels in aqueous environments. One of them is the zeta potential measurement of the surface. Contact angles against water may provide information on the surface structure in contact with water. Based on the results obtained by these measurements, it has previously been concluded that there exists some semi–soluble PVA chains or loops extruding from the material surface into the outer aqueous phase (*1*). This is called a diffused surface (*2*). The results in Table I seem to support the diffuse structure model, since almost zero zeta potential means that the slipping plane is shifted toward the outer aqueous phase by the tethered, semi–soluble PVA chains. They should also interfere with protein adsorption to the surface through the steric hindrance effect. The surface structure of hydrogels with microcrystallites is schematically represented in Figure 13. It is highly possible that the number density and size of the microcrystallites are different among the materials even if their crystallinity is similar with each other.

It seems probable that larger microcrystallites are formed with the increased addition of glycerol because of its promotion effect on PVA crystallization. At the same time, wider amorphous regions should be formed with the increasing glycerol concentration, leading to a more complete covering of the surface with many semi–soluble PVA chains, when compared at a similar crystallinity. The glycerol effect is compared in the surface model of Figure 13(c) and (d). If the diffuse surface can prevent the protein adsorption, platelet adhesion would be also reduced because it is well known that platelet adhesion is promoted by adsorbed proteins such as IgG and fibrinogen. Figures 5 and 6 give evidence for this assumption.

In conclusion, it may be summarized that the annealing of PVA film in the presence of plenty of glycerol makes it possible to produce a diffuse layer on the PVA surface without any chemical modification. In addition, the resulting PVA hydrogels have low water contents such as 30 %. Finally, it should be stressed that blood interactions of PVA are greatly influenced by the history of the PVA sample preparation.

In the future, we need to accumulate more information on the phase structure of PVA hydrogel, especially in the surface region, although it is very difficult because of its highly hydrated structure.

REFERENCES

1. Ikada,Y.;Iwata,H.;Horii,F.;Matsunaga,T.;Suzuki,M.;Taki,W.;Yamagata,S.; Yonekawa,Y.;Handa,H. *J. Biomed. Mater. Res.* 1981, *15*, 697.
2. Ikada,Y. *Adv. Polym. Sci.* 1984, *57*, 103.
3. Ikada,Y.;Suzuki,M.;Tamada,Y. In *Polymers as Biomaterials*; Shalaby,S.W., Ed.; Plenum Press, N.Y., 1984, pp 51.
4. *Polyvinyl Alcohol;* Finch,C.A., Ed.; John Wiley & Sons Ltd., 1973.

5. Boyer,R.F.;Spencer,R.S.;Wiley,R.M. *J. Polym. Sci.* 1946, *1*, 249.
6. Sakurada,I.;Nukushina,Y.;Sone,Y. *Polymer Chem.* (in Japanese), 1955, *12*, 506.
7. Uyama,Y.;Inoue,H.;Ito,K.;Kishida,A.;Ikada,Y. *J. Colloid. Inter. Sci.* 1991, *41*, 275.
8. van Wagenen R.V.;Andrade,J.D. *J. Colloid. Inter. Sci.* 1980, *76*, 305.
9. *Methods in Immunology; a laboratory text for instruction and research;* Garvey,J.S.;Cremer,N.E.;Sussdorf,D.H., Eds., W.A. Benjamin Inc., 1977.
10. Fujimoto,K.;Minato,M.;Tadokoro,H.;Ikada,Y. *J. Biomed. Mater. Res.* 1993, *27*, 335.
11. Peppas,N.A.;Merrill,E.W. *J. Polym. Sci., Polym. Chem. Ed.* 1976, *14*, 441.
12. Noguchi,T.;Yamamura,T.;Oka,M.;Kumar,P.;Kotoura,Y.;Hyon,S.-H.; Ikada,Y. *J. Appl. Biomaterials,* 1991, *2*, 101.
13. Merril,E.W.;Salzman,E.W.;Wong,P.S.L.;Ashford,T.D.;Brown,A.H.; Austen,W.G. *Physiol.* 1970, *29*, 723.
14. Goosen,M.F.A.;Sefton,M.V. *J. Biomed. Mater. Res.* 1983, *17*, 359.
15. Cholakis,C.H.;Sefton,M.V. *J. Biomed. Mater. Res.* 1989, *23*, 399.
16. Cholakis,C.H.;Zingg,W.;Sefton,M.V. *J. Biomed. Mater. Res.* 1989, *23*, 417.
17. Smith,B.A.H.;Sefton,M.V. *J. Biomed. Mater. Res.* 1993, *27*, 89.
18. Tomita,N.;Tamai,S.;Shimaya,M.;Mii,Y.;Ikeuchi,K.;Ikada,Y. *Biomed. Mater. Eng.* 1992, *2*, 71.

RECEIVED May 25, 1993

Chapter 21

# Phase Transition's Control of Collagenous Tissue Growth and Resorption Including Bone Morphogenesis

David Gilbert Kaplan

4215 North Lost Springs Drive, Calabasas Hills, CA 91301

Collagenous structures of living organisms are able to respond to mechanical stress and strain with tissue growth and to the lack of mechanical stimulation with tissue resorption. Mechanical stress and strain affect collagen by driving it to the crystalline or ordered state in accordance with the phenomenon of stress induced crystallinity. The extent to which this occurs has been assessed by applying a uniaxial tensile stress to native collagen fibers. An increase of at least $4^o$ C. in melting temperature confirms that mechanical force can alter the conformation in which collagen exists. This ordering increases steric hindrance to reaction and makes collagen resistant to enzymatic degradation.

Native collagen exists in a triple helical conformation (crystalline state) which is subject to a reversible phase transition to an amorphous state[1]. The conformation of collagen is dependent on its chemical environment. Notably, urea, a product of protein metabolism, drives collagen to the amorphous state[2]. The conditions under which this phase transition of collagen occurs could also be affected by mechanical stress applied to collagen in accordance with the phenomenon of stress induced crystallinity. Stress induced crystallinity results when mechanical forces drive the molecules of an amorphous material to overcome repulsive forces and fall into configurations of decreased potential energy forming ordered states or crystals[3].
　Chemically modified collagen has been demonstrated to be subject to stress induced crystallinity[4]. Moreover, a differential activity to enzymatic degradation exists for amorphous and crystalline collagen[5][6].

0097–6156/94/0540–0243$06.00/0
© 1994 American Chemical Society

This difference in reactivity can be understood since the transitions from the helical to the amorphous state exhibits a positive volume change, the crystalline state being the more densely packed state[4]. Extra volume results in a reduction in steric hindrance to chemical reactivity and thus increased enzymatic degradation of collagen in the amorphous state compared to the crystalline state.

Extensive data exists on the temperature of the phase transition of collagen from the crystalline state to the amorphous state[2]. Yet, living organisms subject their connective tissues to stress and strain. A simple experimental procedure analagous to stressing connective tissue in living organisms could be the application of tensile stress to native tendon by means of a weight. Investigations on the influence of uniaxial force on the transition temperature of collagen would help in our understanding the mechanism by which applied mechanical force controls the physical state of native collagen and thereby the susceptibility of collagen to enzymatic degradation. An increase in the temperature at which this phase transition occurs would reflect the effects of stress induced crystallinity and support the contention that normal metabolism is based in part on the ability of collagen to maintain a crystalline structure[7] [8].

**Experimental:** Samples of collagen fibers were obtained from the tails of eight-month- old male Wistar rats (Simonsen Labs, Gilroy, California.) Fibers were removed from the distal end and washed in distilled water. Samples of approximately three centimeters in length were cut from the fibers and one end of the sample was clamped in the upper tendon fiber holder. A lead weight was crimped at a distance of 1.8 centimeters on the remaining free end. The sample was then suspended from the upper clamp in distilled water at room temperature. The temperature of the water was measured by a mercury thermometer whose bulb was placed next to the fiber sample being tested. The temperature of the distilled water was gradually raised at a rate of $1^{\circ}$ C. per minute. The sample was observed for any visual changes in physical shape including shortening of the sample or loss of opalescense to denote the onset of the phase transition[9].

**Results:** Fiber samples to which no weights had been added maintained a straight geometry until the water bath reached a temperature as low as $53.0^{\circ}$ C. Beginning at this temperature the sample fibers would begin to bend at points along its length. Such motions were sometimes repeated several times as the temperature increased. These bending motions were not associated with the visual loss of opalescence. At a temperature of as low as $60.9^{\circ}$ C., the samples shortened, increased in diameter and the opalescence was seen to give way to a translucent appearance. All of these observations are evidence of melting. All visually apparent changes in these samples were completed at 65.0° C. Samples to which weight were attached remained straight and did not begin to evidence melting until a temperature of 65.1° C. was attained. At this temperature a shortening of the sample length and an increase in sample diameter was observed as the opalescence of the crystalline state was lost. These results are presented in Table I.

TABLE I.  Effect of Temperature on Collagen Fibers

| Animal | Added Weight (g.) | Temp. at initial bending (° C.) | Temp. at initial shrinkage & loss of opalescence (° C.) | Temp. at which changes were completed (° C.) |
|---|---|---|---|---|
| A | 0 | 56.4 | 62.9 | 65.0 |
|  | 2.01 | -- | 67.0 | 67.1 |
|  | 7.64 | -- | 67.0 | 67.0 * |
| B | 0 | 53.0 | 60.9 | 62.0 |
|  | 2.29 | -- | 65.1 | 67.0 |
|  | 7.92 | -- | 66.0 | 66.0 * |

\* Sample broke

The data shows that native collagen responds to moderate uniaxial tensile stress by maintaining the crystalline state to a temperature of at least 4° C. higher than the phase transition of native collagen to which no stress has been applied.

**Discussion:**  A long-recognized phenomenon is that use of a collagenous structure by a living organism induces tissue growth, whereas disuse results in tissue resorption.  For example, Wolff's Law holds that bone, a crosslinked collagenous network into which minerals have been deposited, remodels itself to fit its function.  Unfortunately, explanations for this behavior are not compelling[10].  They are unable to rationalize the seemingly contradictory observations that collagenous tissues respond to both compressive and tensile stresses in an identical manner, both stresses inducing increased growth, and the ultimate mechanism relating bone growth and resorption to mechanical forces has not been satisfactorily elucidated.

An approach based on the traditional concepts of stress induced crystallinity and the differential enzyme activity on crystalline and amorphous collagen offers a relatively direct explanation for tissue growth and resorption.

Living organisms subject their connective tissue to mechanical stress and strain, which induces crystalline or ordered states in almost all materials. Significantly, the formation of crystalline structure is induced by this mechanism whether the stress is compressive or tensile[4]. Further, the presence of enzymes which degrade collagen provides the element required to explain the resorption of collagenous structures. When collagen is exposed to mechanical force, whether tensile or compressive, it is driven to an ordered state where it is resistant to enzymatic degradation due to the increased steric hindrance of the collagen. Should these forces be absent, the collagen reverts to a less ordered state resulting in increased degradation by enzymes and ultimately resorption.

The deposition of collagen is based on the production by the body of collagen precursors. These precursors are deposited and maintain a state of enzyme resistance by the exposure of these precursors to mechanical stress and strain. This mechanism can be used to explain the behavior of non-mineralized collagen wherein the application of mechanical stress results in the formation of such bodily protective structures as calluses. Moreover, mechanical stress and strain, i.e. exercise, can be considered to be a stimulus for the growth and maintenance of the integrity of collagen.

In summary, it is proposed that the healthy metabolism of collagen is based on the reversibility of the phase transition of collagen between the crystalline to the amorphous state. This phase transition could be the critical step in determining whether collagenous structures remain intact or are degraded. Living organisms may "recognize" amorphous collagen by degrading it while maintaining crystalline collagen in its relatively inert form. The facility of this phase transition could be an active component of wound healing and may be related to diseases such as the auto-immune disease, arthritis, periodontal disease, artherosclorosis and osteoporosis.

The loss and formation of crosslinks in the collagen matrix may control the facility of the phase transition[11]. In other words, crosslinks stabilize the crystalline state making collagen more resistant to enzymatic degradation[12]. Areas of tissue with sufficient levels of crosslinks would be stable to degradation. Should crosslinks be lost, tissue would be degraded. In this manner the formation or loss of crosslinks could be a factor in the control of the shape of collagenous tissue (including bone morphogenesis) in healthy states and could contribute to the deleterious effects of the aging of connective tissue.

The growth and development of calcified tissues can be categorized into two aspects, 1) the growth response to applied mechanical forces and

2) the formation of the innate shape of bones. In the first instance, bone response to mechanical stress and strain results in bone growth. As for example, left-handed tennis players would find their tennis playing arm to have increased bone mass over that of their non-playing side[13]. The biochemical and mechanical properties of bone are unchanged following prolonged exercise. The bones from exercised animals demonstrate significantly increased cross-sectional area. Exercise results in an increase of bone mass rather than a change in bone composition[14].

The second aspect of bone growth is the control of bones' innate shape. Murray and Huxley showed in 1925 that a fragment of the limb bud of a chick embryo grafted into the chorioallantois of another, older embryo, will develop into a recognizable femur with a head even though there is no pelvis with which it could articulate[15]. Niven showed that a patella would develop into a recognizable patella even in vitro[16]. The independence of bone growth from mechanical loading can be appreciated in the human skull. The top of the human skeleton is normally only lightly loaded due to muscle action and need be relatively thin to handle such mechanical stress and strain. The human skull is actually thick enough to withstand the occasional sharp blow to which the skull may be subjected. A mechanism must be in place for the skull to be over-designed to enable it to withstand infrequently applied forces of large magnitude. To maintain a skull of excess thickness, the reversion of collagen from the crystalline state to the amorphous state must be made less facile. Crosslinks could serve such a function. As crosslinking levels in collagen are increased, the tendency to revert to the amorphous state is decreased, allowing a designed safety margin of thickness to be built into bone structure. Conversely, the loss of crosslinks could contribute to the disease of osteoporosis and aging in general. As an animal reaches maturation and ages, crosslinks begin to be lost[17, 18]. The loss of crosslinks facilitates the reversion of collagen molecules to the amorphous state, subjecting bones to increased enzymatic degradation and loss of mechanical strength.

The study of the aging of proteins has been an active area of research, while the relationship of the degradation of biomolecules to the ultimate aging process remains elusive. The aging of protein molecules has been related to oxidation - supporting the free radical theory of aging which contends that aging is the result of the formation of free radicals that subsequently react to degrade normal tissues[19]. Recently, Stadtman has shown that the oxidation of proteins is a critical step in protein turnover and thus, the accumulation of oxidized proteins[20]. Many common proteases degrade oxidized proteins more rapidly than the native form. The degradation of endogenous proteins in liver and heart mitochondria and red blood cells is greatly accelerated by exposure of cells to oxygen generating systems[21, 22]. The principle that proteins in the crystalline state are relatively inert to reaction with chemical species has previously been stated. Oxidation of proteins would most likely form less stable crystalline lattice structures, allowing proteins to become more susceptible to enzymatic degradation. The inability to form a stable crystalline lattice structure would also interfere with proteins ability to perform their normal physiological and mechanical functions. These concepts may explain the report by Laibovich and Weiss that normal human collagen is not susceptible to pepsin degradation, while rheumatoid arthritic collagen is readily degraded by this enzyme[23].

We are well aware that the prolonged immobilization of the body as in bedrest or loss of the freedom of movement of bodily appendages results in collagenous tissue resorption[24]. A similar condition is approached in space flight. Reports are that both bone mineralization and collagen metabolism are impaired in growing animals during space flight within a few days after launch[25, 26, 27]. These results are predicted by Wolff's Law which holds that bone remodels itself to fit its function. In microgravity, bones have no supportive function and will be resorbed. We must consider that life on Earth evolved while being subjected to Earth's gravitational influence - driving proteins to ordered states in accord with the phenomenon of stress induced crystallinity. It is proposed that in the absence of the normal influence of Earth's gravitational pull during space flight, collagen molecules revert to the amorphous state and are degraded.

**Literature Cited:**

1.	Lawson, N. W., Giles, W. M. & Pierce, J. A., "Hydrothermal Shrinkage and the Ageing of Collagen," Nature, Vol. 212, 720-22, Nov. 12 (1966).

2.	Stryer, L., Biochemistry, 2nd Edition, W. H. Freeman and Co., New York, N.Y. (1981).

3.	Aklonis, J. J. & MacKnight, W. J., Introduction to Polymer Viscoelasticity, Second edition, John Wiley and Sons, New York, N.Y. (1983).

4.	Gekho, K. & Koga S., "The Effect of Pressure on Thermal Stability and In Vitro Fibril Formation of Collagen," Agri. Biol. Chem., 47 (5) 1027-1033 (1983).

5.	Cheung, D. T., Tong, D., Perelman, N., Ertl, D., & Nimni, M. E., Mechanism of Crosslinking of Proteins by Glutaraldehyde IV: In Vitro and In Vivo Stability of a Crosslinked Collagen Matrix, Connective Tissue Research, Vol. 25, 27-39 (1990).

6.	Goldberg, H. A. & Scott, P. C., "Isolation From Cultured Porcine Gingiviva Explants of a Neutral Proteinase with Collagen Teleopeptidase Activity," Connective Tissue Research, Vol. 15, 209-219 (1986).

7.	Kaplan, D. G., "Bone Response to Mechanical Forces, (A Proposed Explanation for Wolff's Law)", presented at the 198th American Chemical Society Meeting, Miami Beach, Florida, Sept. 1989.

8.  "Bone Theory #1532," Dimensions in Science, produced by Stephen M. Tansey of the American Chemical Society, Washington, D.C., July, 1990.
9.  Kaplan, D. G., Furtek, A. B., & Aklonis, J. J., Crosslinking in Collagen (rat tail tendon), Journal of the Society of Cosmetic Chemists, Vol 82, 163-169 (May/June 1987).
10. Currey, J., The Mechanical Adaptation of Bones, Princeton University Press, Princeton, New Jersey (1984).
11. Kaplan, D. G., Studies on the Viscoelastic Behavior of Connective Tissue, Ph.D. dissertation, University of Southern California, 1972.
12. Flory, P. J., and Spurr, O. K., Melting Equilibrium for Collagen Fibers under Stress. "Elasticity in the Amorphous State" J. Amer. Chem. Soc., 83, 1308 (1961).
13. Jones, H. H., Priest, J. D., Hayes, W.C., Tichenor, C.C., and Nagel, D.A., Humeral Hypertrophy in Response to Exercise. Journal of Bone and Joint Surgery 59A:204-8 (1977).
14. Woo, S. L-Y., Kuei, S.C., Amiel, D., Gomez, M.A., Hayes, W.C., White, F.C., and Akeson, W.H. , The Effect of Prolonged Physical Training on the Properties of Long Bone: A Study of Wolff's Law. Journal of Bone and Joint Surgery, 63A:780-87 (1981).
15. Murray, P.D.F., and Huxley, J.S., Self-differentiation in the Grafted Limb-bud of the Chick. Journal of Anatomy, 59:379-84 (1925).
16. Niven, J.S.F., The Development In Vitro and In Vitro of the Avian Patella. Roux' Archiv, 128:480-501, (1933).14
17. Davison, P., The Contributon of Labile Crosslinks to the Tensile Behavior of Tendons, Connective Tissue Research, Vol. 18, pp. 298-305 (1989).
18. Allain, J.S. etal, Isometric Tension Developed During Heating of Collagenous Tissues, Relationships with Collagen Crosslinking, Biochemica et Biophysica Acta, 533:147-255 (1978).
19. Harmon, D., Aging: A Theory Based on Free Radical and Radiation Chemistry, Journal of Gerontology VII, 298 (1956).
20. Stadtman, E.R., Protein Oxidation and Aging, Science, Vol. 57, 1220-1224, 28 August 1992.
21. Marcillat. O, Zhang, Y, Lin, S.W., Davies, K.J.A., Mitochondria Contain a Proteolytic System Which can Recognize and Degrade Oxidatively-Denatured Proteins, Biochemical Journal, 254, 677 (1988).
22. Davies, K.J.A. and Goldberg, A.L., Proteins Damaged by Oxygen Radicals are Rapidly Degraded in Extracts of Red Blood Cells, Journal of Biological Chemistry, 262, 8227 (1987).
23. Laibovich, J. and Weiss, J.B., Failure of Human Rheumatoid Synovial Collagenase to Degrade Either Normal or Rheumatoid Arthritic Polymeric Collagen, Biochim. Biophys. Acta, 251 (1971) 109-118.
24. Minaire, P., Meunier, P., Edourd, C., Bernard, J, Coupron, P. and Bourret, J., Quantitative Histological Data on Disuse Osteoprosis. Calcified Tissue Research 17:57-73 (1976).

25.   Mechanic, G., Arnaud, S., Boyde, A., Bromage, T., Buckendahl, R., Elliott, J., Katz, E., Durnoua, G., Regional Distribution of Mineral and Matrix in the Femurs of Rats Flown on Cosmos 1887 Biosatellite, FASEB J. 4:34-40 (1990).

26.   Patterson-Buckendahl, P., Arnaud, S., Mechanic, G., Martin, B., Grindeland, R., Cann, C., Fragility and Composition of Growing Rat Bone After One Week in Spaceflight, J. Physiol. 252 (Regulatory Integrative Comp. Physiol. 21): R240-R246 (1987).

27.   Morey, E. R. & Baylink, D. J., "Inhibition of Bone Formation During Space Flight," Science, 201, 1138-41 (1978).

RECEIVED July 30, 1993

# Chapter 22

# pH-Sensitive Hydrogels Based on Hydroxyethyl Methacrylate and Poly(vinyl alcohol)—Methacrylate

Y. J. Wang, F. J. Liou, S. W. Tsai, and G. G. C. Niu

Institute of Biomedical Engineering, National Yang-Ming Medical College, Taipei, Taiwan, Republic of China

Hydrogels, derived from the copolymerization of HEMA and PVA-MA with low concentrations of TMPTMA, were prepared. By increasing the amount of PVA-MA in the hydrogel, the water content increased, but the mechanical strength decreased instead. The PVA-MA containing polymeric hydrogel swelled in responding to the pH changes from 3 to 7. The linear swellability can be increased either by increasing PVA-MA content or by decreasing TMPTMA concentration. While both NaCl and CaCl$_2$ had little effects on the swelling, urea suppressed the pH induced gel expansion significantly.

Poly(vinyl alcohol) (PVA) is one of the hydrogels most often used in biomaterial applications.[1-4] Because of the presence of excessive hydroxyl groups, PVA contained a significant amount of water. PVA was also claimed to have good mechanical strength.[5,6] Another hydrogel, poly(2-hydroxyethyl methacrylate) (poly-HEMA), is well known for its excellent biocompatibility.[7,8] The versatile biomedical applications of poly-HEMA are demonstrated by its uses in contact lenses, vitreous humor replacements and suture materials. To explore a new formulation and other usages of these two polymers, we have copolymerized HEMA with PVA-MA (PVA esterified with maleic anhydride). the preparation and properties of this copolymer are discussed in this article.

## Experimental

**Chemicals**   Poly(vinyl alcohol), 2-hydroxyethyl methacrylate and maleic anhydride were perchased from Merck-Schuchardt (Germany). The average molecular weight of poly(vinyl alcohol) used is about 72,000. The cross-linker, 1,1,1-trimethylol propane trimethacrylate; and initiator, 2,2-diethoxyacetophenone, were obtained from TCI (Kasei, Tokyo, Japan). All chemicals used in this experiment are of reagent grade.

**PVA-MA Preparation.**   Ten grams of PVA was mixed with 22.4 g of maleic anhydride in 100 ml dimethyl sulfoxide. The esterification reaction was proceeded at 60$^{\circ}$C with continuous stirring. After a reaction period of 1, 3, 5 or 7 hours, the

0097–6156/94/0540–0251$06.00/0

reaction product was isolated by the method similar that described by Chiang et al.[9] Briefly, the resulted PVA-MA was purified by repeated solubilization in methanol and precipitation in acetone. The approximated extent of esterification was calculated by using the following equation:

$$E = m * W_a * 72,000 / W_t * W_b$$

where E is the extent of esterification,
    m is the moles of NaOH needed to titrate PVA-MA to the end point,
    $W_a$ is the total weight of PVA-MA recovered after esterification,
    $W_b$ is the total weight of PVA used for esterification,
    $W_t$ is the weight of PVA-MA used for titration.

The purified PVA-MA was dissolved in methanol and stored at $4^oC$ until use.

**Thin Films of Hydrogels.**   Thin films of poly-HEMA, poly(HEMA-co-(PVA-MA)) and poly(PVA-MA) were prepared with compositions listed in Table 1.   The crosslinking reactions of hydrogels were carried out by UV irradiation using 2,2-diethoxyacetophenone as an initiator as described previously.[10]   The Young's modulus of the gel film was determined by the slope obtained from the plot of initial stress versus strain.  The length of the gel in solution was  measured by using a caliper. The expansion of the gel was represented by $(L - L_0)/L_0$, where L is the gel length  at equilibrium and $L_0$, the original gel length. The original dimensions of the gel tested for swelling are about 1 cm long, 0.5 cm wide and 0.25 mm thick.  Each sample was measured in triplicate and the standard deviation obtained was about 15 %.

## Results and Discussion

**Esterification of PVA with maleic anhydride.**   Maleic acid has two dissociation constants with pKa values of 1.9 and 6.3.  On the other hand, PVA-MA has only one pKa (3.2) with a titration end point of pH 6.5.  The extent of esterification increased from 12 % to 40 % by increasing the reaction time from 1 hour to 3 hours.  Further increase in reaction time has no effect on the extent of esterification (Figure 1). A reaction period of 3 hours was therefore chosen in preparing PVA-MA.

**Characterization of poly(HEMA-co-(PVA-MA)) hydrogel.**   The IR spectra of Poly(HEMA-co-(PVA-MA)) is shown in Figure 2.  A strong absorption band at 1720 $cm^{-1}$, corresponding to the stretching mode of carbonyl group of ester, could be found.  In addition, a broader absorption band around 3500 $cm^{-1}$ appeared due to the presence of carboxylic acid in PVA-MA.
With a low  (0.6 %) TMPTMA content,  poly-HEMA contains about 70 % water.  By including PVA-MA (weight ratio, PVA-MA/HEMA = 1/9) in the polymeric network, the water concentration increased to 80 %.  Without the addition of HEMA, the water content of poly(PVA-MA) reached 99 %.  Apparently, the ionized carboxylic acids in the polymeric network formed significant number of hydrogen bonding with water molecules.  In addition, the negatively charged groups repelled each other, leaving space for water molecules in the polymeric network.  This is turn, weakened the mechanical strength of the copolymer.  The Young's modulus of poly-HEMA and poly(HEMA-co-(PVA-MA)) were determined to be 0.45 MPa and 0.1 MPa, respectively.  The decrease in strength is not due to the effect of PVA molecule since poly(HEMA-co-(PVA-AA)) with  a Young's modulus of 1.4 MPa  was stronger than poly-HEMA.
The existence of the negatively charged carboxyl ions in the copolymeric network structure enabled the gel to respond to pH changes.  Figure 3 shows the expansion of the hydrogel in responding to pH changes.  The gel was first equilibrated in  a citrate-

Table 1. Compositions of Poly(HEMA), Poly(HEMA–*co*–(PVA–MA)) and Poly(PVA–MA)

| Item | HEMA | PVA–MA | TMPTMA | DEAP | Weight Ratio, PVA–MA/HEMA |
|---|---|---|---|---|---|
| 1 | 100 | 0 | 0.6 | 0.4 | 0 |
| 2 | 100 | 0 | 6.0 | 0.4 | 0 |
| 3 | 99.5 | 0.5 | 0.6 | 0.4 | 0.005 |
| 4 | 97 | 3 | 0.6 | 0.4 | 0.031 |
| 5 | 95 | 5 | 0.6 | 0.4 | 0.053 |
| 6 | 90 | 10 | 0.6 | 0.4 | 0.111 |
| 7 | 90 | 10 | 2.4 | 0.4 | 0.111 |
| 8 | 97 | 3 | 6.0 | 0.4 | 0.031 |
| 9 | 95 | 5 | 6.0 | 0.4 | 0.053 |
| 10 | 90 | 10 | 6.0 | 0.4 | 0.111 |
| 11 | 90 | 10 | 7.2 | 0.4 | 0.111 |
| 12 | 90 | 10 | 9.6 | 0.4 | 0.111 |
| 13 | 0 | 100 | 0.6 | 0.4 | — |

NOTE: Number represents part (in weight) for each component. Abbreviations: HEMA is 2-hydroxyethyl methacrylate, TMPTA is 1,1,1-trimethylol propane trimethacrylate, TEA is triethanol amine, DEAP is 2,2-diethoxyacetophenone.

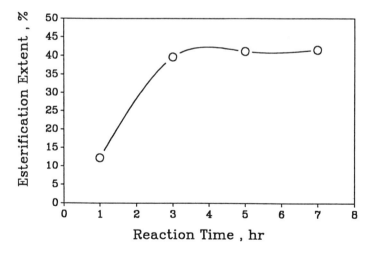

Fig. 1 The effect of reaction time on the esterification extent of PVA-MA.

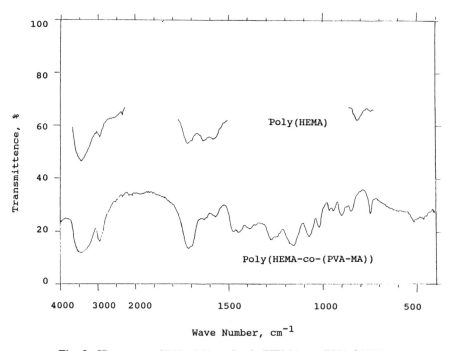

Fig. 2   IR spectra of PVA-MA and poly(HEMA-co-(PVA-MA)).

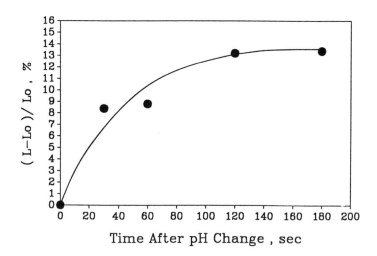

Fig. 3  The expansion of poly(PVA-MA) hydrogel versus time.  The polymer was transfered from citrate-phosphate buffer of pH 3 to that of pH 5.

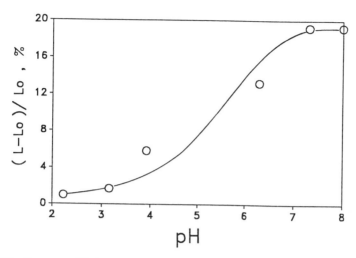

Fig. 4 The linear equilibrium swellability of hydrogel in responding to pH changes.

phosphate buffer ( 4 mM citrate and 20 mM phosphate) at pH 3.0. The gel was then removed from the solution, blotted dry and immersed into citrate-phosphate buffer at pH 7.0 . The gel expanded immediately upon transfer and reached a constant size in 2-3 minutes. The effect of pH on the linear equilibrium swellability of the hydrogels of poly(HEMA-co-(PVA-MA)) is shown in Figure 4. As the pH values of the immersion buffer increased from 2 to 7, the equilibrium length of the hydrogel increased. accordingly  The resulting plot is similar to the acid-base titration curve. The corresponding "pK" value (the pH value at which the gel swelled to half its maximal expansion length) of this pH-volume titration  curve is about 5.5, which is compatible with the $pK_2$ (5.64) of succinic acid. This data confirmed that the maleic acid was converted into succinic acid after polymerization.  Figure 5 shows that the extent of the linear equilibrium swellability increased with increasing content of PVA-MA in the hydrogel film.  Increasing the cross-linking,  on the other hand, lowered the hydrogel swellability in response to the pH changes (Figure 6).
Figure 7 showed that neither NaCl nor $CaCl_2$ affected the swellability of poly(PVA-MA) at pH 5 and below.  However, NaCl suppressed the expansion significantly between pH 5 and 7.  It has been reported that urea enhanced the pH-induced swelling of poly-HEMA hydrogel doped with methacrylic acid.[11] However, we have found that urea suppressed the expansion of poly(PVA-MA).
Other investigators have shown that the incorporation of -COOH or $-NH_2$ containing molecules into hydrogel [12,13] and nylon membrane[14] rendered the dimension of polymeric materials susceptible to pH changes. In this paper, we have demonstrated the use of PVA-MA as  the functional group in pH regulated dimension changes.  In our system, the hydrogel expanded most significantly with pH changes between 3 and 7.  This  corresponds to the pH change in the GI tract when material migrates from stomach to intestine.  Potentisal biomedical applications of this pH-responsive hydrogel include the formulation of drug designed to be released in the intestine.

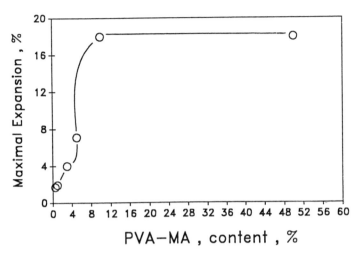

PVA–MA , content , %

Fig. 5  The effect of PVA-MA content on the maximal linear swellability of
poly(HEMA-co-(PVA-MA)). The concentration of TMPTMA is 0,6 %w.

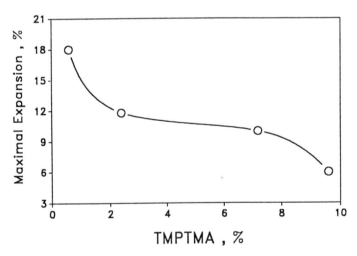

TMPTMA , %

Fig. 6  The effect of TMPTMA content on the maximal linear swellability of
poly(HEMA-co-(PVA-MA)). The weight ratio of PVA-MA to HEMA
is 0.111.

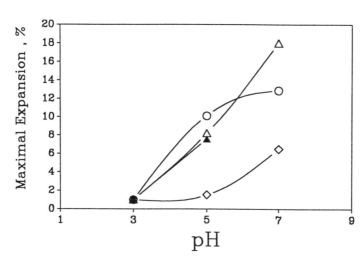

Fig. 7  The effect of NaCl, CaCl$_2$ and urea on the linear swellability of
poly(PVA-MA).  Gel expansion in citrate-phosphate buffer ( ——△—— ),
in citrate-phospate buffer containing 1 M NaCl ( ——○—— ),
in citrate-phosphate buffer containing 0.1 M CaCl$_2$ ( ——▲—— ),
and in citrate-phosphate buffer containing 1 M urea ( ——◇—— ).

References

1.  Peppas,N.A.,  Hydrogels in Medicine and Pharmacy, 1987, CRC Press,
    Cleveland, Ohio, vol. 1, 2 &3.
2.  S. D. Bruck, Biomed. Dev. Artif. Organ, vol.1, p.79, 1973
3.  B. D. Ratner, A. S. Hoffman, S. R. Hanson, L. A. Harker and J. D. Whiffen,
    J. Polym. Sci. Polym. Symp., vol.66, p.313, 1979
4.  P. nathan, E. J. Law, B. G. Macmillan, D. F. Murphy, S. H. Ronel, M. J. D'Andrea
    and R. A. Abrahams, Trans. Am. Soc. Artif. Intern. Organs, vol.22, p.30, 1976
5.  P. Molyneux, Water soluble synthetic polymers, chapter 4, CRC Press, vol.1,
    1983
6.  T. Hirai, Y. Asada, T. Suzuki and S. Hayashi, J. Appl. Polym. Sci., vol. 38,
    p. 491, 1989
7.  M. S. John and J. D. Andrade, J. Biomed. Mater. Res., vol. 7, p.509, 1973
8.  A. S. Hoffman, J. Biomed. Mater. Res., vol. 5, p.77, 1974
9.  W. Y. Chiang and C. M. Hu, J. of Appl. Polym. Sci., vol. 30, p. 3895, 1985
10. C. T. Chiang, F. J. Liou, G. C. C. Niu, Y. A. Fu and Y. J. Wang, J. of Appl.
    Biomater., vol. 1 p. 321, 1990
11. L. Pinchuk, E. C. Eckstein and M. R. Van de Mark, J. of Appl. Polym. Sci.,
    vol. 19, p. 1749, 1984
12. L. B. Peppas and N. A. Peppas, J. of Control. Rel., vol. 8, p.267, 1989
13. J. Kost, T. A. Horbett, B. D. Ratner and M. Singh, J. of Biomed. Mater. Res.,
    vol. 19, p. 1117, 1985
14. Y. Okahata, H. Noguchi and T. Seki, macromolecules, vol. 20, p.15, 1987

RECEIVED April 22, 1993

# Chapter 23

# Kinetic Model for Degradation of Starch–Plastic Blends with Controlled-Release Potential

Liu Zhang[1], John J. Harvey[2], and Michael A. Cole[1]

[1]Agronomy Department and [2]Department of Chemistry, University of Illinois, 1102 South Goodwin Avenue, Urbana, IL 61801

Starch is rapidly degraded when introduced into soil or water, but blending it with water-immiscible plastics delays its degradation and the release of the controlled compound. A kinetic model describing starch-plastic blends as potential controlled release formulations was presented. Blends of cornstarch with plastics were made as model materials and their degradation with α-amylase (E.C. 3.2.1.2) was examined. The results indicated that the kinetics of the starch degradation and release were regulated by: (a) kinetic properties of amylase, including reversible reactions and product inhibition, (b) properties of the plastics, and (c) starch content and component geometry of the polymer blend. The model is applicable to materials which yield incompatible polymer blends and in which the controlled substance is adsorbed to or covalently bonded to the starch component.

**Degradation of Starch Embedded in Plastic Matrices.** Several starch-plastic blends have been developed as degradable plastics (1). The primary purpose of making such blends was to replace plastics derived from petrochemicals with inexpensive and renewable polymers derived from plants. Since starch is readily biodegradable, its loss is the initial step in the overall degradation process of such materials (2). In some cases, the starch in the blends was not digested as easily as anticipated because of the poor accessibility to biotic and abiotic attack and the lack of connectivity of the starch granules each other. Cole (3) reported that mixing gelatinized starch with polyethylene (LDPE) + ethylene acrylic acid (EAA) greatly decreased the starch degradation rate in soil when compared to free starch. Similarly, Allenza et al. (4) reported that blends of intact starch grains with PE were relatively slowly degraded by purified enzymes.

0097–6156/94/0540–0258$06.00/0

**Starch-Based Controlled Release Pesticides.** Starch has been used as a base for controlled release pesticide formulations. Several starch encapsulated herbicides have been developed by the U.S. Department of Agriculture. The purpose of making starch encapsulated herbicides was to control their release and possibly to reduce pesticide leaching in soils. A probable limitation to the wide utility of such materials is the rapid degradation of starch in soil; complete release of starch-encapsulated atrazine and alachlor has been reported to occur in only 5 to 10 d (5). The rapid release rate is consistent with other work which indicates that starch is degraded in soil in about 28 days (6). One would prefer slower degradation and pesticide release so that full-season (160-220 d) pest control could be achieved. Rapid starch degradation is primarily the result of cell growth, not enzyme induction, and the largest microbial increase is in the soil actinomycete population (7). The combination of rapid starch digestion due to enhanced enzyme activity and a large increase in the population of a microbial group with a wide range of pesticide-degrading activities suggested that starch or its derivatives may not be a completely satisfactory carrier for controlled release formulations because the population of pesticide-degrading organisms would likely have an adverse impact on pesticide residence time, thereby decreasing the effective pest-control period.

**The Potential of Starch-Plastic Blends as Controlled Release Formulations.** Since incorporating starch into a hydrophobic matrix retards the starch degradation rate (2), it should be possible to use starch-plastic blends in lieu of the starch alone as controlled release formulation of pesticides. The release of pesticides which are either adsorbed to (Figure 1a) or covalently bonded to (Figure 1b) the starch could be controlled primarily by the rate of starch degradation.

The biodegradability of starch in the plastic matrix mainly depends on the accessibility of starch to microbes and on the connectivity of starch particles each other. Wool and Cole (8) described a simulation model based on percolation theory for predicting accessibility of starch in LDPE to microbial attack and acid hydrolysis. This model predicted a percolation threshold at 30% (v/v) starch irrespective of component geometry and other influential factors. Two critical aspects, the bioavailability and the kinetics of the starch hydrolysis in the plastic matrix must be examined before such blends could be applied as controlled release formulation of pesticides. The goal of this work was to develop a kinetic model describing the degradation and release of starch blended with hydrophobic plastics.

## THEORY

**Flux of Molecules within a Porous Plastic Matrix.** The flux of starch digestion products and enzymes in a porous plastic matrix can be treated as analogous to solute movement in a soil matrix. The transport of starch-digestion products and the enzyme in the plastic matrix and the solutes of soil solutions are both examples of solute movement in a porous body. Solute movement in soil

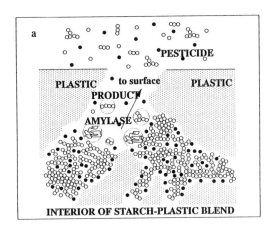

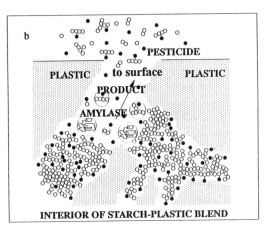

**Figure 1.** Enlargement of starch-filled pore showing space requirements for enzyme and product diffusion. Open and solid circles represent glucose and pesticide molecules, respectively. Dotted circles indicate the effective molecular radius due to molecular rotation. **a**: pesticide is adsorbed to the starch, **b**: pesticide is covalently bonded to the starch.

solution is usually governed by the combination of convection, diffusion, and hydrodynamic dispersion (9), which can be stated qualitatively as:

[combined solute flux] = [flux due to convection]
+ [flux due to diffusion]
+ [ flux due to hydrodynamic dispersion]    (1)

**Convection.** The convection (or mass flow) of water carries with it a convective flux of solutes $J_c$ which is proportional to their concentration $c$, the average apparent velocity $v$, and the water content $\theta$ in the matrix:

$$J_c = v \, \theta \, c \qquad (2)$$

$J_c$ is given in terms of mass of solute passing through a unit cross-section area of the matrix body per unit time (g $cm^{-2}$ $sec^{-1}$ ). Once the starch-plastic blend has been hydrated and the pore space occupied by starch is water-saturated (Figure 1), the convective term becomes unimportant.

**Diffusion.** Diffusion occurs when solutes are not distributed uniformly throughout a solution. Solutes tend to diffuse from where their concentration is higher to where it is lower. The rate of diffusion $J_d$ is related by *Fick's first law* to the gradient of the concentration $c$:

$$J_d = - D_d \, dc/dx \qquad (3)$$

in which $D_d$ is the diffusion coefficient for the solute diffusing in bulk water and $dc/dx$ is the concentration gradient. A diffusion gradient is established within the blend as low molecular weight products are formed during starch-amylase contact periods (Figure 1).

**Hydrodynamic Dispersion.** Hydrodynamic dispersion is a movement process of solutes to even out the concentration or composition of the solutes throughout the solution, which results from the microscopic non-uniformity of flow velocity in the matrix conducting pores. Hydrodynamic dispersion is a process that differs from diffusion in its mechanism but which tends to produce an analogous or synergetic effect to diffusion:

$$J_h = - D_h \, dc/dx \qquad (4)$$

where $J_h$ is the rate of solutes transport by hydrodynamic dispersion; $D_h$ is the dispersion coefficient; and $dc/dx$ is the concentration gradient.

**Kinetic Model for Starch Digestion and Product Release.** Before the comprehensive mathematical model could be constructed, some assumptions were made from prior research on polymer degradation (10): (a) the diffusion of both the enzyme and the products in the plastic matrix obeyed Fick's first law, (b) the diffusion coefficient was constant throughout the matrix during the reaction, (c) the hydrolytic reactions took place inside the hydrophobic plastic matrix, and (d) the reaction between the enzyme and the substrate was a modified Michaelis-Menten type (11) and the product (P), would competitively inhibit enzyme activity. The pertinent reactions and rate constants are:

$$S + E \overset{K_1}{\underset{K_2}{\rightleftharpoons}} ES \overset{K_3}{\underset{K_4}{\rightleftharpoons}} P + E$$

With these assumptions, the following differential equations are formed:

$$dC_{ES}/dt = K_1 C_E C_S + K_4 C_1 C_E - (K_2 + K_3)C_{ES} \tag{5}$$
$$dC_1/dt = K_3 C_{ES} - K_4 C_E C_1 - D(C_1 - C_2) \tag{6}$$
$$dC_2/dt = D(C_1 - C_2) \tag{7}$$

where $C_{ES}$ is the concentration of enzyme-substrate complex inside the plastic matrix, $C_1$ is the concentration of starch-digestion products in the solution inside the plastic matrix, $C_2$ is the concentration of starch-digestion products in the solution outside the matrix, $C_E$ is the enzyme concentration during steady state, $C_{Eo}$ is the initial enzyme concentration in solution, $C_S$ is the concentration of available substrate (starch), $D$ is apparent diffusion coefficient, and $t$ is time.

Integration of equations (5), (6), (7) yields the following equation:

$$P(t) = M ( 1 - e^{-rt} )[ 1/ab + e^{-at}/a(a - b) - e^{-bt}/b(a - b) ] \tag{17}$$

where P(t) is the cumulative product release from the matrix as a function of time, $t$; $M$ is a composite constant for interactions among amylase, matrix, and component geometry; $r$ is a constant to correct for the lag period during the early stages of digestion; and $a$ and $b$ are complex diffusion coefficients whose forms are:

$M = K_1 K_3 D C_S C_{Eo} V / ( K_1 C_S + K_2 + K_3 );$

$r = K_1 C_S + K_2 + K_3;$

$a = [( 2D + K_4 C_E ) - [( 2D + K_4 C_E )^2 - 4D K_4 C_E ]^{1/2} ] / 2;$ and

$b = [( 2D + K_4 C_E ) + [( 2D + K_4 C_E )^2 - 4D K_4 C_E ]^{1/2} ] / 2.$

Full derivation of the equation is provided in the Appendix.

**EXPERIMENTAL**

**Materials.** Blends of granular starch (GS) + Parafilm (PF) or polyethylene (LDPE) containing 10-70% starch (w/w) were prepared by mixing cornstarch (Sigma) with Parafilm (American National Can) or LDPE followed by compression molding. Both mixing and molding were conducted at about 150°

C. Blends of GS-PF or GS-PE containing more than 70% (w/w) cornstarch could not be made because of the low strength of the materials. Sheets of uniform 350 $\mu$m thickness were prepared in all cases and cut into 1 x 2 cm pieces for digestion.

Starch flakes (FS) were prepared by gelatinizing starch at 100° C for 5 min., spreading onto glass plates, air drying followed by 24 h drying at 50° C to produce films. Films were scraped from the plates and ground to yield flakes about 5 to 10 $\mu$m thick and with dimensions ranging from 30 to 700 $\mu$m on a side.

Flakes were blended with Parafilm (PF) as described for granular starch. Because of large volume of air-filled voids in the materials, blends containing more than 50% starch flakes (w/w) could not be made.

**Enzyme Assays.** Starch digestion from blends by porcine $\alpha$-amylase (Sigma) was determined by measuring soluble product formation by the phenol-sulfuric acid method (12). Incubations were conducted in 20 mL of 0.05 M phosphate buffer (pH 7.0) containing 8 pieces of 1 x 2 cm starch-plastic blend and 2 $\mu$L mL$^{-1}$ merthiolate to inhibit microbial catabolism of digestion products (13). Merthiolate did not inhibit enzyme activity or interfere with product assays. Sufficient enzyme was added to give a 100 to 200 units mL$^{-1}$ solution. Incubation temperature was 35° C. Mixtures were shaken at 50 to 70 rpm on a rotary shaker.

For different purposes, two sets of data were attained by using two different sampling methods: (1) Solution was replaced by fresh buffer and enzyme periodically until the starch in the blends was exhausted. Replacement was necessary to minimize the inhibition from products and progressive loss of enzymatic activity, and (2) Solution was not replaced, but 50-200 $\mu$L samples were taken from the bulk solution periodically during a 35 h incubation period.

**Soil Degradation Studies.** Soil burial studies with starch-PE blends were conducted by mixing about 100 mg of blend with 100 g of moist soil in a 250 mL flask equipped with a $CO_2$ trap (14, 15). Since PE is not biodegradable, evolution of $CO_2$ was a measure of starch degradation when corrected for $CO_2$ produced from degradation of soil organic components. Flasks were incubated at 25° C.

Amylase synthesis in soil was measured by adding 1 g of cornstarch to 100 g of moist soil and incubating at 30° C. Samples were removed periodically and assayed for amylase activity (7).

**Estimating of Parameters in the Mathematic Model.** Equation (17) was fitted to the data obtained by sampling method 2. Using a minimization technique, a least squares fit was obtained and parameters estimated for the model.
The minimization protocol was:

$$\text{minimize } f(X) = f\{ M, r, a, b, t_i \} = \sum_{i=1}^{n} [ P(t_i) - P_i ]^2, ( i = 1, 2, 3, ...n )$$

subject to    $0 < M < A$
              $0 < a < b < B$

where $t_i$ is the time after digestion started, $P_i$ is the amount of product released at time $t_i$, $A$ is a constraint constant whose value depends on the starch contents of the specific starch-plastic blends, and $B$ is an empirical constant to constrain the value of $a$ and $b$ during the fitting process. The values of $A$ were typically between 2.5 to 500 for granular starch-plastic blends, and 4.0 to 1400 for starch flakes-plastic blends. The $B$ values usually ranged between 2.0 and 8.0. A computer program, based on the Complex Method (16), was written to perform the minimization.

## RESULTS AND DISCUSSION

**Biodegradability of the Starch in the Blends.** The availability of starch to contact with amylase in starch-plastic blends is a prerequisite for starch digestion. The release rate of non-bioavailable starch granules which were completely occluded by plastic is simply zero, unless the plastic matrix is broken by other means. Hence, the discussion about the kinetics of starch degradation in the following sections refers to that portion of starch which was biodegradable.

The degradability of the starch from the starch-plastic blends is reflected by the starch digestion threshold (17). A critical point which represents the digestion threshold occurs when plotting the percentage of released product as a function of percentage of starch in the starch-plastic blends. From such a plot (Figure 2), one can see that the starch digestion threshold is determined primarily by starch content of the blends. However, other work by the present authors (L. Zhang and M. A. Cole, in preparation) indicated that substantial variations in threshold values can be obtained by varying starch geometry, using plastics with different physical properties, chemical properties, and mechanical strength, and using different preparation methods for the blends. For example, the data in Figure 2 indicate that the digestion threshold of starch flakes-Parafilm blends (FS-PF) was much lower than that of granular starch-Parafilm blends (GS-PF).

**Kinetic Model for Starch Degradation.** Because of the difficulties of measuring degradation of complex mixtures in soil, degradation studies with purified amylase were done to determine the primary factors controlling the degradation rate of starch in blends with hydrophobic materials. Mixtures of granular starch and Parafilm (GS-PF) were chosen to test the model because PF is relatively easier to process than polyethylene (PE), and PF has much less "skin effect" (17) which severely reduces the biodegradability of the starch in PE-containing blends (L. Zhang and M. A. Cole, in preparation).

**Correlation of Model Parameters to Degradation of Starch in the Blends.** Sensitivity analysis of Equation (17) components indicated that degradation rate was very sensitive to enzyme properties ($K_1$, $K_2$, $K_3$, $K_4$) and concentration ($C_{Eo}$, $C_E$), starch content ($C_S$), and matrix factors such as starch geometry, apparent diffusion coefficient ($D$), and properties of the non-starch component which are represented by the model parameters $M$, $r$, $a$, and $b$ in Equation (17). An example of the impact of two factors, starch content and properties of the non-starch component, is the difference in degradation threshold for starch-PE blends of about 55% (v/v) starch and only 20% (v/v) starch in starch-PF blends (Figure 2). The only difference between these two kinds of blends is that they are made from plastics with different mechanical properties and nearly identical elemental compositions. Enzyme concentration was one of the major factors identified as highly influential. With an insoluble substrate and a slowly diffusing catalyst, the probability of contact between substrate and enzyme will be controlled primarily by enzyme concentration. The diffusion rate of enzyme and starch digestion products had little effect on the rate of product formation in high starch blends. Product diffusion rate was influential only at low starch contents, probably because of the "skin effect" and the small path dimensions in low percentage starch materials. The results suggested that amylases of different origin and similar kinetic properties, but with different molecular weights, should be equally effective catalysts. Therefore, the specific microbe in soil which is degrading the starch may not have a major impact on degradation rate.

Experimentally, the initial rate of starch hydrolysis was a function of starch content for both granular starch (Figure 3a) and starch flakes (Figure 3b) in the blends. The use of starch flakes decreased the starch content needed for rapid degradation to occur from 20% (v/v) for granular starch to less than 13% (v/v) for starch flakes (Figure 2). The time required for exhaustion of available substrate was significantly less with starch flakes than with granular starch, with complete digestion achieved in 8 hours for starch flakes compare to 48 hours for granular starch. Overall, the data indicate that substantial control of starch degradation rate is possible by varying the geometry of the starch component and the starch content of the blend. Such control makes it possible to produce blends with high rate or low release rate of pesticides associated with the starch.

**Estimation of Parameters for the Mathematic Model.** Estimated values of the equation parameters for the model are showing in Table I and Figure 4. The estimated parameter $M$ shows good agreement with the starch content and geometry in the blends and with the probability of contact between substrate (starch) and enzyme. The apparent diffusion coefficient $D$ was calculated from the parameters $a$ and $b$. The results indicate that the high values of estimated apparent diffusion coefficient $D$ correspond to low starch content in the blends (Figure 4). The probable reason for this phenomenon is that in very low starch content blends, only those starch granules on the surface of materials were biodegradable (L. Zhang and M. A. Cole, in preparation). The products formed from starch granules on the surface of low starch content materials were not subject to diffusional limitations, because they were digested and dispersed

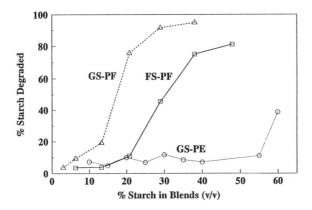

**Figure 2.** Efficiency of starch removal as a function of starch content. Triangles: starch flakes-Parafilm blends, Squares: starch granules-Parafilm blends, Circles: starch granules-polyethylene blends.

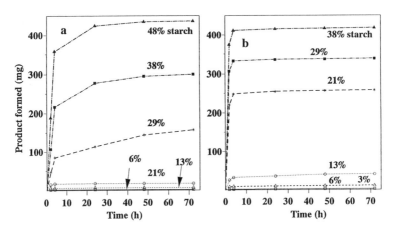

**Figure 3.** Kinetics of starch digestion from **a:** starch granules + Parafilm blends, and **b:** starch flakes + Parafilm blends. Numbers associated with lines indicate starch content (v/v) in blends. Results are typical of three separate experiments.

Table I. Parameters Estimation for the Mathematic Model

Granular starch-Parafilm blends

| Parameter | Percentage of granular starch in blends (v/v) | | | | | |
|---|---|---|---|---|---|---|
| | 6.4 | 13.3 | 20.8 | 29.0 | 38.0 | 47.9 |
| $M$ | 2.9 | 11.5 | 32.5 | 123.5 | 287.0 | 506.0 |
| $a$ | 0.3479 | 0.2269 | 0.1712 | 0.0717 | 0.0667 | 0.0637 |
| $b$ | 11.5944 | 7.5610 | 5.7064 | 2.3906 | 2.5058 | 2.1242 |
| $D$ | 5.612 | 3.660 | 2.762 | 1.158 | 1.077 | 1.026 |

Starch flakes-Parafilm blends

| Parameter | Percentage of starch flakes in blends (v/v) | | | | | |
|---|---|---|---|---|---|---|
| | 3.1 | 6.4 | 13.3 | 20.8 | 29.0 | 38.0 |
| $M$ | 3.3 | 18.1 | 69.3 | 612.5 | 1102.3 | 1360.5 |
| $a$ | 0.3438 | 0.2837 | 0.1139 | 0.0710 | 0.0703 | 0.0672 |
| $b$ | 9.8677 | 6.8023 | 2.8721 | 2.4343 | 2.3897 | 2.3942 |
| $D$ | 4.749 | 3.246 | 1.374 | 1.179 | 1.157 | 1.161 |

directly into the surrounding solution. On the other hand, hydrolysis products in high starch content materials had a long way to move from within the hydrophobic matrix to the surrounding solution, which resulted in low $D$ values. The diffusion coefficient is proportional to cross-section area of the diffusion channel and inversely proportional to the length of diffusion path. This deduction is supported by the evidence that the $D$ values of GS-PF blends of higher than 40% (w/w) starch are almost constant (Figure 4) and the turning points of apparent $D$ values correspond to the starch digestion threshold in Figure 2. Peanasky (18) also concluded that the enzyme molecules did not need to diffuse into the low starch content blends to react and the starch digestion products did not need to diffuse out of the blends to be assayed. Since the data in Figure 5a and 5b show that the lag time was very short, the term, $1 - e^{-rt}$, was ignored in the best fitting process.

The mathematic model provides a reliable means to predict the starch release kinetics of controlled release formulations made from starch plastic blends. The predicted release of starch digestion products from the blends as a function of time was plotted in Figure 5a and Figure 5b. Correspondence between model-derived values and experimental results was excellent in nearly all cases.

**Relationship between Kinetic Model and Soil Degradation Results.** When starch is added to soil, amylase activity increases within 2 days and rises rapidly (Figure 6). Based on 5 day activity in Figure 6, starch was being degraded as about 3.1 g $d^{-1}$ kg soil$^{-1}$, and all compounds adsorbed to the starch would be released in only a few days. In contrast, the maximum starch degradation rate of the blends shown in Figure 7 was about 8 mg starch degraded $d^{-1}$ kg soil$^{-1}$. Since starch degradation rate in blends was controlled in part by enzyme concentration, large increases in soil amylase activity are undesirable because starch degradation would be dramatically increased. Since the starch availability is low in the blends, relatively little response of the soil population to starch blends was expected.

If starch is blended with hydrophobic components like polyethylene and ethylene acrylic acid (19) and processed to yield a laminate structure similar to starch flakes + Parafilm blends, degradation of the starch during burial is still relatively rapid (Figure 7). The degradation kinetics were not exponential as seen in Figure 6, which indicated that the soil microbial population did not increase as it did when free starch was added to soil (Figure 6). The time required to achieve full starch degradation was increased as starch content increased. Release of bioactive compounds such as pesticides adsorbed to the starch would be regulated primarily by the rate of starch digestion and the data indicate that nearly linear release over a 21 to 32 d period would be possible.

The degradation threshold for starch + PE + EAA blends was lower than the threshold for starch flakes + Parafilm blends. In both materials, a higher percentage of the starch was accessible to enzymatic attack at about 20% (v/v) starch. The 21% value is similar to conductivity thresholds reported for electrically conductive plastic-metal blends (20). The similarity in degradation

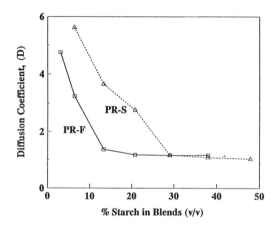

**Figure 4.** Estimated apparent diffusion coefficients as a function of starch content. Triangles: starch granules-Parafilm blends, Squares: starch flakes-Parafilm blends.

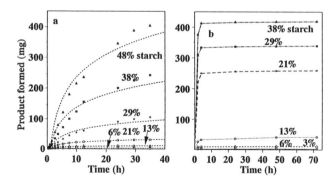

**Figure 5.** Comparison of product release based on the mathematical model (dashed lines) and experimental data (symbols). **a:** starch granules + Parafilm blends, and **b:** starch flakes + Parafilm blends. Numbers associated with lines indicate starch content (v/v) in blends.

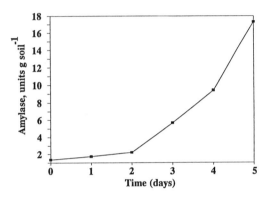

**Figure 6.** Amylase synthesis in starch-amended soil. Each data point is the mean of six analyses. One unit of amylase catalyzes formation of 1 $\mu$mol reducing sugar $d^{-1}$ g soil$^{-1}$.

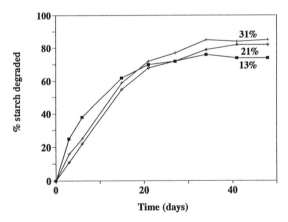

**Figure 7.** Degradation of starch from granular starch + PE + EAA blends during soil burial.

behavior between soil and purified amylase systems suggests that the proposed kinetic model and digestion of blends with purified enzymes can serve as simple and useful means of assessing the degradation rate of potential controlled release materials containing starch or other enzymatically hydrolyzable polymers.

## CONCLUSIONS

The work presented here constructed a general model of starch-plastic blends as potential controlled release formulations. This model provided a practical method of predicting the kinetics of the starch digestion and product release from starch-plastic blends, thus the kinetics of pesticide release is predictable if the pesticides are either adsorbed or covalently bonded to the starch. The model was developed for starch-plastic blends. It should be adaptable to other blends of incompatible polymers, so long as one of the polymers is susceptible to enzymatic hydrolysis.

**APPENDIX: DERIVATION OF THE MATHEMATIC MODEL**

Based on the mechanisms of enzyme hydrolysis of starch and the diffusion processes of enzyme and the starch digestion products in the plastic matrix, the following assumptions were made: (a) the diffusion of both the enzyme and the products in the plastic matrix obeys the Fick's first law, (b) the diffusion coefficient is constant throughout the matrix during the reaction, (c) the hydrolytic reactions take place only inside the hydrophobic plastic matrix, and (d) the reaction between the enzyme and the substrate is a modified Michaelis-Menten type and the product (P), will competitively inhibit the enzyme activity:

$$S + E \underset{K_2}{\overset{K_1}{\rightleftharpoons}} ES \underset{K_4}{\overset{K_3}{\rightleftharpoons}} P + E$$

With these assumptions, the following differential equations are formed:

$$dC_{ES}/dt = K_1 C_E C_S + K_4 C_I C_E - (K_2 + K_3)C_{ES} \tag{5}$$
$$dC_I/dt = K_3 C_{ES} - K_4 C_E C_I - D(C_I - C_2) \tag{6}$$
$$dC_2/dt = D(C_I - C_2) \tag{7}$$

where $C_{ES}$ is the concentration of enzyme-substrate complex inside the plastic matrix, $C_I$ is the concentration of starch-digestion products in the inside matrix solution, $C_2$ is the concentration of starch-digestion products in the outside matrix solution, $C_E$ is the enzyme concentration during steady state, $C_{Eo}$ is the initial enzyme concentration in solution, $C_S$ is the concentration of available substrate (starch), $D$ is apparent diffusion coefficient, $t$ is time.

Equation (5) can be rewritten as:

$$dC_{ES}/dt = (K_1 C_S + K_4 C_1)C_E - (K_2 + K_3)C_{ES} \qquad (8)$$

Since $K_1 C_S >> K_4 C_1$ at the beginning of the reaction, $K_4 C_1$ can be ignored and $C_E = C_{Eo} - C_{ES}$, Equation (8) is changed to:

$$dC_{ES}/dt = K_1 C_S (C_{Eo} - C_{ES}) - (K_2 + K_3)C_{ES} \qquad (9)$$

Integrating Equation (9), we get:

$$C_{ES} = [(K_1 C_S C_{Eo})/(K_1 C_S + K_2 + K_3)] (1 - e^{-rt}) \qquad (10)$$

where $r = K_1 C_S + K_2 + K_3$

Since the enzyme concentration in the reaction solution does not change significantly during the steady state of the hydrolysis process, $C_E$ can be treat as a constant. Equation (6) and Equation (7) can be integrated with Laplace Transform Method (21). The boundary conditions are: $C_1 = 0$ and $C_2 = 0$ at $t = 0$. Thus, the Laplace Transforms of Equation (6) and Equation (7), respectively, are:

$$C_1(s) = [(K_3 C_{ES}/s + D C_2(s)]/(s + K_4 C_E + D) \qquad (11)$$
$$C_2(s) = D[C_1(s) - C_2(s)]/s \qquad (12)$$

where $s$ is Laplace operator.

Substituting for $C_1(s)$ from Equation (11) into Equation (12) and rearranging yields:

$$C_2(s) = K_3 D C_{ES}/[s(s^2 + (2D + K_4 C_E)s + D K_4 C_E)] \qquad (13)$$

Rewrite Equation (13):

$$C_2(s) = K_3 D C_{ES}/[s(s + a)(s + b)] \qquad (14)$$

where $a = [( 2D + K_4 C_E ) - [( 2D + K_4 C_E )^2 - 4D K_4 C_E ]^{1/2} ] / 2$
$b = [( 2D + K_4 C_E ) + [( 2D + K_4 C_E )^2 - 4D K_4 C_E ]^{1/2} ] / 2$

The inverse Laplace Transform of Equation (14) is:

$$C_2(t) = K_3 D C_{ES} [1/ab + e^{-at}/a(a-b) - e^{-bt}/b(a-b)] \qquad (15)$$

Substituting for $C_{ES}$ from Equation (10) into Equation (15) yields:

$$C_2(t) = M_1 (1 - e^{-rt})[1/ab + e^{-at}/a(a-b) - e^{-bt}/b(a-b)] \qquad (16)$$

where $M_1 = K_1 K_3 D C_S C_{Eo} / (K_1 C_S + K_2 + K_3)$

Multiplying Equation (16) by V, the volume of the surrounding solution, we get:

$$P(t) = M (1 - e^{-rt}) [1/ab + e^{-at}/a(a-b) - e^{-bt}/b(a-b)] \qquad (17)$$

where $M = K_1 K_3 D C_S C_{Eo} V / (K_1 C_S + K_2 + K_3)$, and P(t) is cumulative products released as a function of time, $t$.

## ACKNOWLEDGEMENT

This work was supported in part by a grant from the Illinois Corn Marketing Board.

## LITERATURE CITED

1. Aeschleman, P.E., *Microbial Colonization of Degradable Plastic Films*, M.S. Thesis, University of Illinois at Urbana-Champaign, September, 1990.
2. Iannotti, G.; Fair, N.; Tempesta, M.; Neibling, H.; Hsieh, F-H.; and Mueller, R. *Corn Utilization Conference III Proceedings*, National Corn Growers Association, St. Louis, Missouri, June 20-21, 1990.
3. Cole, M.A. In *Agricultural & Synthetic Polymers: Biodegradability & Utilization*, Glass, J.E.; Swift, G., Eds., ACS, Washington, DC, 1990, pp 77-95.
4. Allenza, P.; Schollmemeyr, J.; Rohrback, R.P. In *Degradable Materials: Perspectives, Issues, and Opportunities*, Barenberg, S.A.; Brash, J.L.; Narayan, R. and Redpath, A.E. Eds.; CRC Press, Boca Raton, Florida, 1990, pp 357-379.
5. Wienbold, B.J.;Gish, T.J. *J. Environ. Qual.* ,1992,*21*,382.
6. Cheshire, M.V.; Mundie, C.M.; Shepherd, H. *Soil Biol. Biochem.*, 1969, *1*,117.
7. Cole, M.A. *Appl. Environ. Microbiol.*, 1977, *33*, 262.
8. Wool, R.P.; Cole, M.A. In *Engineered Materials Handbook: Engineering Plastics*, ASM International Press, Metals Park, OH, 1988; Vol. 2, pp 783-787.
9. Hillel, D. *Fundamentals of Soil Physics;* Academic Press, 1980, pp 233-245.
10. Suga, K.; Dedem, G.Van; Moo-Young, M. *Biotechnology and Bioengineering*; John Wiley & Sons Press, New York, 1975, Vol. 27, pp 185-201.
11. Stryer, L. *Biochemistry*; 3rd. Edition; W.H. Freedman & Company: New York, NY, 1988, pp 187-195.
12. Dubois, M.; Gilles, K.A.; Hamilton, J.K.; Rebers, P.A.; Smith, F. *Analy. Chem.*, 1956, *28*, 350.
13. Pengra, R.M.; Cole, M.A.; Alexander, N.*J. Bacteriol.*, 1969, *97*, 1056

14. Loos, M.A.; Kontson, A.K.; Kearney, P.C. *Soil Biol. Biochem.*, **1980**, *12*, 583.
15. Cole, M.A. *Abstr. annu. Mtgs. Amer. Soc. Microbiol.* **1989**, *89*, 364.
16. Rao, S.S. *Optimization*, John Willey & Sons, New York, 1984, pp 345.
17. Wool, R.P.; Peanasky, J.S.; Long, J.M.; and Goheen, S.M., In *Degradable Materials: Perspectives, Issues, and Opportunities*, Barenberg, S.A.; Brasch, J.L.; Narayan, R. and Redpath, A.E. Eds., CRC Press, Boca Raton, FL, 1990, pp 515-544.
18. Peanasky, J.S. *Percolation Effects in Degradable Polyethylene/Starch Blends*, M.S. Thesis, University of Illinois at Urbana-Champaign, May, 1990.
19. Otey, F.; Westhoff, R.P.; Doane, W.M. *Ind. Eng. Chem. Prod. Res. Dev.* **1980**, *19*, 592.
20. Reboul, J.-P. In *Inorganic Fillers*; Lutz, J. T., Jr., Ed. Marcel Dekker, Inc. NY, 1989, pp 255-280.
21. Charles, E.R., Jr. In *Ordinary Differential Equations*, Prentice-Hall Press, Englewood Cliffs, NJ, 1979, pp 256.

## NOMENCLATURE

$A$:    restriction constant in best fitting performance
$B$:    restriction constant in best fitting performance
$a$:    constant defined by Equation (14)
$b$:    constant defined by Equation (14)
$C_E$:    enzyme concentration during steady state
$C_{Eo}$:    initial enzyme concentration in solution
$C_{ES}$:    concentration of enzyme-substrate complex inside the plastic matrix
$C_S$:    concentration of available substrate (starch)
$C_1$:    concentration of starch-digestion products in the solution within matrix
$C_2$:    concentration of starch-digestion products in the solution outside the matrix
$D$:    apparent diffusion coefficient
f(X):    objective equation for least square fitting
$K_1$:    reaction rate constant in Michaelis-Menten equation
$K_2$:    reaction rate constant in Michaelis-Menten equation
$K_3$:    reaction rate constant in Michaelis-Menten equation
$K_4$:    reaction rate constant in Michaelis-Menten equation
$M$:    constant defined by Equation (17)
$M_1$:    constant defined by Equation (16)
$n$:    number of experimental data
$P_i$:    experimental data of cumulative products released right after $i$th sampling
$P(t)$:    cumulative products released as a function of time $t$
$r$:    constant defined by Equation (10)
$s$:    Laplace operator
$t$:    time of sampling

RECEIVED June 11, 1993

# Chapter 24

# Cross-Linking and Biodegradation of Native and Denatured Collagen

K. Tomihata, K. Burczak[1], K. Shiraki, and Yoshito Ikada

Research Center for Biomedical Engineering, Kyoto University, 53 Kawahara-cho, Shogo-in, Sakyo-ku, Kyoto 606, Japan

Collagens are known to have wide biomedical applications. They are for hemostatic agent, blood vessel, heart valve, tendon and ligament, burn dressing, intradermal augmentation, drug delivery systems and so on (1). The denatured type collagen, gelatin, has also been used in medicine as plasma expander, wound dressing, adhesives, and absorbent pad for surgical use (2). Recently it has been reported that both collagen and gelatin can be effectively used as materials for nerve conduits (3,4). Furthermore, it should be noted that crosslinked gelatin has been long used as a material for embolization in endovascular surgery (5). The biodegradation rate of collagen-based implants is known to be strongly affected by the crosslinking density (6-8). Crosslinking can also improve the tensile properties of the materials to such a level as they are handled and sutured with ease. The purpose of this work is to compare different methods for crosslinking of native collagen and denatured gelatin. The crosslinking agents to be employed in this study include glutaraldehyde, water-soluble carbodiimide, bisepoxy compounds, UV radiation, and dehydrothermal treatment.

## Experimental

**Materials.**    Soluble collagen [ type I (70%) + type III (30%) ] was obtained from pig skin by HCl extraction. Gelatin ( type I ) which was extracted from bovine bone by alkaline process was supplied by Nitta Gelatin Co. Ltd, Osaka, Japan and used as received. The average molecular weight of gelatin was about 100,000.

**Film preparation.**    Collagen was dispersed in HCl aqueous solution of pH 3.0 to have a concentration of 1 wt%. The dispersion was cast on a Petri dish and allowed to dry at atmospheric pressure and 25°C to obtain collagen film of about 0.1mm thickness.    Gelatin was dissolved in distilled water of pH 5.5 at 40°C to have a concentration of 10 wt%. For the film preparation, the solution was cast on a Petri dish and allowed to dry in air at 25°C to obtain gelatin film of about 0.3mm thickness.

[1]Corresponding author

0097–6156/94/0540–0275$06.00/0

**Glutaraldehyde (GA) crosslinking.**        Collagen or gelatin films were immersed at 4°C in a GA solution from phosphate buffered saline (PBS, pH 7.4) or double distilled water (DDW, pH 6.0), removed from the solution after 24 hr, rinsed with distilled water, and then dried.

**Water-soluble carbodiimide (WSC) crosslinking.**        1-Ethyl-3-(3-dimethylaminopropyl) carbodiimide hydrochloride was used as WSC. Collagen or gelatin films were immersed in PBS solution of WSC at pH 7.4 and 4°C for 24 hr, rinsed with distilled water, and then dried.

**Bisepoxy compound crosslinking.**        As bisepoxy compounds, a commercial product Denacol was used which is poly(ethylene glycol) diglycidyl ether with the chemical structure as

$$CH_2CHCH_2-O-(CH_2CH_2O)_n-CH_2CHCH_2$$
$$\diagdown O \diagup \qquad\qquad\qquad\qquad \diagdown O \diagup$$

The n value of the bisepoxy compounds used in this study is 1 (EX-810) and 9 (EX-830). They were obtained from Nagase Chemical Co., Osaka, Japan and used without purification. After the aqueous solution of Denacol was added to 10 wt% gelatin solution at 40°C, the mixture was poured into a Petri dish following vigorous mixing and allowed to dry at 25°C.

**Crosslinking by UV irradiation.**        Collagen or gelatin films were placed 10 cm apart from a short wave (254 nm) UV lamp (National GL-15 (15W), Matsushita, Osaka, Japan). The films were irradiated in air at room temperature.

**Crosslinking by dehydrothermal treatment (DHT).**        Collagen or gelatin films were placed in a vacuum oven and evacuated to 0.1 torrs. The temperature within the vacuum oven was kept at 110 to 160°C. The heat treatment of films was allowed to proceed for 24 hr.

**Measurement of the water content of crosslinked gels.**        Crosslinked gel films were immersed in PBS solution of pH 7.4 at 37°C for 20 hr. After the weight of wet gels was measured, they were placed in a vacuum oven for 6 hr at 60°C and 0.1 torrs and the weight of dried gels was measured. The water content (water uptake + water moisture ) of the crosslinked gels was calculated from the following equation:

Water Content = [ (wet weight - dry weight ) / wet weight ] x 100 (%)

**In vitro degradation with collagenase.**        Bacterial collagenase from *Clostridium histolyticum* was used for in vitro degradation of gels. Crosslinked collagen or gelatin gels with the known weight were immersed in collagenase solution containing $Ca^{2+}$ with the concentration of 40 units/ml in 0.05M Tris buffer at pH 7.4 and 37°C. After determined intervals of time the samples were taken out, rinsed with distilled water, dried, and weighed. The extent of the in vitro degradation was calculated as the percentage of the weight of the dried gel after and before collagenase treatment.

**In vivo degradation by subcutaneous implantation.**        Crosslinked collagen or gelatin films with the known weight were sterilized with ethylene oxide gas prior to implantation, rinsed with sterilized PBS, and implanted subcutaneously in Wistar rats. After determined intervals of time the animals were sacrificed and the samples were explanted to weigh after drying. The extent of in vivo degradation was calculated as percentage of the weight of dried gels after and before implantation, similar to degradation with collagenase.

## Results

Figure 1 shows the results of crosslinking by GA which is the most commonly used crosslinking agent for collagen and gelatin. As is apparent from Figure 1, gelatin became water-insoluble as a result of crosslinking with GA. The water content of the swollen gelatin gel was decreased from 100 to 60%. On the contrary, the reaction of collagen with GA did not induce any significant change in the water content. Collagen is not water-soluble at pH 7 and room temperature and has a water content of about 70% even without any chemical crosslinking because of its triple helix structure. Although the water content of collagen remained around 70% even when the GA concentration was increased to 200 mM, the mechanical strength was increased to an appreciable extent by crosslinking with GA. The result is given in Figure 2, which supports crosslinking of collagen with GA.

Collagen and gelatin could be also crosslinked with WSC which has been widely used for covalent immobilization of proteins onto polymer matrixes (9). The water content of crosslinked gels is plotted as a function of WSC concentration in Figure 3. In this case, not only gelatin but also collagen showed a decrease in water content as the WSC concentration became higher.

Bisepoxy compounds can render polymers with amino groups water-insoluble through crosslinking (10). Thus, it seems probable that proteins can be also crosslinked with bisepoxy compounds. However, collagen and gelatin films did not become water-insoluble when immersed in Denacol solution of pH 6 at 4°C for 24hr. To make much higher the protein and Denacol concentrations during crosslinking, gelatin solutions containing Denacol were cast on a dish at 25°C and allowed to dry simultaneously with crosslinking. The water content of the resultant gels is given in Figure 4. Two kinds of Denacol with short (EX-810) and long (EX-830) chains were employed for crosslinking. As is seen, EX-810 with n=1 could decrease the water content of the crosslinked gelatin gel to a higher extent than EX-830 with n=9 did.

In the above crosslinking reactions, water was always present in the reaction systems. In other words, crosslinking was carried out through wet processes. In an attempt to crosslink the proteins through dry processes, electron beam, UV radiation, and simple heat treatment were employed as crosslinking means. The results obtained for UV irradiation and heat treatment (dehydrothermal heat treatment, DHT) are shown in Figures 5 and 6, respectively. It is evident that both the dry processes rendered the collagen and gelatin films water-insoluble. Electron beam irradiation gave a similar result. Comparison of Figure 5 with Figure 6 reveals that heat treatment brought about severe denaturation to collagen molecules, since water contents higher than 70% were obtained for the collagen film as a result of heat treatment.

The major purposes of our protein crosslinking include water insolubilization, improvement of mechanical properties, and control of biodegradation. As examples of in vitro degradation, crosslinked gel films were subjected to hydrolysis with collagenase. Figures 7 and 8 show the results of the films crosslinked with GA and Denacol, respectively. Obviously, the weight of crosslinked gelatin films remaining after enzymatic hydrolysis depends on the concentration of the crosslinking agent for both the GA- and Denacol- crosslinking. The results of in vitro degradation for other crosslinked collagen and gelatin films are given in Figures 9 and 10, respectively. As can be seen, although the water content of the crosslinked collagen gels was almost the same, the rate of degradation was different, depending on the crosslinking extent. On the other hand, the rate of degradation was virtually governed by the water content for the crosslinked gelatin gels. It seems that the enzyme concentration in body is not so

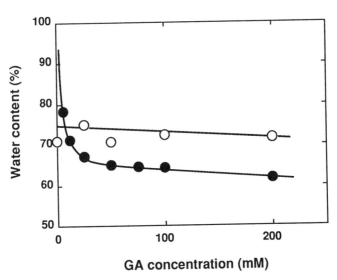

Fig.1 Effect of the GA concentration on the water content of collagen and gelatin films crosslinked with GA in PBS (pH 7.4) at 4°C for 24hr. (O) collagen and (●) gelatin.

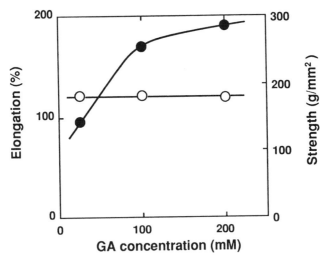

Fig.2 Effect of the GA concentration on the mechanical properties of collagen films crosslinked with GA in PBS (pH 7.4) at 4°C for 24hr. (O) elongation and (●) strength.

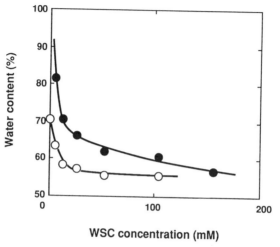

Fig.3 Effect of the WSC concentration on the water content of collagen and gelatin films crosslinked with WSC in PBS (pH 7.4) at 4°C for 24hr. (O) collagen and (●) gelatin.

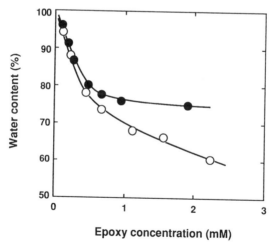

Fig.4 Effect of the Denacol concentration on the water content of gelatin films crosslinked with Denacol in DDW (pH 6.0) at 25°C for 24hr allowing spontaneous water evaporation. (O) EX-810 (n=1) and (●) EX-830 (n=9)

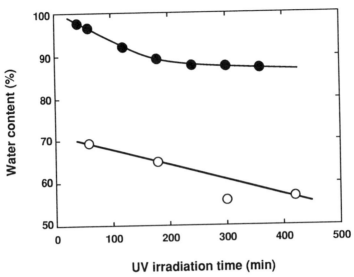

Fig.5 Effect of the UV irradiation time on the water content of collagen and gelatin films crosslinked by UV in a dry state at 25°C (UV lamp ; 254nm and 15W).
(O) collagen and (●) gelatin

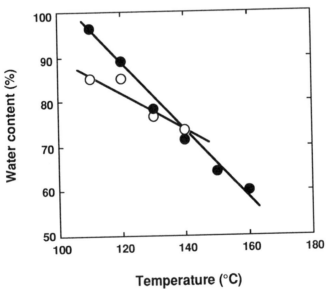

Fig.6 Effect of the treatment temperature on the water content of collagen and gelatin films dehydrothermally (DHT) crosslinked for 24hr at 0.1 torr.
(O) collagen and (●) gelatin

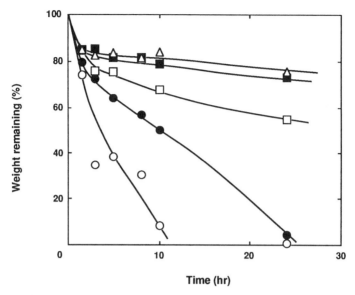

Fig.7 In vitro degradation of GA-crosslinked gelatin films upon hydrolysis in 40units/ml collagenase solution at 37°C and pH 7.4.
(O) 12.5mM, (●) 25.0mM, (□) 50.0mM, (■) 100mM, and (△) 200mM

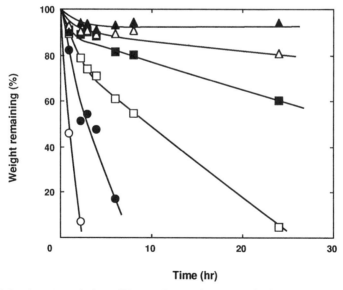

Fig.8 In vitro degradation of Denacol-crosslinked gelatin films upon hydrolysis in 40units/ml collagenase solution at 37°C and pH 7.4 (EX-810).
(O) 0.2mM, (●) 0.4mM, (□) 0.6mM, (■) 0.8mM, (△) 1.0mM, and (▲) 2.0mM

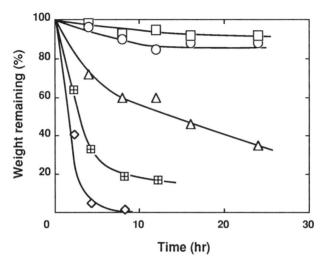

Fig.9 In vitro degradation of various crosslinked collagen films upon hydrolysis in 40units/ml collagenase solution at 37°C and pH 7.4.
(○) GA 200mM, (□) WSC 100mM, (△) UV (5hr), (⊞) DHT (140°C, 24hr), and (◇) non-crosslinked

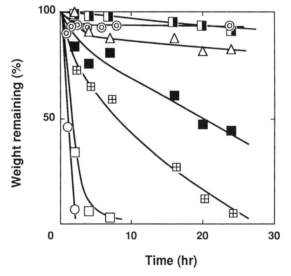

Fig.10 In vitro degradation of various crosslinked gelatin films upon hydrolysis in 40units/ml collagenase solution at 37°C and pH 7.4.
(□) GA 5mM, (■) GA 20mM, (▣) GA 100mM, (△) WSC 50mM, (⊞) DHT (150°C, 24hr), (○) EX-810 0.2mM, and (◎) EX-810 2.0mM

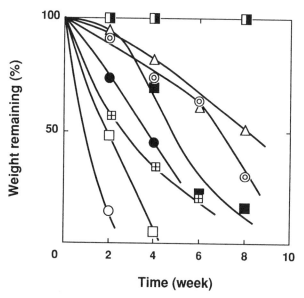

**Time (week)**

Fig.11 In vivo degradation of various crosslinked gelatin films after subcutaneous implantation in rats.
(□) GA 5mM, (■) GA 20mM, (▣) GA 100mM, (△) WSC 50mM, (⊞) DHT (150°C, 24hr), (○) EX-810 0.2mM, (◎) EX-810 2.0mM, and (●) EX-830 2.0mM

high as that used in this in vitro study. The results of in vivo degradation for the crosslinked gelatin gels are shown in Figure 11. The gel films were subcutaneously implanted in rats after sterilization with ethylene oxide gas. It is seen that the rate of in vivo degradation also can be controlled by the crosslinking agent and its concentration. The most important factor governing the biodegradation is again the water content, in other words, the crosslinking extent of gels. Figure 12 shows the film weight after in vivo degradation plotted as a function of water content for the films crosslinked under different conditions.

## Discussion

As demonstrated above, collagen and gelatin can be crosslinked with a variety of crosslinking means in solution, dispersion, and swollen states through both the wet and dry processes. Depending on the biomedical end purpose of the crosslinked gels we can choose the most appropriate method from the above variations. Although the native collagen even without additional chemical crosslinking is insoluble in neutral aqueous solution at room temperature, crosslinking is needed to retard the biodegradation and improve the mechanical strength (*11,12*). In contrast, crosslinking is always required for gelatin unless it is applied in medicine in monodispersed solution. The major difference in physiological activity between the native collagen and the denatured gelatin is the high coagulation activity of collagen toward platelet. Gelatin has no such a platelet activity but maintains a selective affinity to fibronectin similar to collagen. Therefore, it is not always necessary to prepare biodegradable hydrogels from collagen unless the selective affinity to platelet is required. Gelatin is practically

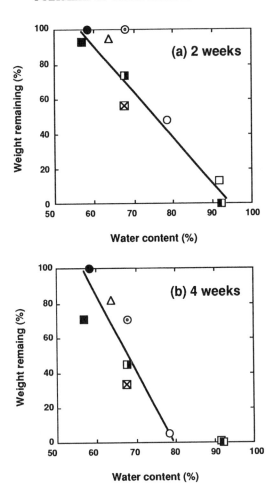

Fig.12 The gel weight remaining after subcutaneous implantation in rats for 2 and 4 weeks plotted against the water content of the crosslinked gelatin gels.
(O) GA 5mM, (☉) GA 20mM, (●) GA 100mM, (△) WSC 50mM,
(☒) DHT (150°C, 24hr), (□) EX-810 0.2mM, (■) EX-810 2.0mM,
(▥) EX-830 0.2mM, and (▦) EX-830 2.0mM.

more convenient than collagen, because a concentrated solution is extremely difficult to prepare from the native collagen.

The mechanism of chemical crosslinkings employed in this study can be proposed as represented in Figure 13. All the crosslinking reactions of collagen and gelatin are likely to utilize their carboxyl and amino groups existing or their molecules. Formaldehyde which has the longest history as the agent for water insolubilization of gelatin may react with the primary amino groups of gelatin.

The most important thing that should be taken into consideration when collagen and gelatin products are applied in medicine is the toxicity of the agent used for crosslinking (13-15). Formaldehyde and glutaraldehyde which have been known to be toxic will be released into the host environment as a result of biodegradation. The

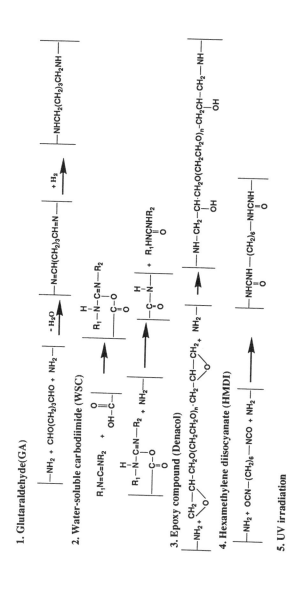

Fig.13 Scheme of crosslinking reaction of protein.

bisepoxy compounds(Denacol) will become also free upon hydrolysis of the crosslinked protein molecules, but are not released with the epoxy groups. They have reacted with amino acid units of proteins and will be released with hydroxyl groups upon hydrolysis. The most interesting crosslinking agent used here is WSC, because this is not incorporated into the crosslinked protein structure but simply changed to a water-soluble urea derivative as shown in Figure 13. Thus one can say that the protein hydrogels crosslinked using WSC is the highest in biosafety, so far as the unreacted WSC and the formed urea derivative have been completely removed from the crosslinked gels. The cytotoxity of the urea derivative was found to be quite low compared with that of WSC (unpublished result).

Physical crosslinking with high-energy radiation, UV, and thermal treatment may not bring new chemical units into the protein molecules, although the crosslinking mechanism is unclear.

## References

1. Piez,K.A. *Collagen;* Mark,H.F.; Bikales,N.M.; Overberger,C.G.; Menges,G.; Kroschwitz,J.I.,eds.; Encyclopedia of Polymer Science and Engineering; Wiley Interscience Publication: 1989, Vol.3; 699- 724.
2. Rose,P.J. *Gelatin;* H.F. Mark,H.F.; Bikales,N.M.; Overberger,C.G.; Menges,G.; Kroschwitz,J.I.,eds.; Encyclopedia of Polymer Science and Engineering; Wiley Interscience Publication: 1989, Vol.7; 488- 513.
3. Lee,G.; Nakamura,T.; Shimizu,Y.; Tomihata,K.; Ikada,Y.; Endo,K. *Program of the 30th Annual Meeting of Japanese Society for Artificial Organs* **1992**, *93.*
4. Li,S.-J.; Archibald,S.Y.; Krarup,C.; Madison,R.D. *Biotechnology and Polymers;* Gebelein,C.G. ed.; Plenum Press: New York, 1991, 281-293.
5. Ikada,Y. *Neurosurgeons*, **1992**, *11*, 236.
6. Weadock,K.; Olson,R.M.; Silver,F.H. *Biomat. Med. Dev. Art. Org.* **1983-84**, *11*, 293.
7. Milthorpe,B.K.; Schindhelm,K.; Roe,S. *Effect of glutalardehyde concentration on collagen cross-linking and absorption;* Heimke,G.; Soltesz,U.; Lee,A.J.C., eds.; Advance in Biomaterials, 9; Elsevier Science Publishers B.V., Amsterdam, 1990, 1-5.
8. Chvapil,M.; Owen,J.A.; Clark,D.S. *J. Biomed. Mater. Res.* **1977**, *11*, 297.
9. Wong,S.S. *Chemistry of Protein Conjugation and Cross-Linking;* CRC Press: Boca Raton, Florida, 1991; 195-199.
10. Piao.,D.-S.; Ikada,Y. *Polymer Preprints, Jpan* **1982**, *31*, 2609.
11. Law,J.K.; Parsons,J.R.; Silver,F.H.; Weiss,A.B. *J. Biomed. Mater. Res.* **1989**, *23*, 961.
12. Kato,Y.P.; Christiansen,D.L.; Hahn,R.A.; Shieh,S.-J.; Goldstein,J.D.; Silver,F.H. *Biomaterials* **1989**, *10*, 38.
13. Speer,D.P.; Chvapil,M.; Eskelson,C.D.; and Ulreich,J. *J. Biomed. Mater. Res.* **1980**, *14*, 753.
14. Eybl,E.; Griesmacher,A.; Grimm,M.; Wolner,E. *J. Biomed. Mater. Res.* **1989**, *23*, 1355.
15. van Luyn,M.J.A.; van Wachem,P.B.; Olde Damink,L.H.H.; Dijkstra,P.J.; Feijen,J.; Nieuwenhuis,P. *Biomaterials* **1992**, *13*, 1017.

RECEIVED May 25, 1993

# Synthetic Bioactive Chain Molecules and Polymers for Controlled Transport of Bioactive Agents

# Chapter 25

# Poly(methacrylic acid) Hydrogels as Carriers of Bacterial Exotoxins in an Oral Vaccine for Cattle

T. L. Bowersock[1], W. S. W. Shalaby[2], M. Levy[1], M. L. Samuels[3], R. Lallone[4], M. R. White[1], D. Ryker[1], and Kinam Park[2]

Schools of [1]Veterinary Medicine, [2]Pharmacy, and [3]Math, Purdue University, West Lafayette, IN 47907
[4]Brookwood Biomedical Institute, Box 26221, Birmingham, AL 35226

Poly(methacrylic acid) hydrogels were tested as a delivery system for oral vaccines in cattle. Hydrogels were absorbed with a vaccine composed of culture supernatants of *Pasteurella haemolytica*, the most common cause of bacterial pneumonia in cattle. Hydrogels absorbed with culture supernatants were administered orally to calves. Calves were then challenged by an intrabronchial instillation of virulent P. haemolytica. For each calf survival time was recorded. All surviving calves were euthanatized 3 days after challenge. A post mortem examination was performed and the lungs of each calf were evaluated for the size and severity of pneumonic lesions. Calves vaccinated orally with culture supernatants had less lung affected by pneumonia, less severe pneumonic lesions, and lived longer than non-vaccinated calves. Results indicate that hydrogels can be used to deliver oral vaccines to calves to enhance resistance to pneumonia caused by *P. haemolytica*.

Pasteurella haemolytica is the most common bacterium that causes death in cattle affected by pneumonia. Numerous vaccines have been developed to prevent this disease with limited success. The newest vaccines to prevent pneumonic pasteurellosis are composed of culture supernatants(1). Culture supernatants (CS) contain the most immunogenic antigens necessary to stimulate protective immunity in calves, and do not cause disease. However, the efficacy of vaccines containing CS has also been questioned (2). As with other vaccines used to prevent pasteurellosis, this vaccine has been used parenterally in feedlots and sales barns where cattle have already been stressed and exposed to multiple pathogens. Herein lies a major reason for poor success of these vaccines: they are not given prior to the time when calves are stressed and infected. Vaccination prior to exposure is desirable i.e., when calves are still on the farm of birth on pasture. This requires extensive handling of animals that is logistically impractical. The most efficient way to vaccinate these calves is through the feed or water.

The oral administration of vaccines to cattle offers many advantages over parenteral inoculations: reduced handling results in less stress to animals; there is a lower labor cost associated with vaccines administered in the feed because animals do not need to be treated one at a time; easier delivery encourages vaccination at a time when calves are less stressed and more likely to respond to a vaccine, e.g., when on pasture with the dam; and there are less injection site reactions that can result in trimming of expensive cuts of meat at slaughter resulting in less profit to packers and producers. Oral administration can also result in the stimulation of mucosal immunity at a time when calves may have (blocking) maternally derived antibodies. Finally, the size and density of hydrogels used to deliver antigens in this experiment are compatible with those of feed pellets. Hydrogels could easily be administered to calves in the feed without any special handling needed.

There is growing interest in the development of vaccines that can be administered orally. Oral vaccination is effective because it results in the stimulation of the gut-associated lymphoid tissue (GALT). Migration of  lymphocytes from GALT results in increased immunity at other mucosal sites including the lung (3). Immunity to a variety of infectious agents has been induced at other mucosal surfaces in laboratory animals and man following the oral administration of antigens (4-7). The oral administration of vaccines is much more difficult to achieve in ruminants. Oral vaccination of cattle requires that the antigen be able to withstand the extreme changes in pH and severe action of proteolytic enzymes in the  upper gastrointestinal tract (GIT).  Cattle have a complex upper GIT composed of 4 stomachs: rumen, reticulum, omasum, and abomasum. The rumen is the first stomach and is the site of cellulolytic microorganisms that initiate digestion of fiber in the ruminant diet. Bypassing the rumen or first stomach is especially demanding since microbial degradation would also be likely to destroy most antigens.  Therefore, a carrier is needed that can protect antigens until they reach the lower gastrointestinal tract.

Hydrogels are crosslinked polymers that swell in the presence of water. The crosslinking creates a meshwork that can trap drugs, chemicals, or antigens. When the hydrogels are hydrated within the body of an animal, the meshwork expands releasing the trapped material. Hydrogels are currently being investigated as a means of oral delivery for drugs, hormones, and peptides. These materials require a controlled release within the body in order to exert a long lived effect with minimal dosing (8-12).

Hydrogels can also be used to protect drugs or antigens from the proteases present in the stomach and lower intestine (13,14). Recently, we demonstrated that orally administered hydrogels can be used to release drugs slowly over time to dogs. Hydrogels were produced that swelled and were retained in the stomach releasing drug for 54 hours (15). Most studies for the delivery of drugs for biomedical purposes use hydrogels that are biocompatible which are broken down within the body by enzymes or pH. Our approach in the present experiment was to use non-absorbable hydrogels made of poly(methacrylic acid) (PMA). This polymer was selected for several reasons: it can absorb and release large proteins efficiently; it does not swell too quickly thereby retaining the density needed to bypass the rumen; it is chemically inert and does not stimulate an immune response; and it is resilient and therefore capable of controlled release of drug in the presence of great frictional activity in the upper GIT of ruminants. This is necessary because if the hydrogel broke down too soon, it would release antigen before leaving the rumen resulting in the proteolytic loss of the antigen. Being non- degradable the PMA hydrogels pass unabsorbed from the body in the feces without causing adverse reactions in the host animal.

We have recently demonstrated that orally administered PMA hydrogels can bypass the rumen to deliver a model antigen to the lower GIT of sheep for 96 hours (16). In the present study we hypothesized that hydrogels could be absorbed with culture supernatants of the bacterium P. haemolytica, administered as an oral vaccine, and stimulate protective immunity to pneumonic pasteurellosis in calves.

## Materials and Methods

Preparation of hydrogels - Poly(methacrylic acid) hydrogels were prepared by polymerizing 40% (w/v) methacrylic acid (Aldrich Chemical Co.) in the presence of 0.8% N, N'- methylenebisacrylamide (Bio-Rad Laboratories), a crosslinking agent. Ammonium persulfate (Polysciences, Inc.) and sodium bisulfite (J.T. Baker Chemical Co.) were used as the initiator and co- initiator, respectively. The solutions were degassed by exposure to a vacuum followed by purging with nitrogen. Polymerization was carried out at $60^{\circ}$ C for 18 hours under nitrogen. The gels were removed from a mold, cut into 5 mm diameter x 3 mm length discs, and washed exhaustively over a 3 day period in distilled deionized water. The gels were then dried at $37^{\circ}$ C for at least one week.

Preparation of culture supernatants - P. haemolytica biotype A1 was grown as previously described (17). Briefly, colonies from a blood agar plate were used to inoculate tryptic soy broth. This culture was incubated at $37^{\circ}$ C for 4.5 hours. This broth was used to inoculate fresh tryptic soy broth which was incubated for 1 hour at $37^{\circ}$ C with 5% $CO_2$ bubbled through it. The broth was stirred as rapidly as possible without foaming. After 1 hour of incubation bacteria were removed from the culture by ultrafiltration (Pellicon, Millipore, Inc.). Culture supernatants free of bacteria were lyophilized and stored dessicated at $-20^{\circ}$ C until loaded into hydrogels.

Loading hydrogels with culture supernatants - Culture supernatants (CS) of P. haemolytica contain many antigens, including a proteinaceous exotoxin (leukotoxin) 102 kd in size. This is the primary component of the CS and an essential antigen necessary in a vaccine. Therefore, it was necessary to determine whether CS could be loaded into and released from PMA hydrogels. The CS were resuspended to a 22% (w/v) solution in distilled deionized water. Poly(methacrylic acid) hydrogels were placed in the CS and allowed to fully absorb for 48 hours at 37° C. Loaded hydrogels (containing 635 ug of total protein of CS per gel) were then dried to a hard glassy consistency by placing them in a 37°C incubator for 48 hours.

Antigen release studies in vitro - To test for release of the CS antigens, 3 loaded hydrogel discs were placed into individual test tubes containing phosphate buffered saline and allowed to hydrate for 48 hours. Eluent was removed daily from each hydrogel for 3 days after the hydrogels were fully hydrated. Fresh saline was placed on the gels after removing the eluent. The eluents were then tested for the presence of leukotoxin, the primary protein antigen present in CS by an enzyme-linked immunosorbent assay (ELISA).

The ELISA was performed by placing 50 ul of each daily sample of eluent into a well on a 96 well polystyrene plate (Immulon 2, Dynatech Laboratories) and allowing the material to bind overnight at 4° C. Three wells were tested for each eluent sample. A polyvalent rabbit antibody made in our laboratory to the 102 kd leukotoxin of P. haemolytica was used to detect the presence of leukotoxin in eluents by incubation at 37° C for 3 hours. A secondary anti-rabbit IgG (Bethyl Laboratories, Inc., Bethyl, TX) conjugated to horseradish peroxidase was placed on the plate and incubated at 37° C for 2 hours. The substrate orthophenyldiamine (Sigma Chemical Company) was added to each well and the plate incubated at room temperature for 30 min. The reaction was stopped by adding sulfuric acid to each well and the plate was read at 490 nm using an EIA spectrophotometer (Molecular Devices Inc.).

Calf vaccine trial - Twelve 4 month old Holstein-Friesian calves, purchased from a local dairy, were divided into 2 groups. These two groups were further divided into 2 groups for 2 separate trials which consisted of 3 experimental and 3 control calves. Experimental calves (CS vaccinates) were given 300 CS- loaded hydrogels per day for 5 days placed in two 15 ml gelatin boluses. Control calves (non-vaccinates) were given 300 plain hydrogels. The hydrogels within the gelatin boluses were administered to the calves by use of a (pill) balling gun. Three weeks after the first day of vaccination each calf was challenged with an intrabronchial inoculation of 25 ml of $10^9$-$10^{10}$ CFU/ml of virulent P. haemolytica. Calves were monitored for clinical signs of disease and euthanatized if in respiratory distress. Calves which survived for seventy-two hours were euthanatized and a post mortem examination performed.

At the post mortem examination, lungs were scored for the percentage of pneumonic lesions, and scored for the severity of gross and histopathological lesions. Lesions were scored on a scale of 0-3 with 0 equal to no lesion noted and 3 equal to a severe lesion. Parameters scored for severity included edema, consolidation, fibrin, cellular infiltration, purulent material, and hemorrhage for both gross and histopathological lesions. The maximum score possible for each calf for gross lesions was 24 and for histopathology was 42. The scoring and determination of areas of pneumonia were performed by the pathologist who had no knowledge of which group a given calf belonged. The data for each calf was ranked (by trial) by survival time and analyzed in conjunction with the percentage pneumonic lung, gross and histopathological lesion score using the Wilcoxon rank sum statistical test.

## Results and Discussion

Antigen release studies - The mean absorbance values for the eluents as tested by ELISA indicated the presence of leukotoxin in the hydrogels for at least 72 hours after elution (Table 1). These results demonstrated that the hydrogels could be loaded with CS including the 102 kd protein antigen and that the leukotoxin was not irreversibly trapped within, but was able to diffuse out of the hydrogels. The fact that the polyclonal antibodies recognized the leukotoxin is important because it indicates that the chemicals and treatments used to load the gels did not adversely affect the epitopes recognized by the polyclonal antibodies. Thus, the leukotoxin retained the epitopes necessary to interact with the immune system. It is likely that other smaller (outer membrane protein) antigens were also present in the CS. We assumed that if the largest antigen of interest could be absorbed into the hydrogels, that smaller antigens would also be absorbed.

Calf challenge studies - Results for the survival time post- challenge in hours, per cent pneumonic lung, gross and histopathological lesion scores for each calf are shown in Table 2. Scores for calves in the first trial were not as sharply defined by group as scores in the second trial due to a difference in the dose of challenge bacteria used in each trial. In trial 1 the challenge dose contained $10^{10}$ CFU/ml of bacteria whereas in trial 2 the challenge contained $10^9$ CFU/ml. The greater challenge dose overwhelmed most of the calves in trial 1 within a few hours. However, it is interesting to note that the severity of the lesions was still greater for the control calves.

In trial 2, when ten-fold fewer bacteria were used in the challenge, all the control calves died and all the CS vaccinated calves survived to the end of the trial. When evaluated in combination with survival time, there was a significantly lower percentage of pneumonic lung, a lower gross lesion score, and a lower histopathological lesion score for CS vaccinated calves compared to sham vaccinates.

Passage of hydrogels through the gastrointestinal tract - Feces were examined during the time calves were being vaccinated and for 5 days after vaccination for the presence of hydrogels. Only five hydrogels, mildly hydrated to non-hydrated, were found in feces. This supports the results of studies performed in sheep in which no intact hydrogels were found in the intestinal contents or the feces over a period of 7 days following the oral administration of hydrogels (16). The hydrogels administered to the sheep were the same size, density, and material as used in the present experiment. Although the presence of leukotoxin was not determined in the hydrogels recovered from the calves, the low number found reflects the high percentage of retention in the reticulum. The reason hydrogels were found in the feces

**Table 1.**  Absorbance values of ELISA indicating the presence of leukotoxin in eluents from hydrogels

| Sample Tested | Absorbance Reading |
|---|---|
| Culture supernatants used to load hydrogels | .421 |
| Tryptic soy broth alone | .207 |
| Eluent 1 | .391 |
| Eluent 2 | .351 |
| Eluent 3 | .468 |

of the calf and not the sheep is most likely attributed to the fact that the calves received a total of 10 times the number of hydrogels the sheep received.  It was far more desirable to have the hydrogels retained in the reticulum over time and for the CS to be released as the gels swelled than to have the hydrogels pass through the gut intact with very little chance of becoming hydrated and releasing the CS.  Figure 1 shows an example of one of the hydrogels recovered from the feces and a hydrogel that had not been administered. The recovered hydrogel was swollen, eroded at the edges, and discolored presumably due to the  absorbance of bile or pigments from ingesta.

At post mortem the entire gastrointestinal tract (rumen to anus) was examined and no intact or recognizable parts of hydrogels were found. This suggests that most hydrogels were probably  eroded in the reticulum due to the extremely coarse consistency of the ingesta. The CS were released by diffusion out of the hydrogels as indicated by in vitro studies shown in Table 1, and as indicated previously by chromium release studies performed in sheep (16).  It is possible that the hydrogels began to  erode from the time they entered the reticulum and this further enhanced the diffusion of CS from the hydrogels.  This hypothesis cannot be proven until further tests are performed. Most importantly, CS were released from the orally administered hydrogels which resulted in the protection of calves challenged by *P. haemolytica*.

## Conclusions

The oral administration of vaccines to prevent respiratory disease has been investigated extensively in laboratory animals primarily for the prevention of influenza in humans (18-19). The oral administration of antigens to prevent respiratory disease in ruminants has been relatively unexplored. In this study, poly(methacrylic acid) hydrogels were absorbed with CS of P. haemolytica containing a mixture of bacterial antigens including the proteinaceous exotoxin  102 kd in size. Calves vaccinated orally with  these hydrogels orally lived longer and had less pulmonary lesions than control calves when challenged with viable P. haemolytica.  The results of this study

Table 2. Gross and histopathological lesion scores, percentage pneumonic lung, and survival time following challenge in CS vaccinated and sham vaccinated calves

| | CS Vaccinates | | | | | Sham vaccinates | | | |
|---|---|---|---|---|---|---|---|---|---|
| Calf # | Percentage pneumonic lung | Gross score | Histopathological score | Survival time (hours) | Calf # | Percentage pneumonic lung | Gross score | Histopathological score | Survival time (hours) |
| *Trial I* | | | | | | | | | |
| •7074 | 0.4 | 6 | 3.5 | 3.5 | •7075 | 53.7 | 15 | 6.2 | 3.5 |
| •7077 | 2.1 | 6 | 5.8 | 3.5 | •7076 | 61.4 | 13.5 | 7.7 | 3.5 |
| •7078 | 25.8 | 12 | 15.8 | 23* | •7080 | 24.2 | 17 | 4.4 | 72** |
| *Trial II* | | | | | | | | | |
| •7082 | 44.2 | 9.5 | 1.2 | 72** | •7081 | 100 | 11 | 3.5 | 12 |
| •7084 | 28.8 | 8.5 | 0.9 | 72** | •7083 | 100 | 10 | 6.8 | 12 |
| •7087 | 31.3 | 10.5 | 1.6 | 72** | •7085 | 100 | 10 | 2.5 | 20 |

\* Calf euthanatized due to severe pneumonia
\*\* Calf euthanatized to end study
Survival plus Percentage Pneumonic Lung P=.040
Survival plus Gross Lesion Score P=.035
Survival plus Histopathology Lesion Score P=.045

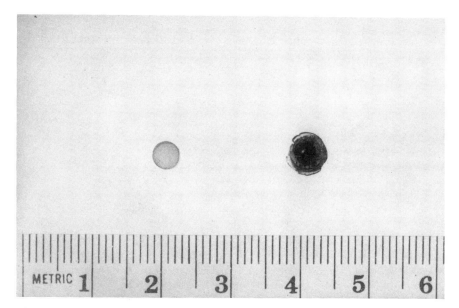

Figure 1. Hydrogel (on right) was recovered from the feces of a CS-vaccinated calf within 5 days after oral administration. Compare the swelling and surface erosion of this hydrogel to that of a hydrogel (on the left) that had not been administered to a calf. The hydrogel on the left was amber in color. The hydrogel on the right was dark green in color presumably due to absorption of bile and pigments from the ingesta in the calf.

are consistent with those of previous studies in which intraduodenal administration of CS of P. haemolytica resulted in enhanced pulmonary antibodies in calves as well as decreased pulmonary lesions in calves challenged by P. haemolytica (17). The decreased lesions in vaccinated calves in the present study show that antigen released by hydrogels was as effective as the direct intraduodenal inoculation of antigen in stimulating protective immunity in the lungs of calves. Oral administration of antigens induces antigen specific lymphocytes that migrate to other mucosal sites and produce antibodies. Recent studies suggest that cellular immunity may also be enhanced by the oral administration of antigens (20). Further studies are needed to determine the immune mechanisms responsible for protection in these calves.

This study showed that hydrogels can deliver antigens orally to ruminants resulting in enhanced immunity at distant mucosal sites. This was demonstrated by reduced pulmonary lesions in calves challenged by a viruluent respiratory bacterium. Studies are underway to determine what other antigens can be loaded into hydrogels and retain their immunogenicity when released into the lower GIT of cattle. Hydrogels provide a practical, safe, economical way to deliver oral vaccines to a large number of animals to prevent diseases which begin at a mucosal surface.

Acknowledgments

This work was supported by Animal Health and Disease Research Funds at the School of Veterinary Medicine, Purdue University, West Lafayette, IN.

## Literature cited

1. Shewen, P.E.;, Wilkie, B.N.; Sharp, A. *Vet. Med.* **1988**,*83*, 1078-1083.
2. Thorlackson, B; Martin, W.; Peters, D. *Can. J. Vet. Res* **1990**,*31*, 573-579.
3. Rudzik, R.; Perey, D.; Bienenstock, J. *J. Immunol.* **1975**, *114*, 1599-1604.
4. McGhee, J.R; Michalek, S.M. *Ann. Rev. Microbiol.* **1981**, *35*, 595-638.
5. Chen, K-S; Strober, W. *Eur. J. Immunol.* **1990**, *20*, 433-436.
6. Nedrud, J.G.; Liang, X.; Hague, N.; Lamm, M.E. *J. Immunol.* **1987**, *139*, 3484-3492.
7. Svennerholm, A.M.; Hanson, L.A.; Holmgren, J.; Lindblad, B.S.; Nilsson, B.; Overeshi, F. *Infect. Immun.* **1980**, *30*, 427-430.
8. Park, K. *Biomat.* **1988**, *9*, 435-441.
9. Shalaby, W.S.W.; Blevins, W.E.; Park, K. *ACS Symp. Ser.:Polym. Drugs and Drug Deliv. Syst.* **1991**, *469*, 237-248.
10. Damge, C.; Michel, C.; Aprahamian, M.; Couvreur, P. *Diabetes* **1988**, *37*, 246-251.
11. Saffran, M.; Jumar, G.S.; Savariar, C.; Barham, J.C.; Williams, F.; Neckers, D.C. *Science* **1986**, *233*, 1081-1084.
12. Peppas, N.A.; Klier, J. *J. Control. Rel.* **1991**,*16*, 203-214.
13. Lee, V.H.L. *J. Control. Release* **1990**, *13*, 213-223.
14. Hageman, M.J. *Proc. Intern. Symp. Control. Rel. Bioact. Mater.* **1992**, *19*, 76-77.
15. Shalaby, W.S.W.; Blevins, W.E.; Park, K. *J. Control. Rel.* **1992**, *19*, 131-144.
16. Bowersock, T.B.; Shalaby, W.S.W.; Blevins, W.E.; Levy, M.; Park, K. *ACS: Polymer Preprints of Proc. 4th Inter. Symp. Polym. Drugs and Drug Deliv. Syst.* **1992**, *33*, 63-64.
17. Bowersock, T.B.; Walker, R.D.; Samuels, M.L.; Moore, R.N. *Can. J. Vet. Res.* **1992**, *56*, 142-147.
18. Chen, K.S.; Burlington, D.B.; Quinnan, G.V., Jr. *J. Virol.* **1987**, *61*, 2150-2154.
19. Bergman, K.; Waldman, R.H.; Tischner, H.; Pohl, W.D. *Intern. Arch. Aller. Applied Immunol.* **1986**, *80*, 107-109.
20. O'Hagan, D.T.; Jeffrey, H.; McGee, J.P.; Davis, S.S.; Rahman, D.; Challacombe,

RECEIVED August 19, 1993

Chapter 26

# Dye-Grafted, Poly(ethylene imine)-Coated, Formed-in-Place Class Affinity Membranes for Selective Separation of Proteins

Y. Li and H. G. Spencer

Department of Chemistry, Clemson University, Clemson, SC 29634-1905

Poly(ethylene imine) (PEI) was adsorbed on a titania microfiltration membrane supported by a porous stainless steel tubular module and crosslinked into a stable layer by multi-functional oxiranes. Cibacron Blue F3GA was immobilized on the activated PEI layer to provide a ligand for specific proteins. Solutions of human serum albumin (HSA) were used as model protein solutions for investigating the affinity separation properties of the membrane. the effects of pH and ionic strength of the protein solution on the binding capacity of the membrane and the influence of the crossflow velocity on the frontal elution profiles were investigated. A procedure for the affinity separation of HSA from human plasma using the membrane was developed and the separation accomplished.

There is an increasing interest, resulting from the development of biotechnology, in the separation of proteins and peptides from biological broths and blood. Affinity separation is a preferred method for such applications because very high specificity is possible. Affinity chromatography is the most used and investigated technique for large scale production but its application has been hampered by the inability of affinity columns to handle high volume flow rates at reasonable ligand utilization efficiencies (1). There are several potential advantages to the use of affinity membranes in place of column chromatography for separation of proteins; for example, large scale separations and a relatively easy operation of the membrane system (2-4).

0097–6156/94/0540–0297$06.00/0
© 1994 American Chemical Society

Formed-In-Place (FIP) microfiltration membranes and their applications in bioseparations have been investigated by the FIP membrane research group at Clemson University in cooperation with Du Pont Separation Systems. A wide range of operating characteristics can be obtained by altering the porous structure of the stainless steel tube, using filter aids, and adding coatings. The FIP microfiltration membranes have been investigated as microfiltration or ultrafiltration devices for food processing, protein separations, and wastewater treatment and as supports for enzyme immobilization to establish continuous process methods and for functional activation for ion-exchange purification procedures (5-6).

This study presents the results of an investigation of the FIP microfiltration membranes as affinity membrane supports. PEI was coated on the membranes to circumvent the problem of unwanted interactions between proteins and the membrane and to provide sites for binding the affinity ligand. Cibacron Blue F3GA, a dye which has previously been used for purifying a variety of proteins from different sources, was grafted on the PEI layer. The amount of dye coupled to the membrane was determined and the efficiencies of the immobilized dye on binding and purification were demonstrated using Human Serum Albumin (HSA) as a representative protein. The effects of operating parameters, pH and concentrations of protein are also described.

**Materials and Methods**

Human serum albumin (HSA) (fraction V) and Plasma (Catalog No. P-9523) were purchased from Sigma Chemical Company, Inc., poly(ethylene imine) (PEI) (MW 600 Da) from Polysciences, Inc., 1,4-butanediol diglycidyl ether (BUDGE) from Aldrich Chemical Company, Inc. and Cibacron Blue F3GA from Fluka. All other reagents were ACS grade.

The FIP microfiltration membranes were permanent microporous titanium dioxide ($TiO_2$) membranes formed on the inside of 1.6 cm I.D. sintered stainless steel tubes with a membrane area of 0.029 $m^2$ supplied by Du Pont Separation Systems, Seneca, SC. Membrane permeability was expressed as permeate flux divided by applied pressure (m $s^{-1}$ $bar^{-1}$). All membranes used in this study had an initial permeability of pure water of at least 9.8 m $s^{-1}$ $bar^{-1}$ at 37°C. The average pore diameters in the titanium dioxide layer are normally about 0.15 μm.

The coating of PEI on the FIP-microfiltration membranes was achieved by a modification of the procedure used by Alpert and Regnier (7). The membrane was soaked in a methanolic solution of PEI (1 to 5% w/v). The solution was degassed and drained from the membrane. The membrane was then dried under a flow of dry nitrogen gas. The second step consisted of contacting the coated membrane with a 5% (v/v) solution of BUDGE in dioxane. After 12 hrs at room temperature, the mixture was allowed to continue reacting at 70°C for another 30 min. The solution was removed from the membrane and membrane rinsed with methanol. Then the membrane was soaked in a methanolic solution of PEI (1% w/v) at 50°C for 1 hr and washed with methanol and water. By varying the PEI concentration in the methanol and the molecular size of PEI, it was possible to create a variety of coating thicknesses. In order to obtain thin coatings, the PEI-coated membrane was washed with methanol after being removed from the coating solution. Methanol washing was omitted in the preparation of the membranes intended to have a heavy coating of PEI.

The immobilization of the dye on the membranes was accomplished in two steps: a one hour dye adsorption at room temperature from a 6% NaCl solution followed by a one hour reaction at 50°C after raising the pH to 10.8 with a carbonate buffer. Subsequent rinsing involved several steps with water and one with methanol.

All the protein solutions were dissolved in the appropriate buffer with low ionic strength and recirculated through the membranes for selected times to accomplish the adsorption. A water flush followed by washing with 1.0 M NaCl was used to remove any non-specifically bound proteins. Then, a 0.5 M NaSCN in 0.01 M phosphate buffer at pH 8.6 was pumped through the membrane to elute the specifically bound proteins into the permeate. The concentrations of eluted proteins were determined spectrophotometrically by absorption at a wavelength of 280 nm using a Bausch & Lomb Spectronic 2000 spectrophotometer. Standard absorbance versus HSA concentration curves were determined with a solution of known concentration and the concentrations of the samples were determined by comparing the absorbance of the sample with the standard curve. The apparent extinction coefficient was $0.531$ mL mg$^{-1}$ cm$^{-1}$. A PAGE (10%) in SDS was used to determine the purity of the HSA separated from human plasma. The gel had no composition gradient and a reducing buffer containing $\beta$-mercaptoethanol was used to prepare the sample. The proteins were visualized by Coomassie blue staining.

**Results and Discussion**

The method used to prepare the PEI coating on the Formed-In-Place microfiltration membranes was a modification of techniques described by Alpert and Regnier (7). It has been shown by Li and Spencer (6) and Chicz (8) that PEI adsorbed to the surface of titania supports can be crosslinked into a stable thin layer or skin on the surface of the support by multi-functional oxiranes or aldehydes. The coating thickness was a function of both molecular weight and concentration of PEI in the coating solution. The BUDGE was chosen as the cross-linking agent because it was considered to be of sufficient length to bridge between adjacent adsorbed PEI molecules and would contribute to the hydrophilicity of the coating. BUDGE can react with the adsorbed polyamine species in two ways: (1) both oxiranes in the crosslinker react with amine groups in a crosslinking reaction and (2) only one oxirane reacts with an amine group and the other oxirane remains free to react with free polyamine molecules in the coating solution applied as a third step to introduce more amine groups and spacers into the coating layer.

The PEI-coated membrane can be used as a support for immobilization of the affinity ligand. The Cibacron Blue F3GA dye was chosen as the affinity ligand and immobilized on the PEI coating layer. In the case of the membrane coated with 1% (v/v) PEI coating solution, the amount of dye coupled on the membrane, roughly estimated by the method described by Champluvier (9), was about 1.0 μmol/mL of the porous layer

The performance of the Dye-PEI membrane was tested using a filtration system described previously(6). The system can be operated in either a crossflow or a deadend configuration, with the selection of the procedure depending on the type of sample being processed. For relative pure solutions, the deadend configuration is preferred, while for solutions containing particle suspensions, the crossflow configuration is favorable. Fig. 1 shows the effect of the crossflow velocity on the frontal elution profiles of HSA. These curves were obtained with a Dye-PEI membrane by circulating 1.0 g/L HSA in 0.01 M phosphate buffer (pH 6.8) at different crossflow velocities. The steady state concentrations of the permeate solution, or passage of HSA, decreased with increasing crossflow velocity. This result was similar to that of the microfiltration of a BSA solution, where at higher crossflow velocity the rejection is higher. As expected, the amount of protein bound on the membrane also slightly decreased with increasing crossflow velocity.

A typical curve of HSA eluted from the membrane is given in Fig. 2. In the sequential separation procedure, the feed HSA solution, 1 g/L in 0.01 M phosphate buffer at pH 6.8, was pumped through the membrane in the deadend manner at a pressure of 50 psi. The HSA, after washing the membrane with water and 1.0 M NaCl solution to remove the non-specifically bound proteins, was eluted with 0.5 M NaSCN in 0.01 M phosphate buffer at pH 8.6. The capacity of protein binding of the Dye-PEI Membrane was about 0.4 g/m$^2$. This corresponds to a capacity of about 4.0 mg/mL of the porous layer, with the volume of the layer roughly estimated from SEM's of the membrane.

The effect of pH on the protein binding is shown in Fig. 3. The amount of protein bound on the membrane was calculated from the frontal elution profiles of 1 g/L HSA solution at different pH's. The capacity of protein binding decreased with increasing pH of the solution. Since the blue chromophore contains both amino and sulfonate groups, an ionic interactions would be expected. It is reasonable to suppose that positively charged residues of lysine $\varepsilon$-CH2-NH$^{3+}$ bind to the negatively charged groups of the dye. This type of interaction might explain the sharp binding decrease at approximately pH 9 because at this range of pH the lysine residues could be deprotonated (*10*).

Although the positive charges of the protein appear to determine the strength of the interaction with the dye, it is clear that electrostatic forces do not by themselves account for the strong binding. However, the ionic effect suggested that the interaction between the albumin and the dye is ionic as well as hydrophobic. A 1 M NaCl solution did not elute the bound HSA from the supported dye, while elution was readily accomplished using a 0.5 M NaSCN solution. The chloride ion in the Hofmeister series tends to disrupt simple ionic binding. Conversely the thiocyanate ion, which is at the other extreme of the series is expected to disrupt the hydrophobic interactions (*11*). These results suggest that both ionic and hydrophobic interactions are responsible for the high affinity binding between the dye and HSA, which is consistent with the extensive studies of the nature of the binding in this system (*12, 13*).

A separation of HSA from human plasma was also performed. A solution of 1 g/L of human plasma at pH 6.8 was circulated through the Dye-PEI membrane. After rinsing with water and 1.0 M NaCl, the bound HSA was eluted with 0.5 M NaSCN at pH 8.6.

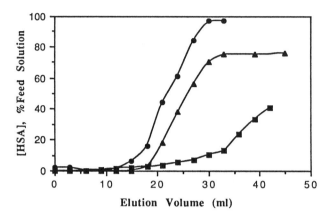

Fig. 1.   Elution profiles of HSA obtained with a Dye-PEI membrane after adsorption of HSA from a 1 g/L solution at different velocities.   The HSA adsorption was accomplished from a 0.01 M phosphate buffer at pH 6.8, 3.3bar (50 psi) and room temperature.   Crossflow velocities: •, 0.00; ▲, 0.53, and ■, 1.5 m/s.

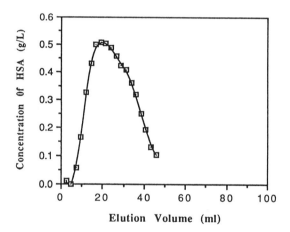

Fig. 2.   Elution profile of HSA obtained with a Dye-PEI membrane by eluting adsorbed HSA with 0.5 M NaSCN in a phosphate buffer at pH 8.6; after equilibrating with the buffer and rinsing with a 1.0 M NaCl solution.

The eluted protein was run on an electrophoresis polyacrylamide SDS gel (10%) to estimate the purity, utilizing the electrophoresis patterns of the whole serum, the permeate during binding and the Cohn fraction V (Sigma) to provide comparisons. Lane 1 in Fig. 4 represents feed human plasma solution, lane 2 permeate solution, lane 3 HSA purified from human plasma, lane 4 HSA (Cohn fraction V). The loadings were 1.6 µL of whole human plasma containing about 40 to 50 µg of HSA per µL in lane 1, 1.6 mL of the permeate of the whole human plasma obtained during the adsorption step of the process, a sample of the protein eluted from the membrane containing 30 µg of protein, and a sample containing 120 µg of fraction V HSA from Sigma. Although the loadings in each lane were not uniform in protein content, qualitative comparisons of the relative HSA content are possible. Comparisons of the pattern in lane 3 with the patterns in lanes 1 and 4 indicate that the HSA has been separated to a high degree from the other proteins contained in the whole plasma.

In the manually operated system used in these experiments, a separation cycle could be completed in less than 30 minutes. An automated process is expected to reduce this time to provide a higher efficiency. The most important result was that the procedure appears promising for obtaining a rather pure target protein in the first step of protein purification. As such it would perhaps appears to have potential for application in perhaps the pharmaceutical industry and in biotechnology for obtaining an expensive component in a single step.

## Conclusions

PEI adsorbed on the surface of an FIP titania microfiltration membrane can be crosslinked into a stable skin on the surface of the support by multi-functional oxiranes. The dye ligand can be grafted on the PEI layer. The affinity membrane separation system can be operated in either crossflow or deadend configurations. For a solution containing particles or having large molecular differences among the components, the crossflow procedure will combine the advantage of both ultrafiltration and chromatography, separating the large molecules from small molecules and purifying a specific protein by binding it with the affinity ligand simultaneously. Crossflow velocity can play a significant role in this procedure. The HSA separated from human plasma using a Dye-PEI membrane was effectively separated, although the efficiency of the system needs to be improved. Further work is to be directed toward the determination of the

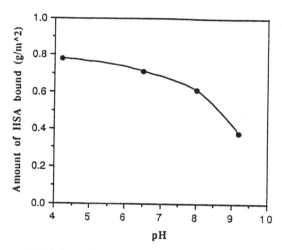

Fig. 3. Amount of HSA bound per unit area of the Dye-PEI membrane versus pH.

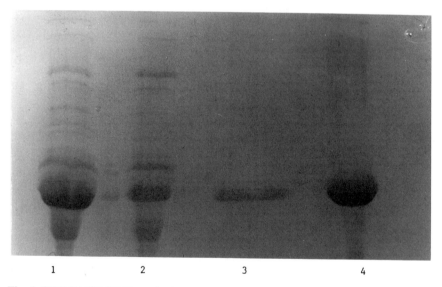

Fig. 4. SDS-PAGE (10%) analysis of HSA purified from Human Plasma. The Proteins were visualized by Coomassie Blue staining. Identification of lanes (with loading): (1) Whole plasma (1.6 µL); (2) whole plasma permeate during adsorption (1.6 µL); (3) HSA purified from plasma (30 µg protein); and (4) HSA,fraction V from Sigma (120 µg protein).

efficiency of the ligand utilization, the determination of the effects of membrane pore size on the process, and the improvement of the efficiency in the purification cycle of the affinity separation process.

## Acknowledgment

The authors gratefully acknowledge the financial support of Du Pont Separation Systems.

References

1. Molinari, R.; Drioli, E. *ICOM'90* **1990**, 1140.
2. Brandt, S.; Coffe, R.A.; Kessler, S.B.; O'Connor, J.L.; Zale, S.E. *Bio/Technology* **1988**, *6*, 779.
3. Klein, E., *Affinity Membranes: Their Chemistry and Performance in Adsorptive Separation Processes*; Wiley: New York, NY, 1991, p. 131.
4. Krause, S.; Kroner, K.H. *ICOM'90* **1990**, 166.
5. Thomas, R.L.; McKamy, D.L.; Spencer, H.G. In *Advances in Reverse Osmosis and Ultrafiltration*; Matsuura, T. and Sourirajan, S., Eds.; National Research Council of Canada, Ottawa, 1989, p. 563.
6. Li, Y. and Spencer, H.G. *J. Biotechnol.* **1992**, *26*, 203.
7. Alpert, A.J.; Regnier, F.E. *J. Chromotogr.* **1979**, *185*, 375.
8. Chicz, R.M.; Shi, Z.; Regnier, F.E. *J. Chromatogr.* **1986**,*359*, 121.
9. Champluvier, B.; Kula, M.R., In *International Symposium on Separations for Biotechnology (2nd)*, Pyle, D.L., Ed.; London, 1990, Vol 2; pp. 295-304.
10. Scopes, R. K. *J. Chromatogr.* **1986**, *376*, 131.
11. Subramanian, S. *Crit. Rev. Biochem.* **1984**, *16*, 169.
12. Gee, A.P.; Borsos, T.; Boyle, M.D.P. *J. Immunol. Methods* **1979**, *30*, 119.
13. Leatherbarrow, R.J.; Dean, P.D.G. *Biochem. J.* **1980**, *189*, 27.

RECEIVED April 22, 1993

# Chapter 27

# Preparation of Insulin-Releasing Chinese Hamster Ovary Cell by Transfection of Human Insulin Gene

## Its Implantation into Diabetic Mice

H. Iwata[1], N. Ogawa[1], T. Takagi[2], and J. Mizoguchi[3]

[1]National Cardiovascular Center, Suita-city, Osaka 565, Japan
[2]Falco Biosystems, Kumiyama-cho, Kuze-gun, Kyoto 613, Japan
[3]Asahi Chemical Industry Company, Ltd., Tahou-gun, Shizuoka 410–23, Japan

The feasibility of gene therapy in the treatment of diabetes was examined. Insulin releasing CHO cells were prepared by transfection of a plasmid that contained human insulin gene under control of the chicken actin promoter. They were implanted into peritoneal cavities of diabetic mice in three different forms. When singly dispersed cells were implanted, only 2 of 11 recipients showed normoglycemia. In the case of implantation of microencapsulated cells, blood glucose levels of all recipients were only transitorily decreased. In the third form of the transplants, that is cells on microcarriers, all of the recipients became normoglycemic and half of them exhibited more than 50 days of normoglycemia. This study demonstrated that insulin releasing cells prepared by genetical modification could normalize glucose levels for a long period if they were implanted in an appropriate form.

If we accept that gene therapy is thought of as *in vivo* protein production and its delivery system is by genetically modified cells, almost all diseases that are currently treated by the administration of protein are candidates for treatment using gene therapy. There have been only a few studies which examined the feasibility of gene therapy in the treatment of diabetes(1,2). Many problems still remain to be examined in detail. For example, recipients were frequently lost by hypoglycemia during these investigations. In the treatment of disease such as diabetes, over-production of protein caused harmful side effects. On the other hand, genetically modified cells sometimes released a sufficient amount of the gene product in *in vitro* culture, but could not provide therapeutic levels of it after they were implanted into animals. The function and fate of genetically modified cells in animals have not been clarified yet. In this study we used xenogeneic cells as the host cell of the insulin gene. We tried to regulate the number of implanted cells by controlling the immune reactions to the implanted cells. We also examined methods to implant cells to effectively draw the functions of the cells *in vivo*. Figure 1 schematically shows the strategy of this research.

## Experimental

Cell; A mutant K1 cell line of Chinese hamster ovary cells (CHO cells) which lacked the enzyme dihydrofolate reductase (DHFR) was used as a host cell (3). It could not grow without supplements, such as thymidine, glycine, and purine. The expression of DHFR was used as a selective maker of genetically modified cells. A culture medium was composed of α–Minimal Essential Medium (Sigma, St. Louis, MO) supplemented with 50 U/L Penicillin and 50U/L Streptomycin (Whittaker Bioproducts, Inc. Md.) and 5% dialyzed fetal bovine serum (Whittaker Bioproducts, Inc. Md.)

Plasmid; The DHFR gene was inserted into the BamHI site of a plasmid which expressed the human insulin gene under control of the chicken actin promoter (4). The resulting construct which is termed by pACT-HIN-DHFR, is shown in Figure 2. It was prepared and purified by a conventional method. Closed circular DNA was purified by equilibrium centrifugation in a CsCl-ethidium bromide density gradient.

Introduction of the plasmid into cells; The plasmid pACT-HIN-DHFR was introduced into CHO cells by a highly efficient calcium phosphate method (5). Cells which were transfected with the plasmid were selected in a culture medium without supplements such as thymidine, glycine, and purine. Selected cells expressed a cloned copy of the DHFR gene and insulin gene. Plasmid DNA introduced into cells was incorporated into chromosomal DNA during culturing. They can release insulin into the culture medium. The ability of insulin release of the obtained cells was amplified by progressive selection of cells resistant to increasing concentrations of methotrexate (MTX) (Sigma, St. Louis, MO).

Implantation of insulin releasing cells into diabetic mice; The ability of the insulin releasing cells to reverse diabetes was examined. Male BALB/c (Japan CLEA, Tokyo, JAPAN) mice were made diabetic by an intraperitoneal injection of streptozotocin(Sigma, St. Louis, MO). Only mice of which plasma glucose concentrations were more than 400 mg/dl were used as diabetic recipients. Three different forms of implants, single cells, microencapsulated cells and cells on microcarriers were examined. The cells cultured on the flask (Corning 25110, Iwaki Glass, Tokyo, JAPAN) were harvested as single cells. They were intraperitoneally implanted into a diabetic mouse. The second form was the cells on the microcarrier. The microcarrier was made of microporous cellulose, about 200 µm in diameter and carrying diethylaminoethyle groups (Asahi Chemical Industry Co., Ltd., Tokyo, Japan). Cells could grow not only on the surface of the carriers, but also inside the pores of the microcarriers. The third one was the implantation of microencapsulated cells. Cells were miroencapsulated into 5% agarose mirocapsules(6). The site of implantation was a peritoneal cavity and cell number was $3 \times 10^7$ in spite of the implant forms. Plasma glucose levels were measured three times a week after implantation. Samples of blood were taken from the subclavian vein. Plasma glucose levels were measured with a Beckman glucose analyzer 2 (Beckman Instruments, Fullerton, CA).

Immunosuppressive therapy; The CHO cell, which is derived from the Chinese hamster, is xenogeneic to a recipient mouse. It is expected that the implanted CHO cells are rejected by the host immune system. A immunosuppressive drug, 15-deoxyspergualin (DSG) (Nippon Kayaku Co., LTD., Tokyo, Japan) was given (2.5 mg/kg body weight) every day. When the non-fasting plasma glucose levels became

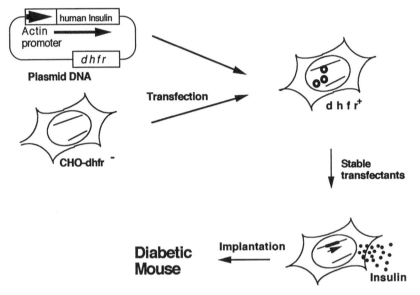

Figure 1. Schematic representation of this experiment.

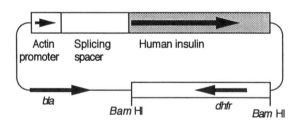

Figure 2. Structure of the pACT-HIN-DHFR recombinant plasmid.

less than100 mg/dl,  DSG treatment was stopped. When  the non-fasting plasma
glucose levels exceeded 200 mg/dl, DSG treatment was started again.
Determination of anti-CHO antibodyi levels:  Antibody levels were determined using
a flowcytometer (Japan Spectroscopic Co., Ltd., Tokyo, Japan).  Recipient serum
was mixed with CHO cells.  The mixture was incubated for 30 min. on ice to form
antibody-CHO cell complexes.  After three washings of CHO cells with Hanks'
balanced salt solution (HBSS), a second anti-mouse IgG and IgM antibody which was
conjugated with fluorescen isothiocyanate (FITC)  was  added and left for 30 min. on
ice.   Cells were washed again with HBSS.  FITC positive cells were measured using
a flowcytometer.  Anti-CHO cells antibody levels in recipient serum were expressed
as  a percentage of FITC positive cells.


**Results**

The cells transfected with plasmid DNA released  insulin  into  a culture medium.
However, the amount of insulin release was not sufficient enough to be used to treat a
diabetic mice.  The copy number of the integrated DNA was increased during
culturing cells under the pressure of MTX.  Figure 3 shows the abilities of insulin
release by CHO cells which was obtained in the culture medium containing a certain
concentration of MTX.  They were increased by a sequential passage of the cells in the
medium containing increasing amounts of MTX.  The cell  line which was obtained
at the 400 nM of MTX released more than 300 $\mu$U insulin per day by $10^5$ cells. The
stability of the amplified gene was examined by subculturing the cell in the culture
medium without MTX for 5 times during 40 days. It was not observed that the cells
lost their ability to release the gene product when the MTX pressure  was  removed.
The cell line which was obtained at 400 nM of MTX was used in the implantation
experiments.
    Singly  dispersed  cells  of  $3.3 \times 10^7$ were  implanted  intraperitoneally.  Blood
glucose  changes  are shown in Figure 4.  Before  implantation  of cells, blood glucose
levels of recipients were much higher than the100 mg/dl blood glucose levels of a
normal mouse.   Only 2  of 11 mice became normoglycemia in response to
implantation of insulin releasing cells for more than 10 days. The blood glucose
levels of one of these two was maintained normoglycemic for more than 50 days.
However,  the other 9 mice could not demonstrate the normal blood glucose levels.
Effects  of implantation of insulin releasing CHO cells were limited when singly
dispersed cells were implanted.  Microencapsulated cells in 5% concentration of
agarose hydrogel were intraperitoneally implanted into each of 5 diabetic mice.
Blood glucose changes are shown in Figure 5. Three of five recipients showed
transitory normoglycemia, but blood glucose reverted to preoperative levels.  The
other  two could not demonstrate normoglycemia. Cells cultured on microcarriers
were also intraperitoneally implanted into each of 4 diabetic mice.  On the contrary to
the implantation of the singly dispersed cells and the microencapsulated cells,  blood
glucose levels of all of recipients  were normalized within several  days as shown in
Figure 6.  Two of these demonstrated more than 50 days normoglycemic periods.
The blood glucose levels of the other mice decreased to normal levels, but it reverted
to preoperative levels after 25 days of observation.  The remaining one was lost by
hypoglycemia.
    The CHO cell is xenogeneic to the recipient mice.  It is anticipated that it
provokes the immune reaction.  The anti-CHO cell antibody formed in the recipients
was monitored.  Figure 7 shows changes in anti-CHO cell antibodies levels in the
blood of recipient mice. Although the immunosuppressive drug,15-deoxyspergualin,

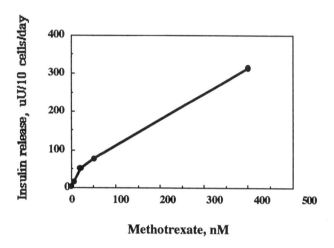

Figure 3. The abilities of insulin release by a cell line obtained at a certain concentration of MTX.

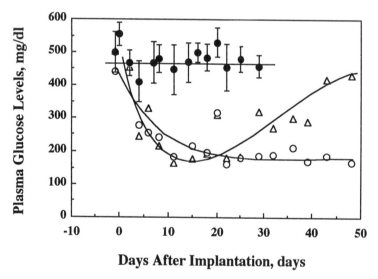

**Days After Implantation, days**

Figure 4. Changes in the non-fasting blood glucose levels of diabetic mice after intraperitoneal implantation of singly dispersed cells.  ● ;n=9, ○ ;n=1, △ ;n=1.

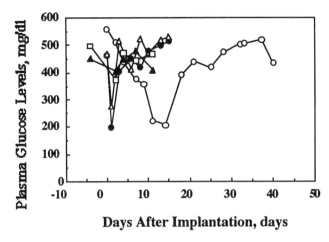

**Days After Implantation, days**

Figure 5. Changes in the non-fasting blood glucose levels of diabetic mice after intraperitoneal implantation of microencapsulated cells in agarose hydrogel.

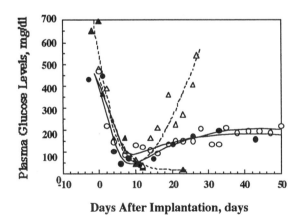

**Days After Implantation, days**

Figure 6 Changes in the non-fasting blood glucose levels of diabetic mice after intraperitoneal implantation of cells cultured on microcarriers.

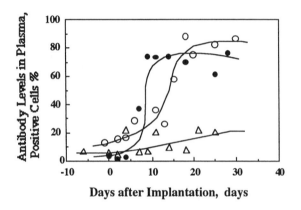

Figure 7. Changes in anti-Chinese hamster antibodies levels in recipient plasma.
O ;Singly dispersed cells, △;Microencapsulated cells,
● ; Cells on micro carriers

was given, anti-CHO cells antibodies were formed when free cells and cells on microcarriers were implanted. On the other hand, when microencapsulated cells were implanted, antibody levels did not increase. The immune reaction was not triggered by microencapsulated CHO cells.

**Discussion**

When singly dispersed cells were implanted, only 2 of 11 recipients showed normoglycemia. The blood glucose levels of the other 9 mice were not normalized. Although an immunosuppresive drug was administered, anti-CHO antibody levels increased after implantation of CHO cells. In the case of implantation of microencapsulated cells, anti-CHO cell antibody levels remained low. Microencapsulated CHO cells could not present antigen to recipient mice or triggerimmune reactions of recipients. Blood glucose levels, however, were only transitorily decreased. Frequently one species possesses natural antibodies for cells of distantly related species. Natural antibodies play an important role in the rejection of xenografts(7). Natural antibodies permeated through the 5% agarose microcapsule and attacked the cells in the microcapsule. In the third form of transplants, that is cells on microcarriers, all of recipients become normoglycemic and half of them demonstrated more than 50 days of normoglycemia, in spite of provocation of immune reactions against cells on microcarriers. Cells on the surface of microcarriers were killed by antibodies against CHO cells, but cells which resided inside the porous microcarriers could proliferate. Dynamic equilibrium between cell killing and cell proliferation were attained and cell numbers was almost kept constant. It resulted in long term normoglycemia.

From this study, it became clear that insulin releasing cells prepared by the gene transfection could control glucose metabolism of the diabetic mouse for a long period if cells were implanted in an appropriate form. Further studies, such as how to control cell growth, how to control the immune reaction and how to remove the implants when adverse effects are observed, are needed to apply gene therapy to clinically curing diabetes.

**References**

1. Selden, R. F., Skoskiewicz, M,J., Russell, P.,S., Goodman, H. M., *N Eng J Med*, 1987, 317, 1067
2. Kawakami, Y., Yamaoka, T., Hirochika, R., Yamashita, K., Itakura, M., Nakauchi, H., *Diabetes*, 1992, 41, 956-961.
3. Kingston, R.E., In *Current Protocols in Molecular Biology* Editors Ausubel, F.M., Brent, R., Kingston, R., Moore,D.D., Seidman, J.G., Smith, J. A., Struhal, K.,Greene Publishing Associates and Wiley-Interscience: New York, NY 1987, Vol.1, pp 9.9.1-9.9.6.
4. Y.Kaneda, K.Iwai, and T.Uchida, *J Biol Chem*, 1989, 264, 12126.
5. Takai, T., Ohmori, H., *Methods in Molecular and Cellular Biology*, 1990, 2,
6. H.Iwata, T.Takagi, H.Amemiya, H.Shimizu, K.Yamashita, K.Kobayashi, T.Akutsu, *J Biomedical Materials Res*, 1992, 26, 967.

RECEIVED July 30, 1993

Chapter 28

# pH- and Ionic-Strength-Dependent Permeation through Poly(L-lysine-*alt*-terephthalic acid) Microcapsule Membranes

Kimiko Makino[1,2], Ei-ichi Miyauchi[1], Yuko Togawa[1], Hiroyuki Ohshima[1], and Tamotsu Kondo[1,2]

[1]Faculty of Pharmaceutical Sciences, Science University of Tokyo, Shinjuku-ku, Tokyo 162, Japan
[2]Research Institute for Biosciences, Science University of Tokyo, Noda-shi, Chiba 278, Japan

The permeability to both anionic and cationic ions, the size, and the electrophoretic mobility of poly(L-lysine-*alt*-terephthalic acid) microcapsules have been measured. The permeation rate and the size are both strongly pH dependent, exhibiting a sharp change in the pH region between 4 and 6, with their minima at pH 4. On the other hand, the electrophoretic mobility measurements show that this membrane has the isoelectric point between pH 2.3 and 2.6. These phenomena are considered to be caused by the shrinkage of the microcapsule membrane, as is seen in polyelectrolyte hydrogel membranes. This shrinkage can be made by an ion-complex formation between fixed charged groups, $-NH_3^+$ and $-COO^-$, located in the polymer chains.

Recently, self-regulating drug delivery systems, in which device itself senses stimuli and releases drug responding to the stimuli, have been developed and many materials have been tested to make devices. Information on the permeability characteristics of microcapsule membranes is important in the design of self-regulating drug delivery systems using microcapsules. Especially when we use microcapsules prepared by an interfacial polymerization method for drug devices, the drug should be loaded into microcapsule core after the water-loaded microcapsules are prepared by polymerization at the interface between aqueous solution(inside microcapsule) and organic solution (*1*). This is because the aqueous material or coproducts remain after polymerization and they have to be removed by washing microcapsule membranes with distilled water, before the usage of the microcapsules. During this process, aqueous drugs are removed by washing, even though the drugs are loaded as an aqueous solution inside microcapsules, before the microcapsule membrane is produced. Therefore, we have to prepare water-loaded microcapsules first, and then to load drugs into the microcapsule core from outside the microcapsules. Also, the drugs once loaded in the microcapsule core should be kept there and released when they are required. To satisfy these needs in loading, storage, and releasing processes, the microcapsule membrane permeability of drugs has to be different at different processes.

For this purpose, we prepared microcapsules with polyelectrolyte membranes that is, poly(L-lysine-*alt*-terephthalic acid) membranes, through which the permeability of solute was responding to the pH and ionic strength of the dispersing medium.

In this paper, we report some experimental results of the studies for the permeability characteristics, particle size, and charge density of poly(L-lysine-*alt*-terephthalic acid) microcapsule membrane. On the basis of such information, we shall discuss the pH- and ionic strength- induced structural change of these microcapsules.

## Materials and Methods

**Preparation of Poly(L-lysine-*alt*-terephthalic acid) Microcapsules.** We prepared water-loaded poly(L-lysine-*alt*-terephthalic acid) microcapsules by an interfacial polymerization technique (2). Mixed organic solvent consisting of cyclohexane and chloroform(3:1 v/v) were prepared. Ten milliliters of 1.5M L-lysine aqueous solution containing 2.5 M sodium carbonate were dispersed in 100 ml of the organic solution containing 1%(v/v) nonionic surfactant, SO-10, to yield a water-in-oil emulsion by stirring at 1360 rpm for 10 min. To this dispersion, 100 ml of 0.1M terephthaloyl dichloride solution in the mixed organic solvent was added and the stirring was continued for more 90 min. We added cyclohexane to stop the reaction, and separated poly(L-lysine-*alt*-terephthalic acid) microcapsules by centrifugation. These microcapsules were washed with cyclohexane, 2-propanol, ethanol, methanol, and distilled water. The microcapsules have been equilibrated in each solution before the usage for every measurement.

**Measurement of Permeability of Solutes through Microcapsule Membranes.** We measured permeation of anionic and cationic electrolyte ions through poly(L-lysine-*alt*-terephthalic acid) microcapsule membranes as a function of pH of the medium at different ionic strengths. The solutes used were 5-sulfosalicylic acid as an anion and phenyltrimethylammonium chloride as a cation. We suspended water-loaded poly(L-lysine-*alt*-terephthalic acid) microcapsules in a buffer solution and mixed this with a 5-sulfosalicylic acid solution or phenyltrimethylammonium chloride solution. The final concentration of the microcapsules is 20 % (v/v). We determined the solute concentration in the suspension medium spectrophotometrically at suitable time intervals after separating microcapsules by centrifugation and filtration through a Millipore filter.

We evaluated the permeability coefficient for solute, P, by the use of the following equation (3)

$$P = - C_f V_m / C_i A t \ \ln(C_t - C_f)/(C_i - C_f)$$

where $C_f$, $C_i$, and $C_t$ are the final, initial and intermediary(at time t) solute concentrations in the suspension medium, $V_m$ is the total volume of microcapsules, and A is the total surface area of microcapsules. The values of $V_m$ and A have been calculated from the precipitation volume and the particle size of microcapsules in each buffer solution. The buffer solutions used were those of KCl-HCl(pH 1.56-2.80), CH3COONa-CH3COOH(pH 3.88-3.95), Na2HPO4-KH2PO4(pH 6.18-8.17), Na2CO3-NaHCO3(pH 10.38-10.44), and Na2HPO4-NaOH(pH 11.37-12.30). The ionic strength of buffer solution was adjusted to the desired values by changing the concentrations of component electrolytes. We have repeated all experiments three times.

**Measurement of Electrophoretic Mobility of Microcapsules.** We measured the electrophoretic mobility of poly(L-lysine-*alt*-terephthalic acid) microcapsules with an automated electrokinetics analyzer (Pen Kem System 3000) in each of the buffer solutions.

**Measurement of Size of Microcapsules.** We measured the size of poly(L-lysine-*alt*-terephthalic acid) microcapsules in various buffer solutions by the usage of a particle sizer (Malvern System 3601). About 15000 microcapsules are dispersed in 1 ml of suspension medium.

## Results and Discussion

The constituent polymers of poly(L-lysine-*alt*-terephthalic acid) microcapsule membranes have both dissociable amino groups and dissociable carboxylic acid groups in their polymer chains. Therefore, these poly(L-lysine-*alt*-terephthalic acid) microcapsule membranes can be either positively charged or negatively charged, depending on the pH of the medium. The permeability to electrolyte ions of poly(L-lysine-*alt*-terephthalic acid) microcapsule membranes is thus expected to change when the pH of the medium is changed.

Figure 1 shows the permeability coefficients of 5-sulfosalicylic acid through the microcapsule membranes in the buffer solutions with various pH and ionic strengths. When the ionic strength of the medium was kept constant, the permeability coefficient was smaller at pH values lower than 4 and was larger at pH values higher than 6. Above pH 6, the permeability values are roughly constant in all alkaline solutions. The increase in ionic strength of the medium caused a rise of the permeability coefficient in the alkaline solution. It is of particular interest that the permeability exhibits a sharp change at around pH 4 in each ionic strength.

Figure 2 shows the permeability coefficient for phenyltrimethylammonium chloride as a function of pH of the medium at different ionic strengths. As in the case of the anion permeation shown in Fig. 1, the permeability coefficient of cations showed a drastic increase in the pH range between 4 and 6 at all ionic strengths when the pH of the medium was raised. In this case, however, the increase in ionic strength of the medium caused a decrease in the permeability coefficient.

The remarkable change in the permeability coefficient observed between pH 4 and 6 can be due mainly to the sharp increase in the microcapsule size in the same pH range, as will be described below.

Figure 3 shows the microscopic photographs of poly(L-lysine-*alt*-terephthalic acid) microcapsules in the buffer solutions of pH 4 and 8 with the ionic strength of 0.154. The size of poly(L-lysine-*alt*-terephthalic acid) microcapsules varies depending on the pH of the dispersing medium. Visual and optical microscopic observations reveal that the microcapsule membranes become nearly transparent in neutral and alkaline media while they can be clearly distinguished from the surroundings in acidic media.

To see the details of the dependence of the size on pH and the ionic strength of the dispersing medium, we measured the size of poly(L-lysine-*alt*-terephthalic acid) microcapsules in the buffer solutions. Figure 4 shows that the size of poly(L-lysine-*alt*-terephthalic acid) microcapsules drastically and discontinuously increase in a narrow pH range between 4 and 6 at all ionic strengths. Although the standard deviation, SD, was less than 20% in each measurement, the SD for the average values in the repeated measurements (3 times) was reduced to less than 1%.

These observations that the membrane becomes transparent and that the size of microcapsules increases in alkaline medium suggest that poly(L-lysine-*alt*-terephthalic acid) microcapsule membranes have a gel-like structure and undergo a volume transition at a pH value between 4 and 6, as usually seen in ionic hydrogels. This pH-driven volume transition is considered to be due to the change of the charge density in the microcapsule membrane.

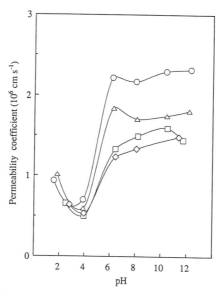

Figure 1. Permeability coefficient of 5-sulfosalicylic acid as a function of pH at various ionic strengths : 0.154(O),0.077(△),0.015(□), and 0.008(◇). SD was within the size of each symbol.

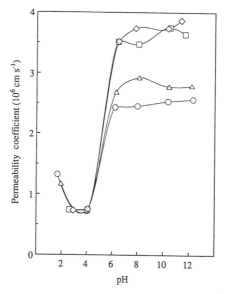

Figure 2. Permeability coefficient of phenyltrimethylammonium chloride as a function of pH at various ionic strengths : 0.154(O),0.077(△),0.015(□), and 0.008(◇). SD was within the size of each symbol.

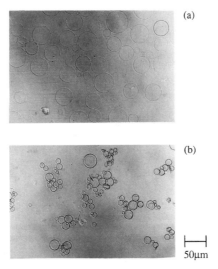

Figure 3. Microscopic photographs of poly(L-lysine-*alt*-terephthalic acid) microcapsules at pH 8 (a) and 4 (b) with the ionic strength of 0.154.

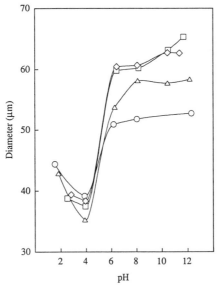

Figure 4. Size of the microcapsules as a function of pH of the medium at various ionic strengths : 0.154(O),0.077(△),0.015(□), and 0.008(◇).

To study the electric properties of these microcapsule membranes, we measured the electrophoretic mobility of poly(L-lysine-*alt*-terephthalic acid) microcapsules. Figure 5 shows the relationship between pH and the electrophoretic mobility of the microcapsules at various ionic strengths. It is observed that poly(L-lysine-*alt*-terephthalic acid) microcapsule membranes have their isoelectric points between pH 2.3 and 2.6 in the dispersing medium with the ionic strengths between 0.005 and 0.12. At the pH values below the isoelectric point, the microcapsule membranes have net positive charges. In this pH region, protonation of the amino groups is complete and dissociation of the carboxylic groups is considerably suppressed. On the other hand, the membranes are negatively charged at pH values higher than the isoelectric point due to decreased protonation of the amino groups and increased dissociation of the carboxylic groups. The mobility showed more negative values in the buffer solution with lower ionic strength when the pH was kept constant.

Suppose that the membrane has its isoelectric point at pH 2.3, the number densities of $-NH_3^+$ and $-COO^-$, both being counted relative to the number density of dissociable amino groups, are calculated as shown in Fig. 6. Here we have also assumed that pK values of carboxylic acid groups and amino groups are located at 3 and 9. The relative number density of $-NH_3^+$ is equal to that of $-COO^-$ at pH 2.3. Above this pH value, the number density of $-COO^-$ is more than that of $-NH_3^+$, whereas below that the number density of $-NH_3^+$ is more than that of $-COO^-$. Also, the net charge density, defined as the absolute value of the relative charge density of $-COO^-$ minus that of $-NH_3^+$, is given in Fig. 7. The net charge density below pH 2.3 is positive because of more $-NH_3^+$ than $-COO^-$, while it is negative above pH 2.3 because of more $-COO^-$ than $-NH_3^+$. Also, the total dissociable number density of carboxylic acid groups is calculated to be 6.0 times more than that of dissociable amino groups. Similar calculations by assuming the isoelectric point be 2.6 result in that the number density of dissociable carboxylic acid groups is 3.5 times more than that of dissociable amino groups and the general features seen in Figs. 6 and 7 are also found for this case. These two limiting values, 3.5 and 6.0 are not unreasonable, because poly(L-lysine-*alt*-terephthalic acid) may have several times more carboxylic acid groups than amino groups. The observation that the number of dissociable carboxylic acid groups are more than that of dissociable amino groups is not unexpected, if the structure of poly(L-lysine-*alt*-terephthalic acid) is taken into account . The following three different situations are possible. (1) If the both terminals of the polymer chain are anchored with lysine molecules, then the polymer chain has only two dissociable amino groups but has dissociable carboxylic acid groups as many as the lysine molecules used in the polymerization procedure. (2) If either one terminal of the polymer chain ends with lysine, the polymer chain has only one dissociable amino group but has one more dissociable carboxylic acid groups than the number of the lysine molecules used in the polymerization procedure. (3) If both terminals of the polymer chain are anchored with terephthalic acid molecules, the polymer chain has no dissociable amino groups. In any case, the polymer chain can contain more dissociable carboxyl groups than dissociable amino groups.

As was described before, both the size and the permeability of poly(L-lysine-*alt*-terephthalic acid) microcapsules show a drastic and discontinuous change in a narrow pH range between 4 and 6 at all ionic strengths. At pH 4, the permeability of poly(L-lysine-*alt*-terephthalic acid) microcapsules is minimum and the microcapsule size becomes smallest by shrinking. The isoelectric point of the microcapsules lies between 2.3 and 2.6. This implies that there are 5.5 times or 3.2 times more $-COO^-$ than $-NH_3^+$ in the microcapsule membranes at pH 4 so that the microcapsule membrane is negatively charged. The deviation of the pH value at

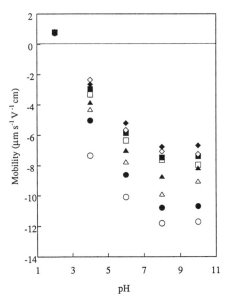

Figure 5. Effects of pH and ionic strengths on the electrophoretic mobility of the microcapsules. Ionic strengths are 0.005(O), 0.01(●), 0.02(△), 0.04(▲), 0.06(□), 0.08(■), 0.1(◇), 0.12(◆).

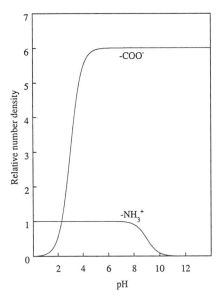

Figure 6. Number densities of dissociated carboxylic acid groups(-COO⁻) and amino groups(-NH₃⁺) relative to the number density of dissociable amino groups (-NH₂ and -NH₃⁺).

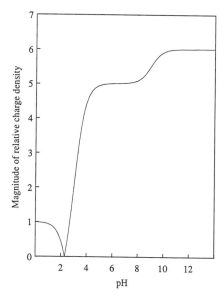

Figure 7. Relative charge density defined as the absolute value of the relative charge density of -COO⁻ minus that of -NH₃⁺.

which the permeability takes its minimum from the isoelectric point can be explained as follows. At pH 4, one -NH₃⁺ group can be neutralized with one of the nearby-located -COO⁻ groups, forming an ion-complex , which makes the polymer network shrunk. At around the isoelectric point, however, -COO⁻ can not always be neutralized with -NH₃⁺, because the number of -NH₃⁺ is equal to that of -COO⁻, giving less probability of forming an ion-complex. Therefore, at this pH, the both of COO⁻ and NH₃⁺ exist independently, without making a complex. At higher pH region than 6, the microcapsule membrane swells by the electric repulsion between -COO⁻ groups because the number density of -COO⁻ becomes much high than that of NH₃⁺. Inversely, at lower pH region than 2, the microcapsule membrane swells by the electric repulsion between NH₃⁺. The structural change of poly(L-lysine-*alt*-terephthalic acid) microcapsule membranes is schematically shown in Fig. 8.

The difference in permeability between cations and anions can be explained as follows. The space in which the solute anion moves along its concentration gradient is expected to be larger in the pH region higher than 4 and poly(L-lysine-*alt*-terephthalic acid) microcapsule membranes swell to a large extent. Therefore, the anion can permeate faster than in acidic media even under the influence of a strong electric field generated by the dissociated carboxylic acid groups.

Increase in the ionic strength of the medium at pH values higher than 6 increased the rate of anion permeation but decreased the rate of cation permeation due to increase in the screening effect of salt ions on the negative charges in the microcapsule membranes. These phenomena are partly due to the electrostatic attraction between the cations and negative charges on the dissociated carboxylic acid groups in poly(L-lysine-*alt*-terephthalic acid) microcapsule membranes. Although this interaction favorably affects the cation permeation through the membranes it is increasingly screened as the ionic strength of the medium increases

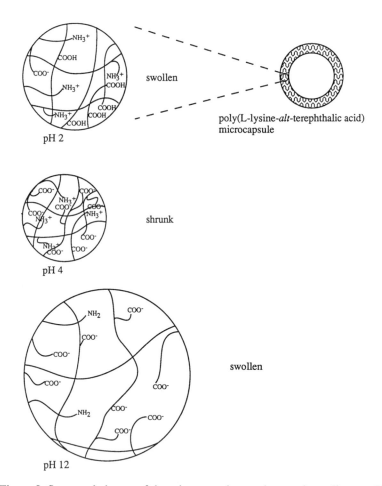

Figure 8. Structural change of the microcapsule membranes depending on pH.

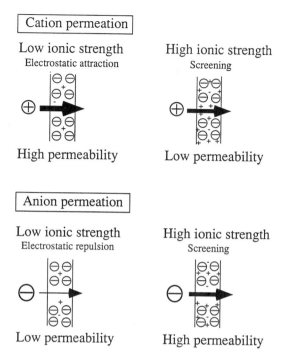

Figure 9. Ionic-strength dependence of solute permeation in alkaline solutions. Large circles with + or - signs stand for permeating solute molecules, small circles with - signs for membrane-fixed charges, and + and - signs for small mobile electrolyte ions.

to lower the rate of cation permeation. Thus, the effect of medium ionic strength on the cation permeation makes a sharp contrast with that on the anion permeation. The above discussion is summarized in Fig. 9.

**Acknowledgments**

This work was supported in part by the Kanagawa Academy of Science and Technology Foundation.

**References**
1.   Makino,K.; Mack,E.J.; Okano,T.; Kim,S.W.*J. Controlled Release,* **1990,** *12,* 235.
2.   Miyauchi,E.; Togawa,Y.; Makino, K.; Ohshima, H.; and Kondo, T. *J. Microencapsulation,* **1992,** *9,* 329.
3.   Takamura, K.; Koishi, M.; and Kondo, T. *Kolloid Z.u.Z.Polymere,* **1971,** *248,* 929.

RECEIVED July 5, 1993

# INDEXES

# Author Index

Amiji, Mansoor, 135
Antonucci, Joseph M., 147,184,191,210
Aronhime, M., 87
Bowersock, T. L., 288
Burczak, K., 275
Cheng, G. W., 191
Code, James E., 147
Cohn, D., 87
Cole, Michael A., 258
Friends, G., 76
Fujimoto, Keiji, 219,228
Harvey, John J., 258
Ikada, Yoshito, 35,228,275
Illinger, J. L., 103
Imanishi, Y., 66
Ito, Y., 66
Iwata, H., 306
Johnson, Russell A., 202
Kaplan, David Gilbert, 243
Karasz, F. E., 103
Kasuya, Y., 219
Kawaguchi, H., 219
Keeny, III, Spurgeon M., 210
Kondo, Tamotsu, 314
Künzler, J., 76
Lallone, R., 288
Langer, R., 16
Levy, M., 288
Li, Y., 297
Liou, F. J., 251
Makino, Kimiko, 314
Minato, Masao, 228
Miyamoto, M., 219
Miyauchi, Ei-ichi, 314
Mizoguchi, J., 306

Niu, G. G. C., 251
Ogawa, N., 306
Ohshima, Hiroyuki, 314
Ozark, R., 76
Park, H., 2
Park, Kinam, 2,135,288
Penhasi, A., 87
Prud'homme, Robert K., 157
Reed, B. B., 184
Rogers, K. R., 116
Roweton, S. L., 58
Ryker, D., 288
Samuels, M. L., 288
Schneider, N. S., 103
Schumacher, Gary E., 147
Shalaby, Shalaby W., 58,116,202
Shalaby, W. S. W., 288
Shiraki, K., 275
Spencer, H. G., 297
Stansbury, Jeffrey W., 171,184,191
Suzuki, K., 66
Takagi, T., 306
Tesk, John A., 210
Togawa, Yuko, 314
Tomihata, K., 275
Trokanski, M., 76
Tsai, S. W., 251
Vacanti, C. A., 16
Vacanti, J. P., 16
Wang, Francis W., 210
Wang, Y. J., 251
White, M. R., 288
Williams, Joel L., 49
Yin, Yu-Ling, 157
Zhang, Liu, 258

# Affiliation Index

American Dental Association Health
  Foundation, 184
Army Research Laboratory, 103
Asahi Chemical Industry Company, Ltd., 306
Bausch and Lomb Inc., 76
Becton Dickinson Research Center, 49

Brookwood Biomedical Institute, 288
Children's Hospital, 16
Clemson University, 58,116,202,297
Falco Biosystems, 306
Geo-Centers, Inc., 103
Harvard Medical School, 16

The Hebrew University of Jerusalem, 87
Japanese Red Cross, 219
Keio University, 35,66,219,228,275
Massachusetts General Hospital, 16
Massachusetts Institute of Technology, 16
National Cardiovascular Center, 306
National Institute of Standards and
    Technology, 147,171,184,191,210
National Institutes of Health, 147

National Yang-Ming Medical College, 251
Princeton University, 157
Purdue University, 2,135,288
Science University of Tokyo, 314
University of Illinois, 258
University of Massachusetts, 103
West China University of Medical
    Sciences, 191

# Subject Index

## A

Absorbable oxalate polymers
    biological performance of transcortical
        pins, 208–209
    chemical composition of polymerization
        charges, 204t
    crystallinity vs. composition, 206,207f
    development, 202–203
    experimental description, 203
    in vitro absorption, 208f
    IR frequencies of functional groups, 206
    polymer characterization procedure, 205
    polymer synthetic procedure, 204
    postpolymerization schedules, 204,205t
    properties, 206,207t
    study procedure, 205–206
Acrylate-containing spiro orthoesters,
    structures, 172,173f
Acrylic monomers and oligomers, multifunc-
    tional hydrophilic and hydrophobic, See
    Multifunctional hydrophilic and hydro-
    phobic acrylic monomers and oligomers
Adhesion, biomaterials, 45,46f
Affinity separation, advantages for protein
    and peptide separation from biological
    broths and blood, 297
Aging, proteins, 247–248
Alkylene oxalate polymers, application, 202
Amine-derivatized poly(ethylene terephthal-
    ate), preparation, 139,141f
Annealing temperature, poly(vinyl alcohol)
    hydrogel property effect, 228–241
Arg-Gly-Asp-Ser, function, 221
Arg-Gly-Asp-Ser-carrying microspheres,
    leukocyte activation, 220–227

Arg-Gly-Asp-Ser-immobilized membrane
    cell adhesion activity, 72–74f
    coupling, 72
    synthesis, 68,72
Ascorbic acid, chemical structure, 148,149f
Ascorbic acid as etchant–conditioner for
    resin bonding to dentin
    experimental procedure, 148–150
    tensile bond strengths, 150,154t
    treatment time vs. surface changes,
        150–153f,155
ASTM test methods, microcellular foam
    characterization, 60
(4-Azido-2-nitrophenyl)polyethylene glycol,
    preparation, 139,142f

## B

Bacterial exotoxins, poly(methacrylic acid)
    hydrogels as carriers, 288–295
Bioactive chemicals, polymerizable, use for
    surface biolization, 66–74
Bioadhesion, definition, 45
Biocompatibility
    bioadhesive surfaces,45–48
    definition, 37
    groups, 37,38f
    interfacial, See Interfacial biocompatibility
    surface property effect, 66
Biocompatible materials, classification of
    approaches, 66,67f
Biodegradability, starch in starch–plastic
    blends, 264,266f
Biodegradable drug delivery devices, 54,56

Biodegradation of native and denatured
collagen
cross-linking agent effect, 283–285
experimental materials, 275
film preparation procedure, 275
in vitro degradation
bisepoxy compound cross-linked films,
277,281*f*
cross-linked collagen films, 277,282*f*
cross-linked gelatin films, 277,282*f*,283
glutaraldehyde cross-linked films,
277,281*f*
with collagenase, 276
in vivo degradation
cross-linked gelatin films, 283*f*
subcutaneous implantation, 276–277
water content effect, 283,284*f*
Biolization, surface, *See* Surface biolization
by grafting polymerizable bioactive
chemicals
Biological responses, classifications, 37,38*f*
Biological surface modification methods,
examples, 136
Biomaterial–living system interactions,
schematic representation, 37,38*f*
Biomedical materials, 35,36*t*
Blood compatibility, surface design, 68–71*f*
Bone
control of innate shape, 247
stress-induced growth, 246–247
synthetic biodegradable polymers, 25–27*f*
Brunauer–Emmett–Teller surface area,
microcellular foam characterization, 61

C

Capillary flow porometry, microcellular
foam characterization, 61
Cartilage, synthetic biodegradable
polymers, 19–27
Cell-adhesive surfaces, biocompatibility,
47–48
Cell–polymer constructs, transplantation
into animals, 21–25
Cell transplantation
history, 17
synthetic biodegradable polymers, 17–19
CH$_2$=CHCONH(CH$_2$)$_5$COOSu, synthesis, 72

Chinese hamster ovary cell, insulin releasing,
preparation by human insulin gene
transfection, 306–313
Chondrocytes, transplantation, 20–23*f*
Coating of tablets, use of polymers, 3,5,6*t*
Collagen
chemical environment vs. conformation, 243
cross-linking and biodegradation, 275–285
stress-induced crystallinity, 243
Collagenous tissue growth and resorption
bone growth and development, 246–247
deposition vs. precursor production, 246
experimental procedure, 244
occurrence, 243–244
phase transition effect, 246
prolonged immobilization vs.
resorption, 248
protein aging, 247–248
stress-induced crystallinity, 245–246
temperature effect on fibers, 244,245*t*
Contact lens materials, properties requiring
optimization, 76
Controlled release, rationale, 49
Controlled release dosage forms, use of
polymers, 5,7–12
Controlled-release formulations, starch–
plastic blends, 259,260*f*
Controlled-release polymeric drug devices
development, 49
forms, 49,51*t*
Convection, description, 261
Conventional dosage forms, use of
polymers, 2–6
Counterion condensation
determination, 161*f*,166*f*
prediction of salt partitioning in gel
and solution, 164
Cross-linking of native and denatured
collagen
bisepoxy compound concentration vs.
water content, 277,279*f*
bisepoxy compound cross-linking procedure, 276
chemical cross-linking, mechanism,
284,285*f*
cross-linking by dehydrothermal treatment, 276
cross-linking by UV irradiation, 276
cross-linking effect, 284–285
experimental materials, 275

Cross-linking of native and denatured
  collagen—*Continued*
film preparation procedure, 275
glutaraldehyde concentration
  vs. mechanical properties, 277,278*f*
  vs. water content, 277,278*f*
glutaraldehyde cross-linking procedure, 276
treatment temperature vs. water content,
  277,280*f*
UV irradiation time vs. water content,
  277,280*f*
water content of cross-linked gels,
  measurement, 276
water-soluble carbodiimide concentration
  vs. water content, 277,279*f*
water-soluble carbodiimide cross-linking
  procedure, 276
Crystallinity, 229–230
Culture supernatants, use in vaccines
  against pasteurellosis, 288
Cyclic acetals, ring-opening dental resin
  systems, 184–190

D

Degradation of starch–plastic blends,
  kinetic model, 258–274
Denatured collagen, cross-linking and
  biodegradation, 275–285
Dental resin(s), fluorescent cure monitoring,
  210–217
Dental resin systems based on cyclic acetals,
  use of ring-opening polymerization for
  formation, 184–190
Dentin
  acid agent pretreatment for bonding,
    148–155
  bonding problems, 147–148
  composition, 147
Diffusion, description, 261
Diffusion-controlled drug delivery systems,
  use of polymers, 7,8*f*,10*t*
Difunctional acrylic monomers,
  synthesis, 191
Dimethyl isophthalate, structure, 203
Dimethyl terephthalate, structure, 203
Dissolution, definition, 9
Dissolution-controlled drug delivery
  systems, use of polymers, 7–10*t*

Dissolution drug delivery devices, 54,56
Drug packaging, use of polymers, 12–13
Dye-grafted, polyethylenimine-coated,
  formed-in-place class affinity membranes
  for selective protein separation
  applications, 303
  cross-flow velocity vs. protein binding,
    300–302*f*
  efficiency, 303
  experimental procedure, 298–300
  future work, 303,305
  human serum albumin elution, 301,302*f*
  human serum albumin separation from
    human plasma, 301,303,304*f*
  hydrophobic interactions vs. protein
    binding, 301
  ionic interactions vs. protein binding, 301
  material selection, 300
  performance testing procedure, 300
  pH vs. protein binding, 301,304*f*

E

Encapsulated drug devices
  examples, 49,50*t*
  polymer selection, 55*t*,56
  rate-controlled membranes, 50
  release rate, 50
  single-component system, 51–53*t*
Encapsulation, biomaterials, 42,44*t*,45
Endothelial cells, use with synthetic biodegradable polymers, 30–31
Entrapment, biomaterials, 39,41
Ethyl (*N,N*-dimethylamino)benzoate, fluorescent cure monitoring of dental resins,
  212–215*f*
Ethylene oxide based multifunctional vinyl
  oligomers, synthesis, 193–195
Excipients, use of polymers, 3,4*t*
Exotoxins, bacterial, poly(methacrylic acid)
  hydrogels as carriers, 288–295

F

Fluorescent cure monitoring of dental resins
  advantages, 210–211
  correlation of fluorescence intensity ratio
    with IR degree of cure, 214,217*f*

Fluorescent cure monitoring of dental resins—*Continued*
curing conditions, 211,213*t*
ethyl (*N,N*-dimethylamino)benzoate, 212–215*f*
experimental description, 211
fluorescence intensity ratio vs. curing conditions, 214,216
fluorescence measurement procedure, 211–212
IR degree of cure vs. curing conditions, 214,216*f*
IR measurement procedure, 212
resin materials, 211
Fluorescent probes, use for in situ monitoring of resin cure kinetics, 210
Fluorosiloxanes, hydrophilic, lipid resistant, *See* Hydrophilic lipid-resistant fluorosiloxanes
Flux of molecules within porous plastic matrix, 259,261
Foams, microcellular, *See* Microcellular foams
Formed-in-place class affinity membranes, dye grafted, polyethylenimine coated, *See* Dye-grafted, polyethylenimine-coated, formed-in-place class affinity membranes for selective protein separation
Formed-in-place microfiltration membranes, application in bioseparations, 298
Froth, morphology, 60

**G**

Gas–polymer melt method, microcellular foam production, 59
Gel–solution systems, polyelectrolyte, salt partitioning, 157–169
Gene therapy, use in diabetes treatment, 306
Glass-transition temperature, water effect, 109–113
Graft polymerization, biomaterials, 42,43*f*
Growth, collagenous tissue, 243–248

**H**

Hepatocyte transplantation, use of synthetic biodegradable polymers, 26,28–30

Highly fluorinated difunctional monomers and multifunctional oligomers, 195–197*f*
Human insulin gene, preparation of insulin-releasing Chinese hamster ovary cell, 306–313
Hydrodynamic dispersion, description, 261
Hydrogels
advantages for oral delivery of drugs, 289–290
pH-sensitive, *See* pH-sensitive hydrogels based on 2-hydroxyethyl methacrylate and poly(vinyl alcohol) esterified with maleic anhydride
poly(vinyl alcohol), *See* Poly(vinyl alcohol) hydrogels
Hydrophilic lipid-resistant fluorosiloxanes
experimental procedure, 77–79
mechanical property optimization, 80–86
oleic acid absorption vs. concentration, 80,81*f*
oleic acid uptake vs. concentration, 80,83*f*,84
physical and mechanical property objectives, 78,80
physical property optimization, 80–86
synthetic procedure, 77–79
total weight loss vs. heating time, 84,85*f*
water contact angle vs. time, 84*f*
Hydrophilic multifunctional acrylic monomers and oligomers, *See* Multifunctional hydrophilic and hydrophobic acrylic monomers and oligomers
Hydrophilic polyurethanes, water solubility, 103–104

**I**

Insulin gene, human, preparation of insulin-releasing Chinese hamster ovary cell, 306–313
Insulin-releasing Chinese hamster ovary cell preparation by human insulin gene transfection
anti Chinese hamster antibody level vs. cell type, 309,312*f*,313
blood glucose levels
after microcarrier-cultured cell implantation, 309,311*f*,313
after microencapsulated cell implantation, 309,311*f*,313

Insulin-releasing Chinese hamster ovary
cell preparation by human insulin gene
transfection—*Continued*
blood glucose levels—*Continued*
after singly dispersed cell implantation,
309,310*f*,313
cell preparation, 307
experimental description, 306–307
future work, 313
immunosuppressive therapy procedure, 307
insulin release by cell line vs. methotrexate
concentration, 309,310*f*
insulin-releasing cell implantation into
diabetic mice, 307
plasmid introduction into cells, 307
plasmid preparation, 307,308*f*
strategy, 307,308*f*
Interfacial biocompatibility
bioadhesive surfaces, 45–48
classification, 37,40*f*
nonstimulative surfaces, 37,39–46
Intestinal mucosa, use of cell–polymer
constructs, 31
Intramedullary stem, methods of fixation, 117
Inverse emulsion foams, morphology, 60
Ion-exchange drug delivery systems, use of
polymers, 9,12
Ionic strength, permeation effect in
poly(L-lysine-*alt*-terephthalic acid)
microcapsule membranes, 314–323

K

Katchalsky's theory, prediction of salt
partitioning in gel and solution, 162–164
Kinetic model for degradation of starch–
plastic blends with controlled-release
potential
biodegradability, 264,266*f*
correlation of model parameters to starch
degradation in blends, 265,266*f*
derivation of mathematical model, 271–273
enzyme assay procedure, 263
experimental materials, 262–263
flux of molecules within porous plastic
matrix, 259,261
kinetic model and soil degradation results,
relationship, 268,270*f*,271
mathematic model parameter estimation
procedure, 263–264

Kinetic model for degradation of starch–
plastic blends with controlled-release
potential—*Continued*
model, 264–269
nomenclature, 274
parameter estimation for mathematical
model, 265,267*t*–269*f*
soil degradation study procedure, 263
starch digestion and product release
model, 261–262

L

Leukocyte activation by
Arg-Gly-Asp-Ser-carrying microspheres
amount of immobilized peptide, 223,224*t*
cytoskeleton inhibitors vs. oxygen consumption, 223,226*f*,227
hypochlorite ion liberation vs. incubation
time, 223,225*f*
leukocyte activation evaluation
procedure, 221
microsphere preparation procedure,
221,222*f*
oxygen consumption vs. incubation time,
223,224*t*
phagocytosis evaluation procedure, 221
superoxide anion liberation vs. incubation
time, 223,225*f*
synergistic mechanism, 227
Light microscopy, microcellular foam
characterization, 60
Lipid-resistant fluorosiloxanes, hydrophilic,
*See* Hydrophilic, lipid-resistant fluorosiloxanes

M

Macroscopically nonstimulative surfaces,
biocompatibility, 37,39–41*f*
Materials for biomedical applications, 135
Matrix devices
biodegradable devices, 54,56
description, 52
dissolution devices, 54,56
nondegradable devices, 54
polymer selection, 55*t*,56
release rates, 52,54

Medicofunctionality, bulk property effect, 66
Mercury porosimetry, microcellular foam characterization, 61
Methacrylate-substituted spiro orthoester, polymerization, 172,173f
Methoxypolyethylene glycol, coupling to polyurethane soft segment, 139,140f
2-Methylene spiro orthocarbonate with pendant methacrylate group
blue light vs. sunlamp effect on methacrylate conversion, 181,182t
$^{13}$C-NMR spectrum of monomer, 174,176f
cyclic carbonate elimination, 179,180f
dental resin formulation preparation, 177
IR spectra of monomer and intermediates, 174,175f
isomers, proportions, 177,178f
isomers of monomer, 177,178f
monomer synthesis, 172,174,175f
polymerization by non-ring-opening and double-ring-opening pathways, 179,180f
polymerization initiators, 174t,177
structure, 172,173f
vinyl ether type group effect on polymerization, 179,181
Methylenebisacrylates, formation mechanism, 191,193,194f
Microcellular, definition, 58
Microcellular foams
applications, 61–62
characterization methods, 60–61
definition, 58
formation methods, 59
morphologies, 59–60
potential biomedical applications, 62–64
types, 58
Microcellular materials, forms, 58
Microspheres, Arg-Gly-Asp-Ser carrying, leukocyte activation, 220–227
Microspheres containing immobilized bioactive molecules, applications, 220–221
Molecularly nonstimulative surfaces, biocompatibility, 39,41–46
Monolithic drug delivery systems, 7,10t
Multicomponent reservoir, 52
Multifunctional acrylic oligomers, synthesis, 191
Multifunctional hydrophilic and hydrophobic acrylic monomers and oligomers
applications, 200

Multifunctional hydrophilic and hydrophobic acrylic monomers and oligomers—
Continued
cyclopolymerization vs. cross-linking, 199,200f
ethylene oxide based multifunctional vinyl oligomer synthesis, 193–195
highly fluorinated difunctional monomers and multifunctional oligomer
solvent effect on synthesis, 199
structure, 199
synthesis, 195–197f
tetramethyldisiloxane-based multifunctional vinyl oligomers, 195,198f,199

N

Native collagen, cross-linking and biodegradation, 275–285
Natural absorbable polymers, limitation, 202
Net charge density, definition, 319
Nondegradable drug delivery devices, 54
Nonstimulative surfaces, biocompatibility, 37,39–46
Nonvinyl cyclic acetal, synthesis, 185,188f
Nylon 12 open-cell foam, microstructure, 62–64

O

Oral vaccination of cattle
advantages, 289
gut-associated lymphoid tissue, stimulation, 289
poly(methacrylic acid) hydrogels as carriers of bacterial exotoxins, 289–295
requirements, 289
Organ transplantation, problems, 16
Orthopedic prostheses
current technology, 116
optimum polymer–bone interface, 117
Osmotic drug delivery systems, use of polymers, 9,11t
Oxalate-based polymers, See Absorbable oxalate polymers
Oxybismethacrylates, formation mechanism, 191,192f

P

*Pasteurella haemolytica*, prevention, 288
Pasteurellosis, vaccines, 288
Peptide separation from biological broths
and blood, affinity techniques, 297
Percutaneous transport, polymeric devices,
49–56
Permeability, microcellular foam
characterization, 60
Permeability coefficient for solute,
calculation, 315
Permeation, pH, and ionic strength effect
in poly(L-lysine-*alt*-terephthalic acid)
microcapsule membranes, 314–323
pH, permeation effect in poly(L-lysine-*alt*-
terephthalic acid) microcapsule membranes, 314–323
pH-sensitive hydrogels based on 2-hydroxy-
ethyl methacrylate and poly(vinyl alcohol)
esterified with maleic anhydride
characterization of hydrogel, 252,254–257*f*
chemicals, 251
hydrogel thin film preparation, 252,253*t*
poly(vinyl alcohol) esterification with
maleic anhydride, 252,253*f*
poly(vinyl alcohol) esterified with maleic
anhydride preparation, 251–252
Pharmaceutical products, role of
polymers, 2–13
Phase transition, control of collagenous
tissue growth and resorption, 243–248
Physicochemical surface modification
methods, examples, 136
Plastic matrices, degradation of starch, 258
Poisson–Boltzmann equation, solution,
158–162
Poly(alkylene oxalate), structure, 203
Poly(*trans*-1,4-cyclohexyldicarbonyl-*co*-
hexa-methylene oxalate), structure, 203
Poly(dimethylsiloxane)
advantages and disadvantages for contact
lens applications, 76–77
wettable contact lens material, development, 77
Polyelectrolyte gel–solution systems, salt
partitioning, 157–169
Poly(ether urethane amide) segmented
elastomers
diacid effect on kinetics, 93,97,98*f*

Poly(ether urethane amide) segmented
elastomers—*Continued*
differential scanning calorimetic thermograms, 97,99*f*
elemental analysis, 93,97*t*
experimental description, 88–89
[1]H-NMR spectrum, 93,96*f*
intrinsic viscosity effect on isocyanate
conversions, 97,98*f*
IR spectra, 93,95*f*
kinetics of reactions, 90–94
model system synthetic procedure, 89
poly(ether urethane amide) synthetic
procedure, 89
polymer analytical procedure, 89
polymerization reaction, 93,95–99
solvent effect on synthesis, 99–101*f*
synthesis, proposed reaction mechanism, 87
Poly(ethylene oxide)
biocompatibility effect, 138–143
conformation in aqueous environment, 136
surface modification of polymeric biomaterials, 135–143
Poly(2-hydroxyethyl methacrylate), biomedical applications, 251
Poly(2-hydroxyethyl methacrylate)-*co*-poly-
(vinyl alcohol) maleic anhydride ester
hydrogel, characterization, 251–257
Poly(L-lysine-*alt*-terephthalic acid) microcapsule membranes
electrophoretic mobility measurement
procedure, 315–316
ionic strength
vs. microcapsule size, 316,318*f*
vs. permeability, 316,317*f*
vs. solute permeation in alkaline
solutions, 321,323*f*
microcapsule preparation, 315
microcapsule size measurement
procedure, 316
net charge density, 319,321*f*
number densities of dissociated amino
groups, 319,320*f*
pH-driven volume transition, 316
pH vs. electrophoretic mobility, 319,320*f*
pH vs. microcapsule size, 316,318*f*
pH vs. permeability, 316,317*f*
pH vs. structural changes, 319,321,322*f*
solute permeability measurement
procedure, 315

Polymer(s) in pharmaceutical products
  applications, 2
  controlled release dosage form usage,
    5,7–12
  conventional dosage form usage, 2–6
  drug packaging usage, 12–13
Polymer selection for drug delivery devices,
  influencing factors, 55*t*,56
Polymeric devices for transcutaneous and
  percutaneous transport
  encapsulated drug devices, 49–53
  matrix devices, 52,54,56
  polymer selection, 55*t*,56
Polymeric materials
  advantages as biomaterials, 135
  surface modification with poly(ethylene
    oxide), 135–143
Polymeric prodrugs, description, 12
Polymerizable bioactive chemicals, use for
  surface biolization, 66–74
Polymerization shrinkage, reduction
  methods, 171
Polymerization shrinkage in dental resin
  composite systems, source of
  problems, 184
Poly(methacrylic acid), advantages for use
  in oral delivery of drugs, 290
Poly(methacrylic acid) hydrogels as carriers
  of bacterial exotoxins
  antigen release, 291–293*t*
  calf challenge results, 292,294*t*
  calf vaccine trial procedure, 291–292
  culture supernatant preparation, 290
  hydrogel loading with culture supernatants,
    290–291
  hydrogel passage through gastrointestinal
    tract, 292–293,295*f*
  hydrogel preparation, 290
Polypropylene open-cell foam, microstructure, 62–64
Polyurethane(s) containing hydrophilic
  block copolymer soft segment–water
  interaction
  differential scanning calorimetric measure-
    ments of sorbed water state, 107–109
  experimental procedure, 104,105*t*
  water uptake
    immersion measurements, 105*t*–107
    vs. temperature, 105*t*–107
    vs. glass-transition temperature, 109–115

Polyurethane elastomer, 87–88
Poly(vinyl alcohol), properties, 251
Poly(vinyl alcohol) film, methods for
  insolubility, 228
Poly(vinyl alcohol) hydrogels
  bulk properties, 231–233*f*
  contact angle measurement procedure, 230
  crystallinity determination procedure,
    229–230
  crystallinity vs. annealing temperature,
    231,233*f*
  ex vivo arteriovenous shunt experimental
    procedure, 231
  ex vivo platelet adhesion, 237,239*f*
  experimental objective, 229
  film preparation, 229
  in vitro platelet adhesion
    vs. crystallinity, 237,238*f*
    vs. incubation time, 234,236*f*,237
    vs. water content, 237,238*f*
  in vitro protein adsorption
    vs. crystallinity, 234,235*f*
    vs. water content, 234,235*f*
  interfacial structure models, 240*f*,241
  platelet adhesion measurement procedure,
    230–231
  protein adsorption determination
    procedure, 230
  sol fraction
    determination procedure, 229
    vs. annealing temperature, 231,232*f*
  studies for biomedical applications, 237
  surface properties, 234*t*
  water contact angle vs. annealing
    temperature, 234*t*
  water content
    determination procedure, 229
    vs. annealing temperature, 231,232*f*
    vs. crystallinity, 231,233*f*
  $\zeta$ potential measurement procedure, 230
  $\zeta$ potential vs. annealing temperature, 234*t*
Prodrugs, polymeric, description, 12
Property–structure relationship, absorbable
  oxalate polymers, 202–209
Protein(s), aging, 247–248
Protein separation, use of dye-grafted, polyethylenimine-coated, formed-in-place
  class affinity membranes, 297–305
Protein separation from biological broths
  and blood, affinity techniques, 297

**R**

Reservoir drug delivery systems, 7,10*t*
Resin bonding to dentin, ascorbic acid as
  etchant–conditioner, 147–155
Resorption, collagenous tissue, 243–248
Reticulate foams, morphology, 60
Ring-opening dental resin systems based on
  cyclic acetals
bulk polymerization procedure, 185,187
diametral tensile strength vs. composition,
  189*t*,190
experimental procedure, 185
free radical ring opening, 184,186*f*
monomers used, 185,186*f*
nonvinyl cyclic acetal synthetic procedure,
  185,188*f*
polymerization pathway, 187–189
solution polymerization procedure, 185,187
vinyl cyclic acetal synthetic procedure,
  185,188*f*
Ring-opening polymerization, 2-methylene
  spiro orthocarbonate with pendant
  methacrylate group, 171–182


**S**

Salt partitioning in polyelectrolyte gel–
  solution systems
counterion condensation determination,
  161*f*–162,166*f*
counterion condensation–electrostatic
  interaction theory, 164–166
experimental procedure, 157
experimental vs. theoretical predictions,
  165–168*f*
importance, 157
Katchalsky's theory, 162–164
pH vs. degree of neutralization, 166,169*f*
Poisson–Boltzmann solution equation,
  158–162
salt absorption vs. gel cross-link density,
  165,168*t*
studies, 157
Scanning electron microscopy, microcellular foam characterization, 60
Self-regulating drug delivery systems, membrane permeability requirements, 314–315

Single-component drug delivery
  description, 51
  release rates, 51–53*t*
Sintering, microcellular foam production, 59
Sol fraction, calculation, 229
Solution–gel systems, polyelectrolyte, salt
  partitioning, 157–169
Starch
  biodegradability in starch–plastic blends,
    264,266*f*
  degradation in plastic matrices, 258
Starch-based controlled-release pesticides,
  function, 259
Starch degradation, kinetic model, 264–269
Starch digestion, kinetic model, 261–262
Starch–plastic blends
  kinetic model for degradation, 259–274
  potential as controlled-release formulations,
    259,260*f*
Steric repulsion
  surface-bound hydrophilic polymers,
    136,137*f*
  surface-bound poly(ethylene oxide), 136
Stress-induced crystallinity, description, 243
Structure–property relationship, absorbable
  oxalate polymers, 202–209
Surface biolization by grafting polymerizable
  bioactive chemicals
  process, 66,67*f*
  surface design for blood compatibility,
    68–71*f*
  surface design of cell adhesive materials,
    68,72–74*f*
Surface-bound hydrophilic polymers, steric
  repulsion, 136,137*f*
Surface-bound poly(ethylene oxide), steric
  repulsion, 136
Surface design for blood compatibility
  polymerizable thrombin inhibitor, 68,71*f*
  poly(vinyl sulfonate), 68–70*f*
Surface design of cell adhesive materials
  Arg-Gly-Asp-Ser coupling, 72
  Arg-Gly-Asp-Ser-immobilized membrane
    synthesis, 68,72
  cell adhesion activity of Arg-Gly-Asp-Ser-
    grafted membrane, 72–74*f*
  $CH_2=CHCONH(CH_2)_5COOSu$
    synthesis, 72
  graft polymerization, 72
Surface modification of biomaterials, 135

Surface modification of polymeric biomaterials with poly(ethylene oxide), steric
repulsion approach, 135–143
Surface phosphonylated thermoplastic
polymers
cytotoxicity, 124–127*f*
cytotoxicity evaluation procedure, 118
hydroxyapatite formation,
124,128*t,f*–133*f*
hydroxyapatite formation procedure,
118–119
surface characterization procedure, 117–118
surface modification, 117,119*t*–124
Sustained release, rationale, 49
Syntactic foams, morphology, 60
Synthetic biodegradable polymers, tissue
engineering, 16–31


T

Tablet coating, use of polymers, 3,5,6*t*
Tablet excipients, use of polymers, 3,4*t*
Tetramethyldisiloxane-based multifunctional
vinyl oligomers, synthesis, 195,198*f,*199
Thermally induced phase separation, 59
Thermoplastic polymers, surface phosphonylated, *See* Surface phosphonylated
thermoplastic polymers
Tissue–adhesive surfaces, biocompatibility,
45–47*t*
Tissue engineering using synthetic biodegradable polymers
advantages, 16–17
bone, 25–27*f*
cartilage, 19–25

Tissue engineering using synthetic biodegradable polymers—*Continued*
cell transplantation, 17–19
endothelial cells, 30–31
hepatocyte transplantation, 26,28–30
Toxic compounds related to polymers,
examples, 35,36*t*
Tracheal epithelium, use of cell–polymer
constructs, 31
Transcutaneous transport, polymeric
devices, 49–56
Transdermal products, examples, 49,50*t*


U

Ultra high molecular weight polyethylene,
use in orthopedic protheses, 116–117
Urethelium, use of cell–polymer constructs,
30–31


V

Vaccination, oral, *See* Oral vaccination of
cattle
Vinyl cyclic acetal, synthesis, 185,188*f*
Vitamin C, ascorbic acid, 148,149*f*


W

Water, interaction with polyurethanes
containing hydrophilic block copolymer
soft segments, 103–115
Water content, calculation, 229

Production:  C. Buzzell-Martin
Indexing:  Deborah H. Steiner
Acquisition:  Anne Wilson

Printed and bound by Maple Press, York, PA

# Bestsellers from ACS Books

*The ACS Style Guide: A Manual for Authors and Editors*
Edited by Janet S. Dodd
264 pp; clothbound ISBN 0–8412–0917–0; paperback ISBN 0–8412–0943–X

*The Basics of Technical Communicating*
By B. Edward Cain
ACS Professional Reference Book; 198 pp;
clothbound ISBN 0–8412–1451–4; paperback ISBN 0–8412–1452–2

*Chemical Activities* (student and teacher editions)
By Christie L. Borgford and Lee R. Summerlin
330 pp; spiralbound ISBN 0–8412–1417–4; teacher ed. ISBN 0–8412–1416–6

*Chemical Demonstrations: A Sourcebook for Teachers,*
*Volumes 1 and 2,* Second Edition
Volume 1 by Lee R. Summerlin and James L. Ealy, Jr.;
Vol. 1, 198 pp; spiralbound ISBN 0–8412–1481–6;
Volume 2 by Lee R. Summerlin, Christie L. Borgford, and Julie B. Ealy
Vol. 2, 234 pp; spiralbound ISBN 0–8412–1535–9

*Chemistry and Crime: From Sherlock Holmes to Today's Courtroom*
Edited by Samuel M. Gerber
135 pp; clothbound ISBN 0–8412–0784–4; paperback ISBN 0–8412–0785–2

*Writing the Laboratory Notebook*
By Howard M. Kanare
145 pp; clothbound ISBN 0–8412–0906–5; paperback ISBN 0–8412–0933–2

*Developing a Chemical Hygiene Plan*
By Jay A. Young, Warren K. Kingsley, and George H. Wahl, Jr.
paperback ISBN 0–8412–1876–5

*Introduction to Microwave Sample Preparation: Theory and Practice*
Edited by H. M. Kingston and Lois B. Jassie
263 pp; clothbound ISBN 0–8412–1450–6

*Principles of Environmental Sampling*
Edited by Lawrence H. Keith
ACS Professional Reference Book; 458 pp;
clothbound ISBN 0–8412–1173–6; paperback ISBN 0–8412–1437–9

*Biotechnology and Materials Science: Chemistry for the Future*
Edited by Mary L. Good (Jacqueline K. Barton, Associate Editor)
135 pp; clothbound ISBN 0–8412–1472–7; paperback ISBN 0–8412–1473–5

For further information and a free catalog of ACS books, contact:
American Chemical Society
Distribution Office, Department 225
1155 16th Street, NW, Washington, DC 20036
Telephone 800–227–5558

# Highlights from ACS Books

*Good Laboratory Practice Standards: Applications for Field and Laboratory Studies*
Edited by Willa Y. Garner, Maureen S. Barge, and James P. Ussary
ACS Professional Reference Book; 572 pp; clothbound ISBN 0–8412–2192–8

*Silent Spring Revisited*
Edited by Gino J. Marco, Robert M. Hollingworth, and William Durham
214 pp; clothbound ISBN 0–8412–0980–4; paperback ISBN 0–8412–0981–2

*The Microkinetics of Heterogeneous Catalysis*
By James A. Dumesic, Dale F. Rudd, Luis M. Aparicio, James E. Rekoske,
and Andrés A. Treviño
ACS Professional Reference Book; 316 pp; clothbound ISBN 0–8412–2214–2

*Helping Your Child Learn Science*
By Nancy Paulu with Margery Martin; Illustrated by Margaret Scott
58 pp; paperback ISBN 0–8412–2626–1

*Handbook of Chemical Property Estimation Methods*
By Warren J. Lyman, William F. Reehl, and David H. Rosenblatt
960 pp; clothbound ISBN 0–8412–1761–0

*Understanding Chemical Patents: A Guide for the Inventor*
By John T. Maynard and Howard M. Peters
184 pp; clothbound ISBN 0–8412–1997–4; paperback ISBN 0–8412–1998–2

*Spectroscopy of Polymers*
By Jack L. Koenig
ACS Professional Reference Book; 328 pp;
clothbound ISBN 0–8412–1904–4; paperback ISBN 0–8412–1924–9

*Harnessing Biotechnology for the 21st Century*
Edited by Michael R. Ladisch and Arindam Bose
Conference Proceedings Series; 612 pp;
clothbound ISBN 0–8412–2477–3

*From Caveman to Chemist: Circumstances and Achievements*
By Hugh W. Salzberg
300 pp; clothbound ISBN 0–8412–1786–6; paperback ISBN 0–8412–1787–4

*The Green Flame: Surviving Government Secrecy*
By Andrew Dequasie
300 pp; clothbound ISBN 0–8412–1857–9

For further information and a free catalog of ACS books, contact:
American Chemical Society
Distribution Office, Department 225
1155 16th Street, NW, Washington, DC 20036
Telephone 800–227–5558